# Geminoid Studies

Hiroshi Ishiguro · Fabio Dalla Libera
Editors

# Geminoid Studies

Science and Technologies for Humanlike
Teleoperated Androids

 Springer

*Editors*
Hiroshi Ishiguro
Osaka University
Toyonaka, Osaka
Japan

Fabio Dalla Libera
Osaka University
Toyonaka, Osaka
Japan

ISBN 978-981-10-8701-1      ISBN 978-981-10-8702-8   (eBook)
https://doi.org/10.1007/978-981-10-8702-8

Library of Congress Control Number: 2018934871

Printed on acid-free paper

This Springer imprint is published by the registered company Springer Nature Singapore Pte Ltd. part of Springer Nature
The registered company address is: 152 Beach Road, #21-01/04 Gateway East, Singapore 189721, Singapore

# Preface

The first part of this book surveys the technologies involved in building a very humanlike robot, or android. The appearance, movements, and perceptions of such an android must mimic those of a human, and thus require unique developmental techniques. Chapters 1 and 2 describe how to develop a very humanlike robot. First, the android must have a very humanlike appearance. Body-part molds are made from a real person using shape-memory foam. Then, soft silicon skins are made for the entire body using these molds. Covering the entire body with soft material is a unique feature that is quite different from that employed in existing humanoid robots. The soft skin not only provides a humanlike texture for the body surface, but also allows safe physical interaction, that is, humanlike interpersonal interactions between real people and the android. The humanlike movements are produced by pneumatic actuators that are driven by highly compressed air. The high compressibility of air provides very flexible motions in the android joints without software control. This also makes the human–android interactions more humanlike and safer. To generate humanlike body movements in the android, the motions of a human performer are measured by means of a motion capture system, and the measured movements are copied to the android. People strongly expect humanlike responsive behaviors when interacting with the android, because it looks very similar to a human. It is, therefore, necessary for the android perceptual system to obtain various information to produce rich interactions. However, cameras and microphones embedded in the androids body are not sufficient, as the range and type of observations are restricted. To overcome this limitation, a sensor network is integrated with the android. Multiple types of sensor (such as camera, microphone, floor sensor, and laser range finder) are placed in the environment surrounding the android, thus endowing the environment with perceptive capabilities. This method allows the android to store a large amount of information on human activities and obtain contextual information. The mental state of the android can be defined based on contextual information, allowing humanlike responsive behavior to be produced. In the process of making the android more humanlike, we require knowledge about human behavior and interpersonal cognition; this is a novel approach to studying human nature by developing the android. We call this research framework android science, as explained in Chap. 2. Another approach for implementing humanlike behavior in the android is teleoperation by a real person. A human operating the android from a distance can play a role in forming perceptions, decision making, and motion and utterance generation. We can study human–android interactions prior to developing a humanlike android using the technology of teleoperation. Teleoperation provides a novel telecommunication effect whereby the person operating the android feels that it is his/her own body, while a second person interacting with the operated android also feels it is possessed by the operator. That is, teleoperation works as a robotic communication media that can transfer the operator's presence (or Sonzai-kan in Japanese) to a remote place. The teleoperation system of the android and its significance in android science are described in Chap. 3.

Robots operating in the real world must be able to understand, communicate with, and interact with real people. However, human–robot communication remains unnatural, and hence ineffective. Therefore, a very humanlike robot–an android–may prove to be the ultimate man–machine communication device. To create realism and avoid the "uncanny valley," this android requires realistic behavior that matches its realistic looks. The second part of this book describes methods for the automatic generation of humanlike motions (including body, facial, lip, and head motions) in the android.

Chapter 4 describes methods for generating body movements. To generate humanlike body movements (gestural movements) in an android, the motions of a human performer are measured by means of a motion capture system, and the measured movements are copied to the android. Here, it is important that the three-dimensional appearance of the performer should be transferred to the android. This is because the kinematics of the android differ from the human musculoskeletal system, and existing methods of copying a human's joint angles to that of an android are insufficient for generating humanlike movements.

Chapter 5 describes a teleoperation system in which the lip motion of a remote humanoid robot is automatically controlled by the operators voice. In the speech-driven lip motion generation method, the lip height and width degrees are estimated based on vowel formant information, so there is no need to create specific models for different languages. Subjective evaluation indicates that the audio-based method can generate lip motion that is more natural than vision-based and motion capture-based approaches. Issues regarding online real-time processing are also discussed in this chapter.

Chapters 6 and 7 describe analysis and generation strategies for head motion. Head motion naturally occurs in synchrony with speech and may carry paralinguistic information such as intention, attitude, and emotion in dialogue communication. With the aim of automatically generating head motions during speech utterances, analyses are first conducted on human–human dialogue interactions to verify the relations between head motions and linguistic and paralinguistic information carried by speech utterances. Chapter 6 describes the analysis of motion-captured data for several speakers during natural dialogues, including intra- and inter-speaker variabilities. Chapter 7 describes head motion (nodding and head tilting) generation models based on rules inferred from these analyses of the relationship between head motion and dialogue acts. Issues regarding eye gaze control during head motion control are also discussed in Chap. 7. Evaluation results show that the rule-based methods can generate more natural motions than conventional methods of directly mapping the original motions to the android.

Almost fifty years ago, Mori proposed the hypothesis of a non-monotonic relation between how humanlike an artificial device appears and people's fondness toward it. As a device becomes similar to our body in its appearance, our affinity gradually rises. However, when the device acquires a certain level of similarity to a real human or human body parts, people suddenly start to feel a strong negative impression toward it. Mori called this sudden change in trend the "Bukimi no Tani" (valley of uncanniness). This hypothesis is very persuasive alongside the examples he presented comparing a very humanlike myoelectric hand and a wooden hand created by a Buddhist image sculptor. Mori's hypothesis attracted interest in various fields, including human–robot interaction researchers, and has resulted in a large number of studies. Despite considerable efforts, until now, no clear evidence has been found to prove Mori's hypothesis. One reason for this is the limitations of the methods used in most studies. Still photographs, such as morphed figures between a real human face and an artificial face, or video recordings of animated characters and robots have mainly been used for testing. In his original paper, Mori described his hypothesis in terms of people's interactions with an artificial device, but few studies have considered such an interactive situation. Another possible reason is the technical limitation that very humanlike artificial devices with interactive capabilities, such as robots, are hard to realize.

Now, how does this "uncanny valley" response of mankind affect the design of robots intended to interact with people in socially functional ways, as we do with other humans? The main purpose of developing a humanoid robot is so that it can interact with people. If the task of a robot is just to fold laundry or clean rooms, the robot need not interact with people, and thus humanlike behavior is not required. Only when helping people to conduct more complicated tasks that require dialogue with verbal and nonverbal communication, as among humans, are humanlike robots necessary. This humanlike appearance and behavior are important, as we are fine-tuned to recognize and understand detailed intentions from observing human behavior. Past studies have found evidence that people react differently toward machine-looking entities and humans. People may be able to infer the other entity's "intention" as well as those of humans, such as with dogs, horses, or factory machines, but this is only after long-term experience and training. It is much easier for us to understand the intention of other humans. Therefore, the appearance (including behavior) of robots is one of the most important factors in designing socially interactive robots. However, given the "uncanny valley" response, how should we design a robot's appearance?

While the mechanism behind the "uncanny valley" remains unclear, the basic system behind such human responses is gradually being revealed by using androids as a device for exploring human nature. In the following chapters, Shimada et al. hypothesize that lateral inhibition, as seen in neuronal responses, may be one cause of the uncanny valley response (Chap. 8).

In Chap. 9, Matsuda et al. compare the brain activities of people watching various actions performed by a person and by robots with different degrees of humanlike-ness. Based on their observations, they suggest that appearance does not crucially affect the activity of the human mirror-neuron system, and argue that other factors, such as motion, should be targeted to improve the humanlike-ness of humanoid robots.

Złotowski et al. report an exploratory study using a live interaction with robots having different types of embodiment, an area that has not been considered in previous studies (Chap. 10).

With advances in robotic technologies, we are now at the stage to explore requirements for complicated tasks such as persuading or instructing people. When robots start to play advanced roles in our lives, such as caring for the elderly, attributes such as trust, reliance, and persuasiveness will be critical for performing appropriate services in roles such as doctors, teachers, and consultants. For robots to live alongside and communicate smoothly with people, and to be relied on to perform critical jobs, robots must convey the assuredness required of people in the same position; that is, to give the impression of reliability and trust. Past studies have shown the importance of humanlike nonverbal channels and the superiority of the physical presence of robots to software agents or computer terminals in everyday conversation. Researchers have been trying to make such interactive robots "humanoid" by equipping them with heads, eyes, and hands so that their appearance more closely resembles that of humans and they can make movements and gestures analogous to humans. These forms are not only efficient from the viewpoint of functionality, but also because people have a tendency to anthropo-morphize; that is, it is the humanlike appearance that best helps us interact with robots. However, to date, scant attention has been paid to how the appearance of service robots should be designed. Appearance has always been the role of industrial designers and has seldom been a field of study. Based on psychological and philosophical results, one dominant factor that rules human behavior is that of the body shaping the mind or, more specifically, the phenomenon of appearance forming an impression or changing people's attitudes. For robots to obtain social attributes, the effect of appearance must by systematically examined.

In the following chapters, people's reactions toward robots, especially android robots with humanlike appearance, are examined from three aspects. In Chaps. 11–13, the initial response of people (especially children) to android robots is examined. Chapters 14–16 explore whether robots can change people's attitudes using different approaches. Furthermore, when robots are teleoperated by a real person, the act of operation, especially conversation, reveals interesting insights into the fundamentals of human nature in terms of personality and individuality. In Chap. 17, exploratory attempts to investigate such properties are described.

Most of the studies described above investigate how people react when facing robots. However, when robots are used as a device for interaction, that is, when robots are teleoperated with people on both sides, those in front of the robots and the teleoperators are affected. Soon after starting to operate the Geminoid (a tele-operated android robot), people tend to adjust their body movements to those of the robot. They talk slowly in order to synchronize with the robot's lip motion and make small movements to mimic those of the robot. Some operators even feel as if they themselves have been touched when others touch the teleoperated android. This illusion, the *Body Ownership Transfer* (BOT) effect of teleoperated robots, is due to the synchronization between the act of teleoperating the robot and the visual feedback of the robot's motion. If BOT can be induced merely by operation without haptic feedback, then applications such as highly immersive teleoperation interfaces and prosthetic hands/bodies that can be used as real body parts could be developed.

In the following chapters, initial studies on BOT are introduced. Starting from the exploration of BOT and its basic nature (Chap. 18), various aspects of BOT are examined. How BOT occurs under different teleoperation interfaces (Chap. 19), how social situations influence BOT (Chap. 20), and the minimal requirements for BOT to occur are investigated through neuropsychological experiments (Chap. 21). It is shown that the will to control the robot (agency) and seeing the robot in motion (visual feedback) are sufficient for BOT, and thus the operator's sense of proprioception is not required.

An interesting conjecture arises here: When the operator feels that the android's body is his/her own, will a feedback phenomenon occur? For example, when the android's facial expression changes on its own, without being controlled by the operator, will the operator's emotion be affected? In the past, psychologists studied a phenomenon known as the "facial feedback hypothesis," which is based on William James famous idea that the awareness of bodily changes activates emotion. If feedback from the teleoperated robot could affect the operator, we may be able to implement a new device that can support the regulation of one's physical or psychological activities. The final two chapters explore some ambitious trials on regulating people through the act of teleoperating android robots. Chapter 22 describes an attempt to regulate people's emotion through facial expression feedback from android robots, while Chap. 23 presents experimental results that indicate the neural activity of the operator can be enhanced through teleoperating android robots with a brain–computer interface.

Most previous studies have focused on laboratory-based experiments. In this book, however, we introduce field studies that exemplify the kind of roles androids might take in the real world. These studies demonstrate how people perceive androids when faced with them in everyday situations. Androids have a similar appearance to humans, but, at the same time, can be recognized as robots. Therefore, it is expected that androids could have a similar presence as humans in the real world. Moreover, androids have the potential to create a different influence to that of actual humans. In this part, we introduce several field experiments that were held in four different situations (a café, hospital, department store, and stage play). In the first experiment, an android copy of Prof. Ishiguro, named Geminoid HI-2, was installed in a café at Ars Electronica in Linz, Austria (Chaps. 24 and 25). The results show that people treated the android as a social entity, not merely a physical object. In Chap. 26, we describe a study on how a bystander android influences human behavior in a hospital. This experiment shows that when the android nods and smiles in synchronization with the patient, the patients impressions toward the doctor are positively enhanced. As a third field experiment, the android tried to sell a cashmere sweater to customers in a department store. The results show that 43 sweaters were sold over 10 days. This indicates that the android could operate as a salesperson in a store (Chaps. 27 and 28). Finally, in Chaps. 29 and 30, we investigate how people perceive an android that reads poetry as a stage play. In this experiment, we create a stage play in collaboration with an artist and ask the audience about their impressions of how the android performs as an actor.

We conclude this preface by thanking the Japan Society for the Promotion of Science (JSPS) for supporting this work through the KAKENHI Grant-in-Aid for Scientific Research (S) Number 25220004.

Toyonaka, Japan                                                                    Hiroshi Ishiguro
April 2017                                                                          Fabio Dalla Libera

# Contents

# Chapter 1
# Development of an Android System Integrated with Sensor Networks

**Takenobu Chikaraishi, Takashi Minato and Hiroshi Ishiguro**

**Abstract** To develop a robot that has a humanlike presence, the robot must be given very humanlike appearance and behavior, and a sense of perception that enables it to communicate with humans. We have developed an android robot called "Repliee Q2" that closely resembles human beings; however, sensors mounted on its body are not sufficient to allow humanlike communication with respect to factors such as the sensing range and spatial resolution. To overcome this issue, we endowed the environment surrounding the android with perceptive capabilities by embedding it with a variety of sensors. This sensor network provides the android with humanlike perception by constantly and extensively monitoring human activities in a less obvious manner. This paper reports on an android system that is integrated with a sensor network system embedded in the environment. A human–android interaction experiment shows that the integrated system provides relatively humanlike interaction.

**Keywords** Humanlike behavior · Humanlike perception · Idling motion
Environment-embedded sensors

## 1.1 Introduction

In recent years, there has been much research and development toward intelligent robots that are capable of interacting with humans in daily life [3, 4]. However, even if robots were able to deal with various tasks, mechanical-looking robots are not

This Chapter is a Modified Version of Previously Published Papers, [1, 2], edited to be comprehensive and fit with the context of this book.

T. Chikaraishi · H. Ishiguro
Graduate School of Engineering, Osaka University, Osaka, Japan

T. Minato (✉)
Asada Project, ERATO, Japan Science and Technology Agency, Osaka, Japan
e-mail: minato@atr.Jp

H. Ishiguro
ATR Intelligent Robotics and Communication Laboratories, Kyoto, Japan

H. Ishiguro and F. Dalla Libera (eds.), *Geminoid Studies*,
https://doi.org/10.1007/978-981-10-8702-8_1

acceptable for tasks that require a humanlike presence. This is because our living space is designed for humans, and the presence of other people provides a kind of assurance to individuals. For a robot to have a humanlike presence, it needs to be recognized as a person, behave like a human, and be able to communicate naturally with people.

Apart from the engineering aspects, the development of such a robot will contribute to studies about the principles of human communication. This is because improving the robot's function to communicate with people step-by-step will lead to a gradual elucidation of the communication function of human beings.

Turning now to interpersonal communication, there are various communication channels, such as language, paralanguage (e.g., tone and voice patterns), physical motion (e.g., gazing and gestures), interpersonal distance, and clothes [5]. To develop a robot that can naturally communicate with people through these communication channels, the following points are required:

- The robot must have a very humanlike appearance and behavior.
- The robot must have a rich sense of perception.

In this study, we use an android called "Repliee Q2" (Fig. 1.1) that closely resembles a human and can produce humanlike motions. To give the android perceptive capabilities, we implement voice recognition, gesture recognition with multiple omnidirectional cameras, and human tracking with floor sensors.

**Fig. 1.1** The developed "Repliee Q2" android

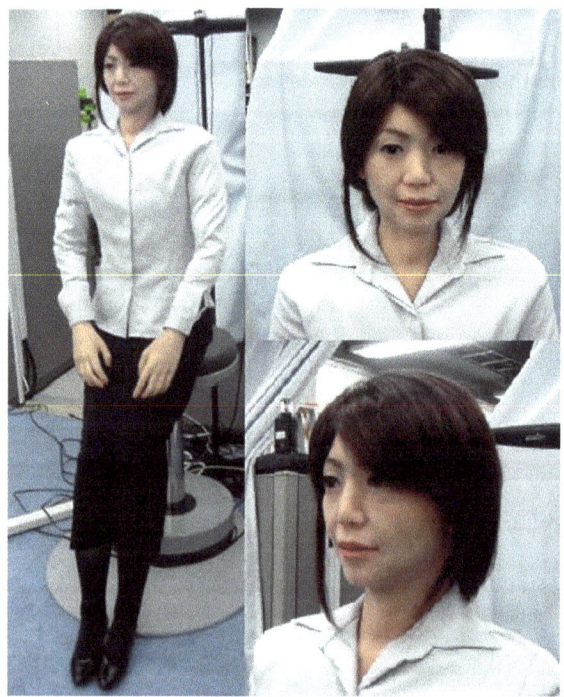

Existing research on communication has improved the communication capabilities of robots through the development of sensors that are similar to human sensory organs (e.g., [6]). However, recognition processes using sensors mounted on the body of the robots place considerable constraints on factors such as the sensing range and spatial resolution. Under these constraints, it is difficult for robots to obtain sufficient external information to allow natural communication with humans.

An alternative method for achieving a rich sense of perception is to embed a variety of sensors in the environment surrounding the robot, that is, to endow the environment with perceptive capabilities [7–9]. This method allows the robot to store a considerable amount of information on human activities and obtain contextual information, because the sensors embedded in the human living space can observe human activities constantly and extensively while remaining less obvious. In this case, it is not necessary to mount all of the sensors on the robot's body. Even if the robot itself senses nothing, it can naturally communicate with people by behaving as if it is doing the sensing itself.

The method used in this study overcomes the limitations of on-body sensors by constructing a sensor network to achieve natural human-robot communication. The sensor network consists of both on-body sensors and as many diverse sensors as possible embedded in the environment. However, the sensory information obtained by this system is completely different from that of the five human senses, which becomes a problem in the development of a perceptive system. The integration and selection of the sensory information necessary for natural communication depends on the situation. The sensors obtain quantitative information, such as a person's location; however, the information that is required for communication is subjective, such as whether the other person is at an appropriate distance for talking, which again depends on the situation. This study prepares situation-dependent scenarios to deal with contexts including subjective information. The scenarios describe which motion module is to be executed, and when it is to be executed, according to the sensory information while assuming a specific situation. The android is thus able to integrate and select the sensory information according to the current situation. In this paper, we report the development of a system to facilitate the description of these scenarios.

The remainder of this paper is organized as follows. The system's hardware organization and software architecture are described in Sect. 1.2. A demonstration of the system at a world expo is reported in Sect. 1.3. To verify that the system achieves natural human–android interaction, we implemented a simple and indirect interaction where people viewed the android in a waiting situation (where it remained in one place). Generally, individuals exhibit various motions depending on their mental state, even when they simply remain in one place. A variety of sensory information is necessary to implement these subconscious motions. In Sect. 1.4, we define the android's mental state according to the person's state as observed by sensors and implement a scenario in which the android chooses her motion module according to her mental state. In Sect. 1.5, the implemented android is compared with an android whose motion module is randomly selected according to the subjects' impression, and the effectiveness of the system is verified.

## 1.2 Hardware Organization and Software Architecture

This section presents the hardware organization and software architecture (Figs. 1.2 and 1.3). The system must process a large amount of data from a variety of sensors to achieve humanlike perception. Therefore, the sensor processing is distributed to a number of computers. The android's motion modules contain datasets of desired values for each degree of freedom (DOF), and their recognition results trigger the execution of motions in the android. Briefly, this involves the following steps:

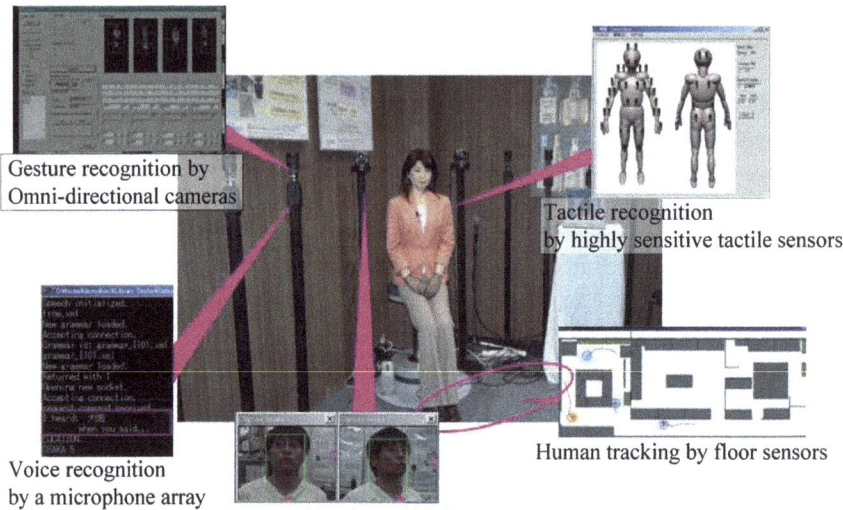

**Fig. 1.2** Hardware organization

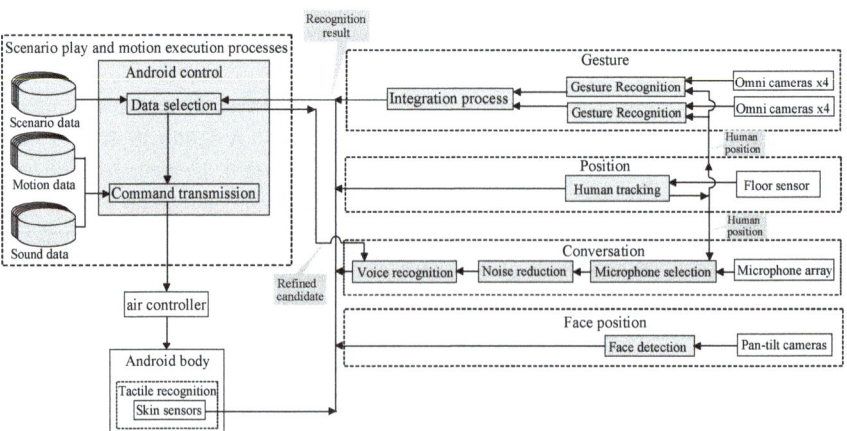

**Fig. 1.3** Software architecture

```
<?xml version="1.0"?>
<scenario>
        <event eventno="1" sensor="UNUSE" motion="G001.dat">
                <result><UNUSE nextevent="2"/></result></event>
        <event eventno="2" sensor="UNUSE" motion="G002.dat">
                <result><UNUSE nextevent="3"/></result></event>
        <event eventno="3" sensor="UNUSE" motion="G003.dat">
                <result><UNUSE nextevent="4"/></result></event>
        <event eventno="4" sensor="VOICE" motion="G004.dat" grammar="grammar_G004.xml">
                <result><YES nextevent="5"/>
                        <NO nextevent="8"/>
                        <ERROR nextevent="16"/>
                </result></event>
        <event eventno="5" sensor="UNUSE" motion="G005.dat">
                <result><UNUSE nextevent="6"/></result></event>
        <event eventno="6" sensor="UNUSE" motion="G006.dat">
                <result><UNUSE nextevent="7"/></result></event>
        <event eventno="7" sensor="UNUSE" motion="G007.dat">
                <result><UNUSE nextevent="11"/></result></event>
        <event eventno="8" sensor="UNUSE" motion="G008.dat">
                <result><UNUSE nextevent="9"/></result></event>
        <event eventno="9" sensor="UNUSE" motion="G009.dat">
                <result><UNUSE nextevent="10"/></result></event>
        <event eventno="10" sensor="UNUSE" motion="G010.dat">
                <result><UNUSE nextevent="11"/></result></event>
        <event eventno="11" sensor="UNUSE" motion="G011.dat">
                <result><UNUSE nextevent="12"/></result></event>
        <event eventno="12" sensor="FLOOR" motion="G012.dat">
                <result><REGION1 nextevent="13"/>
                        <REGION2 nextevent="14"/>
                        <REGION3 nextevent="15"/>
                        <OUTFLOOR nextevent="13"/>
                        <ERROR nextevent="13"/>
                </result></event>
        <event eventno="13" sensor="UNUSE" motion="G013.dat">
```

**Fig. 1.4** Example of a scenario

(1) Obtain recognition results from computers where sensor processes are running.
(2) Use the recognition results to select the motion module's ID from the scenario.
(3) Execute the android's motion and play a voice file.
(4) Return to step (1).

Repeating these steps determines the context of the situation in order to execute human–android communication. Determining the context of the situation has the additional advantage that the system can refine recognition candidates. For example, in a greeting situation, the system can restrict greetings to such candidates as bowing, shaking hands, or waving a hand, and exclude other candidates. In addition, because these modules are described in XML, they are easily created by people without programming skills (Fig. 1.4). Combining various sensors has another advantage in that recognition by one sensor can help another type of sensor. For example, position data from floor sensors can help identify the region of interest for camera images to recognize a gesture, and gesture recognition can help voice recognition to refine recognition candidates.

### 1.2.1   Repliee Q2's Body

Repliee Q2 was modeled on an actual person and constructed with medical silicon rubber. Hence, it has a very humanlike appearance and artificial skin that is as soft as human skin. It has 42 air actuators in its body, including nine on the left and right arms, two on the left and right hands, four on the waist, three on the eyes, one on the eyebrows, one on the eyelids, one on the cheek, seven on the mouth, and three on the neck. These enable Repliee Q2 to generate rich facial expressions (Fig. 1.5). The face, neck, arms, and hands are covered with silicone skin, and 42 highly sensitive tactile sensors are mounted beneath the skin.

(a) *Pneumatic controller and air compressor*: The reactions of the actuators are very natural because they incorporate air dampers against external force. This achieves a much safer interaction with humans than other actuators such as oil pressure actuators and DC servomotors. The actuators can represent the unconscious breathing movements of the chest with a silent drive, because the air compressor can be located at a point distant from the android.

### 1.2.2   Sensor Network and Data Processing System

Seven computers (Pentium IV 3 GHz) are used to process the sensor data.

(a) *Floor tactile sensor units*: Floor tactile sensor units (Vstone VS-SS-SF55) are used for human tracking (Fig. 1.6). The floor sensor units consist of densely arranged tactile sensors (at 5 mm intervals) to detect human position by gravity. Covering the floor of a room with these units enables human tracking. The specifications of the sensor units are as follows. Each floor sensor unit measures $500 \times 500 \times 15$ mm,

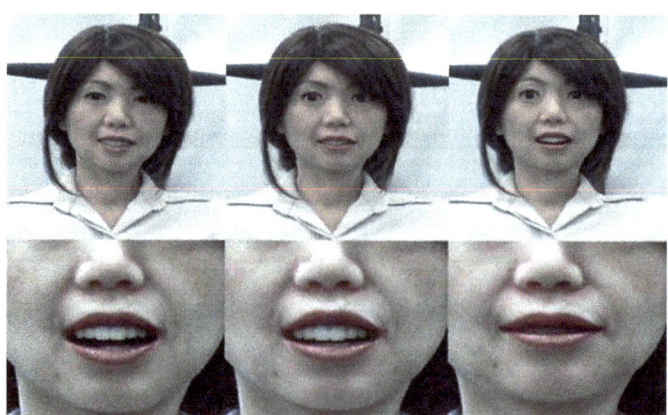

**Fig. 1.5**   Repliee Q2 facial expressions

**Fig. 1.6** Floor tactile sensor unit

with detection blocks measuring $100\,\text{mm} \times 100\,\text{mm}$, and has a sensitivity of 200–250 $\text{g/cm}^2$. The resolution ability is $10 \times 10\,\text{cm}$, which is enough to detect human positions, because most adult feet are 20–30-cm long. In this study, $4 \times 4$ units ($4\,\text{m}^2$) are placed in front of the android in the region where we expect human–android communication to occur.

The floor sensors are connected through a serial port. After processing the data from the connection, an android control computer obtains the data via TCP/IP. A gesture recognition computer and voice recognition computer also obtain the data to improve their recognition accuracy. When a person walks, he or she activates multiple detection blocks that cause a many-to-one problem for tracking. Therefore, we utilize "Human Tracking Using Floor Sensors Based on the Markov Chain Monte Carlo Method" by Murakita et al. [10].

(b) *Omnidirectional cameras*: Eight omnidirectional cameras (Vstone VS-C42N-TR) are embedded into the environment to recognize human gestures, including "bowing" and "hands up." The cameras are connected through an image capture board. The specifications of the omnidirectional cameras include an imaging angle of 15° above the horizontal plane and 60° below the horizontal plane, with $768 \times 494$ effective pixels.

In this research, we utilize "VAMBAM: View and motion based aspect models for distributed omnidirectional vision systems" by Nishimura et al. [11, 12], which enables the system to recognize human gestures independent from the location of the person relative to the cameras. However, the method applies a subtraction image to determine the region of interest, but this is not robust against optical noise. For this reason, we use human position data from floor sensors to define the detection region in the images from the omnidirectional cameras.

(c) *Microphone array*: Eight non-directional microphones mounted in the lower part of the omnidirectional cameras are used for voice recognition. The microphone-equipped omnidirectional cameras are embedded into the environment to obtain speech sounds. For voice recognition, it is necessary to obtain sound containing the least amount of noise possible. Thus, we use the following three steps. First, the microphone that is closest to the sound source is selected using human position information obtained from the floor sensors. Second, background noise is subtracted using the input average of several past frames below a certain threshold. Third, the voice recognition process receives the refined sound data. In this research, we use Microsoft's SpeechSDK [13] as a voice recognition engine.

(d) *Pan/tilt/zoom cameras*: The system must obtain human face positions, because the android should face the person when talking. For that purpose, two CCD pan/tilt/zoom cameras (Sony EVI-D100) are positioned on either side of the android. The camera specifications include an angle of pan/tilt of $\pm100°$ (horizontal) and $\pm25°$ (vertical), and a maximum pan/tilt speed of 300°/s (horizontal) and 125°/s (vertical). The cameras are controlled via a serial port, and the pan/tilt movement covers the view in front of the android. According to the face position data, the system modifies the command values to the actuator on the android's neck, so the android can turn to face the person. To obtain the human face position and direction, we used the "Study on Extraction of Flesh-Colored Region and Detection of Face Direction" by Adachi et al. [14].

(e) *Highly sensitive tactile sensors*: Highly sensitive tactile sensors can perceive bodily contact. A total of 42 tactile sensors are mounted beneath the android's silicon skin. The skin sensors consist of piezoelectric devices that are sheathed by a silicon film layer. The sensors detect deformation speed as physical contact.

## 1.3 Demonstration at a World Exposition

To verify that the system works in a non-laboratory environment, we exhibited it at a prototype robot exhibition at the 2005 World Exposition held in Japan and demonstrated it to ordinary Expo visitors. This confirmed that our implementation works and can be operated by a non-expert. The prototype robot exhibition portrayed a city, residential area, and park in the year 2020 and demonstrated the coexistence of prototype robots developed for practical applications with people. Our system was exhibited in a cafe in the city area for eleven days (Fig. 1.7). In this demonstration, a microphone held in the android's hand was used instead of the microphone array, because the surrounding noise was too high. The scenario portrayed the android interviewing a person visiting the cafe. The android asked questions such as "Where are you from?" and "Which exhibit was the most interesting?," and the person answered verbally or with gestures (Fig. 1.8).

**Fig. 1.7** Demonstration at the 2005 World Exposition, Aichi, Japan

**Fig. 1.8** The android interviewing a visitor

## 1.4 Implementing Natural Behavior in the Android by Defining Mental States

We verified that the android system integrated with the sensor network is effective for natural interaction by developing the waiting behavior of the android. We assumed a scene in which the android waits while another person stands nearby, as shown in Fig. 1.12. Generally, an individual's motions change according to his or her mental state, even when simply waiting (remaining in one place). For example, the frequency of eye blinking increases, breathing is stronger, and gazing becomes unstable when a person feels stressed. It is assumed that the android's motion changes according to sensor information, causing people to attribute mental states to the android, resulting in a more humanlike impression. Additionally, the android should not simply stand still, because a past study has found that a perfectly still android has a less humanlike appearance [15]. To define the scenario, we set up the same situation as in Fig. 1.12, but replaced the android with a person (subject B). We investigated the mental state of subject B while recording the motion of the other person (subject A) using the sensor network (the tactile sensor data of subject B was simulated from the observations). We classified the sensor data on subject A's motion relative to subject B's mental states. We also prepared the android's motion modules to make them similar to the motions of individuals with the same mental states as subject B. We then defined the scenario using the motion modules and the sensor data pattern. In Sect. 1.5, the android behavior for specific scenarios is compared with that of an android with a randomly selected motion module.

### 1.4.1 Observation of a Waiting Person

We conducted an experiment to observe the situation shown in Fig. 1.9. We observed the motions of a waiting person and asked about her/his mental state after the observation. The experimenter informed the subject (subject B) that this experiment was to investigate natural human movement and asked her/him to sit on a chair for 20 min without standing up. The experimenter then informed another subject (subject A) of the purpose of the experiment (the same as above) and asked her/him to behave freely within the 2 m × 2 m area shown in Fig. 1.9. This space was covered with floor sensors. Two people played subject B (one male in his 20 s and one female in her 20 s), and two people played subject A (two males in their 20 s). All of the subjects were university students. The scene was videotaped, and common motions among subjects A and B were extracted from the video. After the recording, subject B freely reported his/her mental state.

The six common motions listed in Table 1.1 were extracted from subject B.

**Fig. 1.9** The experimental setup

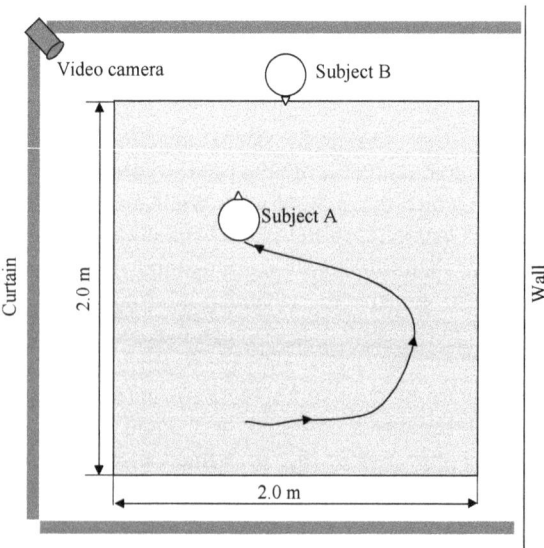

**Table 1.1** Observed behavior of subject B

| Behavior of subject B |
| --- |
| Look in the direction subject A is looking |
| Look or glance at subject A |
| Turn to her/his front without looking at subject A |
| Throw a glance at subject A |
| Look at the ground |
| Follow the movement of subject A |

**Table 1.2** Classification of the feelings of subject B toward subject A. Subject B was later replaced with the android in the experiment

| Position ID of A | Behavior of A | Feeling of B toward A | Mood of B |
| --- | --- | --- | --- |
| 2, 5 | Stare at B | Unpleasant | Disgust |
| 8 | Stare at B | Unpleasant + oppressive feeling | |
| 11 | Stare at B | Very unpleasant + oppressive feeling | |
| 1–6 | Not stare + walk around | Not so concerned about the action of A | Acceptance |
| 1, 3, 4, 6 | Stare at B | Harassed | Fear |
| 7–12 | Not stare + walk around | Anxiety | Apprehension |
| 8, 11 | Look somewhere | Anxiety | |
| 1–12 | Stare + walk around | Anxiety | |
| 7, 9, 10, 12 | Stare at subject B | Anxiety | |
| 1–7, 9, 10, 12 | Look somewhere | Worry about where A is looking | Interest |

## 1.4.2 Mental States of the Waiting Person

We classified the feelings of subject B toward subject A based on the location of subject B and the interpersonal distance between them. The results are presented in Table 1.2. The position ID (first column) denotes the digits in Fig. 1.10, which correspond to the square space of 2 m × 2 m in Fig. 1.9. For example, the first row means that when subject A is standing in position 2 or 5 staring at subject B, subject B felt unpleasant. The last column denotes the mood of subject B following Plutchik's model of emotions [16].

The emotions of subject B were classified into five categories. Each emotion is believed to have been elicited for the following reasons.

(a) *Disgust*: Subject B reported that they did not want subject A to stare at them, as they felt oppressed. This implies that subject B wanted to remove an object that could

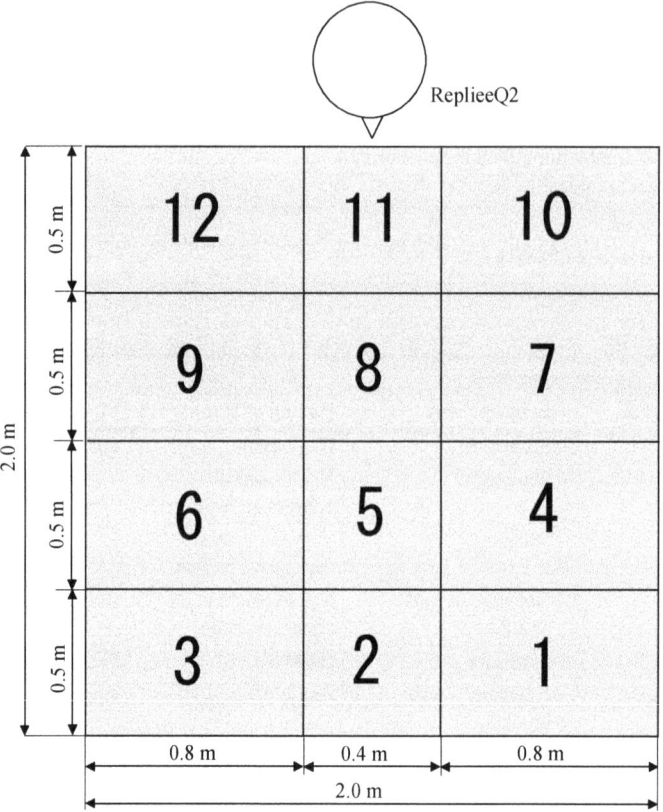

**Fig. 1.10** ID of the position of subject A

be harmful to them; that is, the emotional function of "rejection" was activated. In this case, subject B's emotion is "disgust."

(b) *Acceptance*: Subject B reported that they did not care because subject A did not care about them. This implies that subject B accepted the situation; that is, the emotional function "affiliation" was activated. In this case, subject B's emotion is "acceptance."

(c) *Fear*: Subject B reported that they felt mental distress from being stared at by subject A. This implies that subject B wanted to protect themselves from suffering; that is, the emotional function is one of "protection." In this case, subject B's emotion is "fear."

(d) *Apprehension*: The emotion "fear" occurred because subject B reported that they felt mental distress from being stared at or being too close to subject A. Moreover, they reported that they worried about subject A's intentions. This implies that the emotional function of "exploration" was also activated and the emotion of "anticipation" occurred. As a result of this mixture, the emotion of "apprehension" occurred.

(e) *Interest*: The emotion of "anticipation" occurred because subject B reported being worried about what subject A was looking at. However, they did not feel disgust; that is, they accepted it (the emotion was "acceptance"). As a result of this mixture, the emotion of "interest" occurred.

In this situation, a person's mental state is defined according to the other person's motion and position. The sensor network that we developed obtains this kind of information more robustly than on-body sensors.

### 1.4.3 Motion Selection of the Android According to its Mental State

Based on the above analysis, we defined a process to recognize the behavior of subject A (see Fig. 1.11). The first step is to check for the presence of a person (subject A) within the area covered by the floor sensors. If subject A is in the area, the sensor network obtains her/his position. Next, the movement of subject A is checked. For example, if subject A does not move, the face direction (i.e., looking direction) is measured. Furthermore, if subject A is moving toward subject B (i.e., the android), the duration of her/his gaze is measured, and the behavior of subject A is specified accordingly. "Neutral" in Fig. 1.11 denotes that subject A does nothing. The android system is able to detect when it is touched and the direction of a sound source; however, no behavior that requires the sensor network to be recognized was observed in

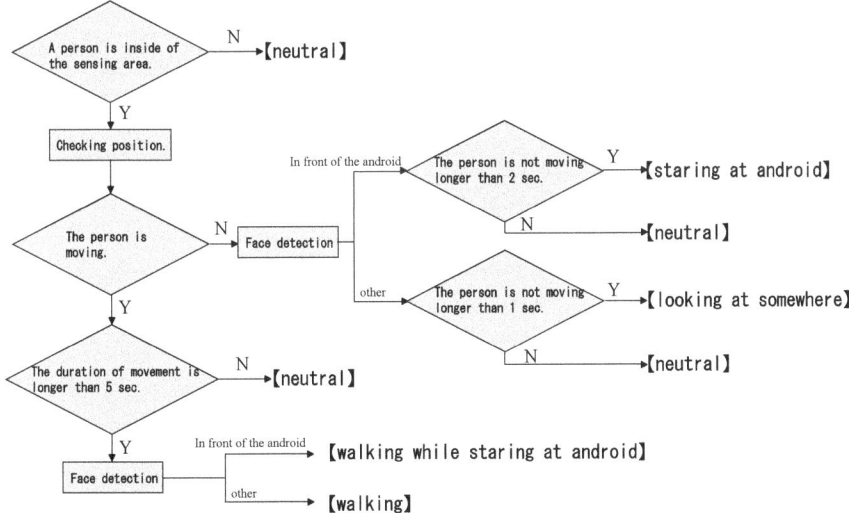

**Fig. 1.11** Procedure for recognizing a person's behavior

the above experiment. The android's mental state is determined by the recognized behavior according to Table 1.2 (where subject B is replaced with the android).

In this scenario, the system repeats the following steps.

(1) Recognize a person's behavior.
(2) Determine the mental state of the android according to Table 1.2.
(3) Select and execute the motion module according to the procedure of Fig. 1.11.

The remainder of this section describes the android motion features to be regarded when creating motion modules. We intuitively designed a motion module for each mental state while referring to the motions of subject B in the above experiment. In the following description, the frequency of motion is standardized for the situation in which nobody is around the android (hereafter, the normal situation).

The motion of subject B when feeling disgust had the following common features. The motion modules of the android for this emotion were designed by taking these features into consideration.

• Does not make an eye contact with a person.
• Bends backward to create some distance from the person.
• Increases blinking rate.

The motions for fear, apprehension, and interest had the following common features. The motion modules for these emotions were designed by taking these features into consideration.

• The frequency of looking at the person is high, but no staring occurs.
• The frequency of blinking is high.
• The movement of the eyes is quick.

The motion modules for acceptance were designed by taking the following features into consideration.

• The frequency of looking at the person is low, and no staring occurs.

The frequency of looking at the person, blinking, and eye movement for each emotion was designed in the following order:

$$normal = acceptance < interest < apprehension < disgust < fear$$

## 1.5 Evaluation

### 1.5.1 The Experimental Setup

To verify the effectiveness of the developed system, we conducted an experiment to compare the android in the above scenario with an android whose motion module is randomly selected with respect to the subjects' evaluation of humanlikeness for the androids. Hereafter, the android that follows the above scenario is called the *scenario android* and the android with randomly selected motion is called the *random android*. A subject interacts with both androids in the experiment. The subject was not informed of the difference between the two androids. To counterbalance the order of the interaction, we designed two conditions:

Condition (1)    The subject interacts with the random android for 3 min (first session) and then interacts with the scenario android for 3 min (second session) after a 1 min break.

Condition (2)    The subject interacts with the scenario android for 3 min (first session) and then interacts with the random android for 3 min (second session) after a 1 min break.

Each subject participated in one condition.

The experiment was conducted in the environment shown in Fig. 1.12. Figure 1.13 shows a scene from the experiment. Initially, the subject was informed that the experiment is to evaluate the humanlikeness of the android. The subject was then told to move around each android freely for a 3 min period. After the interactions, the subject was asked to score the humanlikeness of the androids in the first and second sessions. The scores were assigned on a scale of seven, ranging from $-3$ (not humanlike) to 3 (very humanlike). The subjects were 16 university students. Eight subjects participated in condition 1, and the other eight participated in condition 2.

**Fig. 1.12** Setup of the experimental room

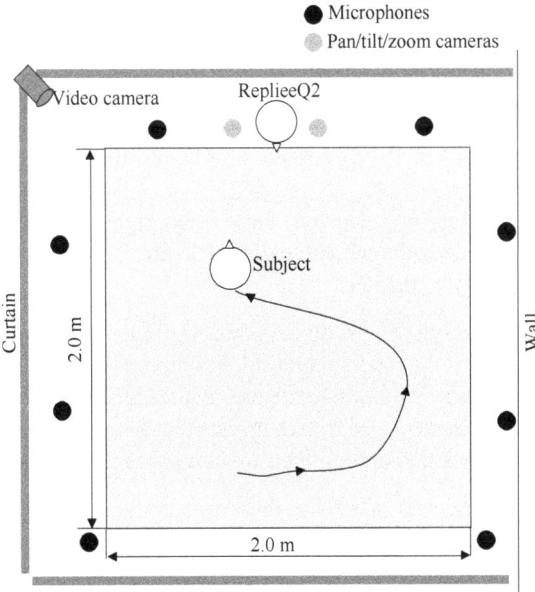

**Fig. 1.13** Experimental scene

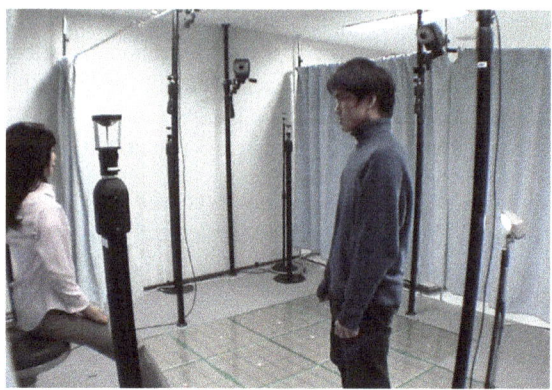

**Table 1.3** Experimental results

| Condition | Android | Averaged score |
|---|---|---|
| (1) (Random → Scenario) | Random (R1) | −0.63 |
| (2) (Scenario → Random) | Scenario (S1) | 0.75 |
| (1) (Random → Scenario) | Scenario (S2) | 0.63 |
| (2) (Scenario → Random) | Random (R2) | 0.5 |

## 1.5.2 Results

The averaged scores are presented in Table 1.3. To consider the order of interaction, the averaged scores of the first and second sessions are given separately.

(a) A T test revealed that there is a significant difference between the random android in the first session (R1) and the scenario android in the first session (S1) ($p < 0.05$).

(b) A T test revealed that there is no significant difference between the random android in the second session (R2) and the scenario android in the second session (S2) ($p = 0.452$).

From result (a), the humanlikeness of the scenario android is significantly greater than that of the random android. This means that the android's behavior in response to the subject's motion made the android more humanlike. Moreover, the difference between results (a) and (b) suggests that there is an order effect.

To check this order effect, we compared the scores with respect to the order.

(c) A T test revealed that there was a significant difference between the random androids in the first and second sessions ($p < 0.1$).

(d) A T test revealed that there was no significant difference between the scenario androids in the first and second sessions ($p = 0.438$).

These results suggest that the interaction with the scenario android in the first session influenced the impression of humanlikeness toward the random android in the second session. In other words, previous interaction with the scenario android made the random android more humanlike. In summary, we have seen that:

- The android behavior generated by the developed system makes the android more humanlike.
- The experience of interacting with the scenario android in advance influences the interaction with the random android.

In addition to the humanlikeness scores, the subjects were asked to freely report their impressions. Many subjects reported that the humanlikeness of the scenario android was derived from:

- The motion of looking at the subject.
- When the subject came close to the android, it frequently looked around. This gave the impression that the android was nervous.
- The android changed its posture depending on its gaze direction.
- When the subject came close to the android, it blinked more frequently.

Some subjects reported that the android was less humanlike because it looked around too frequently. On the contrary, some subjects reported that the random android was humanlike when it averted its gaze after momentary eye contact. For both androids, many subjects reported that involuntary movements, such as slight mouth and shoulder movements, were humanlike. In addition, some subjects had a positive impression of the eye and gaze movements, whereas others had a negative impression.

Next, we consider the order effect. Even when the android's motion is randomly selected, the subjects seem to feel as if the motion is sometimes responsive or contingent on their own motion. The subjects believed this occurred by chance when they first interacted with the random android. However, once the subjects think that the android is humanlike in the scenario android session, the chance response seems to help maintain the android's impression of humanlikeness.

The experimental results reveal that the developed system can augment the naturalness of human–android interaction.

## 1.6   Considerations and Future Work

The developed system cannot achieve the exact behavior of humans; however, subjects felt a sensation of humanlikeness in the android. This suggests that the human recognition process strongly changes its target and resolution depending on the context. In other words, it is suggested that highly accurate recognition is not a requirement of achieving natural human–android interaction if the android is able to obtain the context of the situation. The same can be said for the humanlikeness of the android motions. In contrast, some contexts require accurate recognition. For example, some subjects reported that the android looked around too frequently to be humanlike. The act of looking around makes the android more humanlike; however, it makes the android less humanlike in a situation where the subjects expect the android to make eye contact with them. An on-body sensor has limitations in obtaining the context of a situation with respect to factors such as the sensing range and spatial resolution; however, a sensor network is a promising method because it can monitor human activities constantly and extensively.

People's behavior is sometimes inconsistent with the context. In this case, a method depending on contextual information will frequently fail. It is, therefore, necessary to recognize certain factors without contextual information and to determine the context according to the recognized result. A version of the chicken-and-egg (context-and-recognition) problem needs to be solved in future work.

## 1.7 Conclusion

This paper proposed a system in which an android is integrated with a sensor network to achieve a robot that has a humanlike presence and the ability to naturally communicate with people. In the experiment, we designed the android's mental states using sensory information from a sensor network and achieved natural human–android interaction based on these mental states. The proposed system was compared with a method in which the android motion was randomly selected, and its effectiveness was verified. In a series of experiments, the overall context was given in advance by defining the scenario. The sensor network is a powerful tool for obtaining contextual information because it can monitor human activities constantly and extensively.

## References

1. Takenobu Chikaraishi, Takashi Minato, and Hiroshi Ishiguro. (2010). Development of an android system integrated with sensor networks. In *Robotics 2010 Current and Future Challenges*, ed. Houssem Abdellatif. ISBN: 978-953-7619-78-7, InTech. http://www.intechopen.com/books/robotics-2010-current-and-future-challenges/development -of-an-android-system-integrated-with-sensor-networks.
2. Takenobu Chikaraishi, Takashi Minato, and Hiroshi Ishiguro. (2008). Development of an android system integrated with sensor networks. In *2008 IEEE/RSJ international conference on intelligent robots and systems (IROS 2008)*, 326–333.
3. Takayuki Kanda, Hiroshi Ishiguro, Tetsuo Ono, Michita Imai, Takeshi Maeda, and Ryohei Nakatsu. (2004). Development of Robovie as a platform for everyday-robot research. *Electronics and Communications in Japan* (translated from Denshi Joho Tsushin Gakkai Ronbunshi), Part 3, 87 (4).
4. Kajita, S. (2002). Research of biped robot and humanoid robotics project (HRP) in Japan. In *The fourth international conference on machine automation (ICMA'02)*, 1–8.
5. Daibo, I. (1993). Psychology of intimacy and function of communication (In Japanese with English abstract). Technical Report of IEICE (The Institute of Electronics, Information and Communication Engineers), HC-93-52, 33–40.

6. Hiroshi G. Okuno, Kazuhiro Nakadai, Tino Lourens, and Hiroaki Kitano. (Oct 2001). Human-robot interaction through real-time auditory and visual multiple-talker tracking. In *Proceedings of IEEE/RSJ international conference on intelligent robots and systems (IROS-2001)*, 1402–1409. Maui, Hawaii: IEEE.
7. Morishita, H., K. Watanabe, T. Kuroiwa, T. Mori, and T. Sato. (2003). Development of robotic kitchen counter: A kitchen counter equipped with sensors and actuator for action-adapted and personally-fit assistance. In *Proceedings of the 2003 IEEE/RSJ international conference on intelligent robots and systems IEEE robotics and automation society and robotics society of Japan*, vol. 10, 1839–1844 .
8. Taketoshi Mori, Hiroshi Noguchi, and Tomomasa Sato. (2005). Daily life experience reservoir and epitomization with sensing room. In *Proceedings of workshop on network robot systems: Toward intelligent robotic systems integrated with environments*.
9. Hiroshi Ishiguro. (1997). Distributed vision system: A perceptual information infrastructure for robot navigation. In *International joint conference on artificial intelligence (IJCAI-97)*, 36–41.
10. Murakita, T., T. Ikeda, and H. Ishiguro. (Aug 2004). Human tracking using floor sensors based on the Markov chain Monte Carlo method. In *Seventeenth international conference on pattern recognition (ICPR)*, 917–920.
11. Takuichi Nishimura, Takushi Sogo, Shinobu Ogi, Ryuichi Oka, and Hiroshi Ishiguro. (2001). Recognition of human motion behaviors using view-based aspect model based on motion change. *Transactions of the IEICE*, J84-D-II (10): 2212–2223.
12. Ishiguro, H., and T. Nishimura. (2001). VAMBAM: View and motion based aspect models for distributed omnidirectional vision systems. In *Proceedings international joint conference artificial intelligence*, 1375–1380.
13. http://www.microsoft.com/japan/msdn/accessibility/speech/.
14. Adachi, Y., A. Imai, M. Ozaki, and N. Ishii. (2001). Study on extraction of flesh-colored region and detection of face direction. *International Journal Knowledge-Base Intelligent Engineering Systems*, 5–2.
15. Minato, T., M. Shimada, S. Itakura, K. Lee, and H. Ishiguro. (2005). Does gaze reveal the human likeness of an android? In *Proceedings of the 4th IEEE international conference on development and learning*, 106–111.
16. Plutchik, R. The emotion: Facts, theories and new model, New York: Random House.

# Chapter 2
# Building Artificial Humans
# to Understand Humans

**Hiroshi Ishiguro and Shuichi Nishio**

**Abstract**  If we could build an android as a very humanlike robot, how would we, humans, distinguish a real human from the android? The answer to this question is not simple. In human–android interactions, we cannot see the internal mechanism of the android, and thus, we may simply believe that it is a human. This means that humans can be defined in two ways: by their organic mechanism and by their appearance. Further, the current rapid progress in the development of artificial organs makes this distinction confusing. The approach discussed in this paper is to create artificial humans based on humanlike appearances. The developed artificial humans, an android and a geminoid, can be used to understand humans through psychological and cognitive tests. We call this new approach to understanding humans "Android Science."

**Keywords**  Robot · Android · Geminoid · Cognitive science

**Field of research**  Artificial human

This chapter is a modified version of a previously published paper [1], edited to be comprehensive and fit with the context of this book.

H. Ishiguro
Department of Systems Innovation, Graduate School of Engineering Science,
Osaka University, Osaka, Japan

H. Ishiguro · S. Nishio (✉)
Advanced Telecommunications Research Institute International,
Keihanna Science City, Kyoto, Japan
e-mail: nishio@botransfer.org

H. Ishiguro and F. Dalla Libera (eds.), *Geminoid Studies*,
https://doi.org/10.1007/978-981-10-8702-8_2

## 2.1 Introduction

Why are people attracted to humanoid robots and androids? The answer is simple: because human beings are attuned to understanding or interpreting human expressions and behaviors, especially those that exist in their surroundings. Infants, who are supposedly born with the ability to discriminate various types of stimuli, gradually adapt and fine-tune their interpretations of detailed social clues from other voices, languages, facial expressions, or behaviors [2]. Perhaps due to this functionality of nature and nurture, people have a strong tendency to anthropomorphize nearly everything they encounter. This is also true for computers or robots. In other words, when we see PCs or robots, some automatic process inside us attempts to interpret them as human. The media equation theory [3] was the first to explicitly articulate this tendency. Since then, researchers have been pursuing the key element that makes people feel more comfortable with computers or creates an easier and more intuitive interface to various information devices. This pursuit has also spread to the field of robotics. Recently, the focus of robotics has shifted from traditional studies on navigation and manipulation to human–robot interaction. A number of researches have investigated how people respond to robot behavior and how robots should behave so that people can easily understand them [4–6]. Many insights from developmental or cognitive psychologies have been implemented and examined to see how they affect the human response or whether they help robots produce smooth and natural communication with humans.

However, human–robot interaction studies have neglected one issue: the "appearance versus behavior" problem. Empirically, we know that appearance, one of the most significant elements in communication, is a crucial factor in the evaluation of interaction (see Fig. 2.1). The interactive robots developed so far have very mechanical outcomes that appear as "robots." Researchers have tried to make such interactive robots "humanoid" by equipping them with heads, eyes, or hands so that their appearance more closely resembles that of humans and to enable them to make analogous human movements or gestures such as staring, pointing. Functionality was considered the primary concern in improving communication with humans. In this manner, many studies have compared robots with different behaviors. Thus far, scant attention has been paid to the robots' appearance. Although there have been many empirical discussions on very simple static robots such as dolls, the design of a robot's appearance, particularly to increase its human likeness, has always been the role of industrial designers; it has seldom been a field of study. This is a serious problem for developing and evaluating interactive robots. Recent neuroimaging studies show that a certain brain activation does not occur when the observed actions are performed by non-human agents [7, 8]. Appearance and behavior are tightly coupled, and there are strong concerns that the evaluation results might be affected by appearance.

In this chapter, we introduce *android science,* an interdisciplinary research framework that combines two approaches, one in robotics for constructing very humanlike robots and androids, and another in cognitive science that uses androids

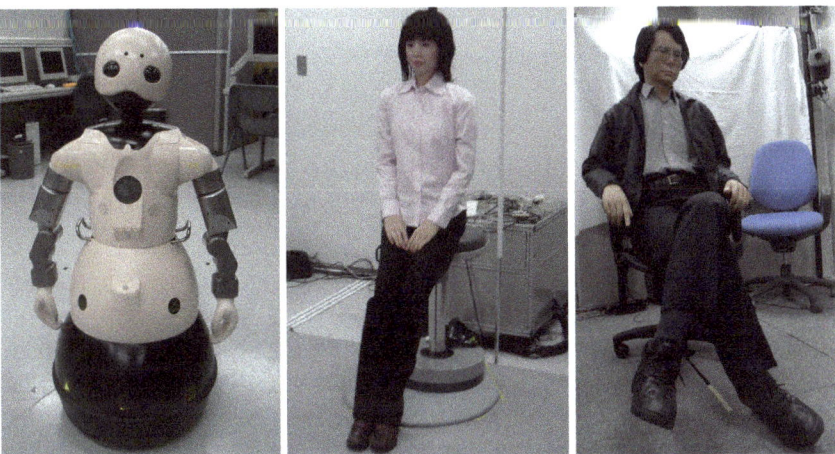

**Fig. 2.1** Three categories of humanlike robots: humanoid robot Robovie II (left: developed by ATR Intelligent Robotics and Communication Laboratories), android Repliee Q2 (middle: developed by Osaka University and Kokoro Corporation), geminoid HI-1 (right: developed by ATR Intelligent Robotics and Communication Laboratories)

to explore human nature. Here, androids serve as a platform to directly exchange insights from the two domains. To proceed with this new framework, several androids have been developed. The development of android systems and several results is described. However, we encounter serious issues that sparked the development of a new category of robot called *geminoid*. The concept and development of the first geminoid prototype are described. Preliminary findings to date and future directions of study with geminoids are also discussed.

## 2.2  Android Science

Current robotics research uses various findings from the field of cognitive science, especially in the area of human–robot interaction, in an attempt to adopt findings from human–human interactions to make robots that people can easily communicate with. At the same time, cognitive science researchers have also begun to utilize robots. As research fields extend to more complex, higher-level human functions such as seeking the neural basis of social skills [9], robots will be expected to function as easily controlled devices with communicative ability. However, the contribution from robotics to cognitive science has not been adequate, because the appearance and behavior of current robots cannot be handled separately. As traditional robots look quite mechanical and very different from human beings, the effect of their appearance may be too strong to ignore. As a result, researchers cannot clarify whether a specific finding reflects the robot's appearance, its movement, or a combination of the two.

**Fig. 2.2** Framework of
*Android Science*

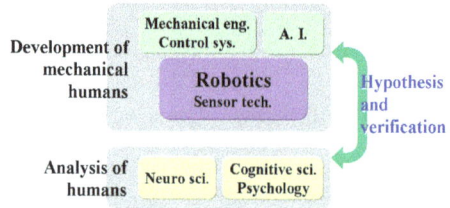

We expect to solve this problem using an android whose appearance and behavior closely resembles that of a human. The same thing is also an issue in robotics research, as it is difficult to clearly distinguish whether the cues pertain solely to the robot's behavior. An objective, quantitative means of measuring the effect of appearance is required.

Androids are robots whose behavior and appearance are highly anthropomorphized. Developing androids requires contributions from both robotics and cognitive science. To realize a more humanlike android, knowledge from human sciences is also necessary. At the same time, cognitive science researchers can exploit androids to verify hypotheses regarding human nature. This new, bidirectional, interdisciplinary research framework is called *android science* [10]. Under this framework, androids enable us to directly share knowledge between the development of androids in engineering and the understanding of humans in cognitive science (Fig. 2.2).

The major robotics issue in constructing androids is the development of humanlike appearance, movements, and perception functions. A further issue in cognitive science is "conscious and unconscious recognition." The goal of android science is to realize a humanlike robot and identify the essential factors for representing human likeness. How can we define human likeness? Further, how do we perceive human likeness? It is common knowledge that humans have conscious and unconscious recognition. When we observe objects, various modules are activated in our brain. Each of them matches the input sensory data with human models to affect reactions. A typical example occurs when, even if we recognize a robot as an android, we react to it as a human. This issue is fundamental both for engineering and scientific approaches. It will be an evaluation criterion in android development and will provide cues for understanding the human brain's mechanism of recognition.

To date, several androids have been developed. Repliee Q2, the latest android [10], is shown in the middle panel of Fig. 2.1. Forty-two pneumatic actuators are embedded in the android's upper torso, allowing it to move smoothly and quietly. Tactile sensors, which are also embedded under its skin, are connected to sensors in its environment, such as omnidirectional cameras, microphone arrays, and floor sensors. Using these sensory inputs, the autonomous program installed in the android can make smooth, natural interactions with nearby people.

Even though current androids have enabled us to conduct a variety of cognitive experiments, they are still quite limited. The bottleneck in terms of interaction with humans is the lack of ability to conduct a long term conversation. Unfortunately, because current artificial intelligence (AI) technology for developing humanlike brains is limited, we cannot expect humanlike conversation with robots. When meeting humanoid robots, people usually expect humanlike conversation. However, the technology lags way behind this expectation. AI progress takes time, and AI that can make humanlike conversation is our final goal in robotics. To arrive at this final goal, we need to use currently available technologies and gain a deeper understanding of what it is to be a human. Our solution for this problem is to integrate android and teleoperation technologies.

## 2.3   Developing Androids

Up to now, several androids have been developed. Figure 2.3 shows Repliee R1, the first android prototype, and Repliee Q2, the latest android [10]. As stated above, engineering issues in creating androids involve the development of humanlike appearance, movements, and perception. Here, we describe our approach to resolving each of these issues.

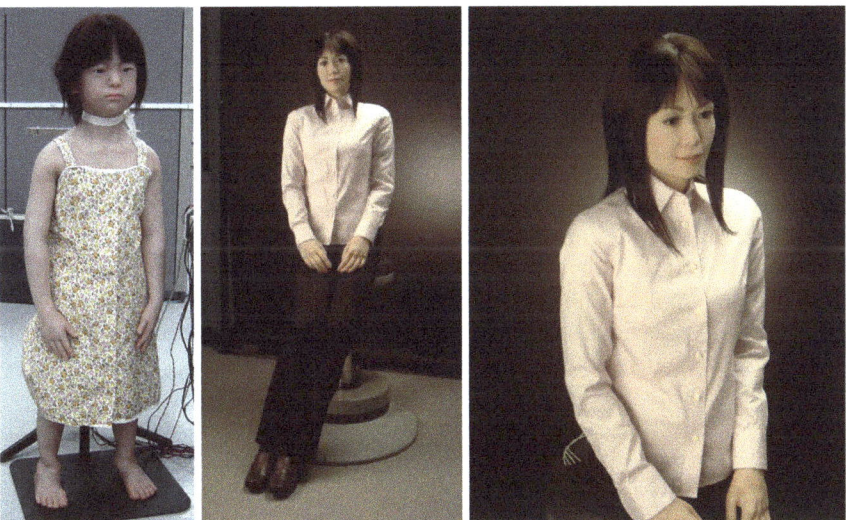

**Fig. 2.3** First android, Repliee R1 (left; developed by Osaka University), and the latest android, Repliee Q2 (right; developed by Osaka University and Kokoro Corporation)

## 2.3.1  Humanlike Appearance

The main difference between conventional robots and androids is in their appearance. To create a very humanlike robot, we began by copying the surface of the human skin.

First, body part molds were made from a real human using the shape-memory foam used by dentists. Then, plaster human-part models were made from the molds. A full-body model was constructed by connecting these plaster models. Again, a mold for the full-body model was made from the plaster model, and a clay model was made using the mold. Professionals in formative art modified the clay model to recover the details of the skin's texture. The human model loses its form in the first molding process, because human skin is soft. After that modification, a plaster full-body mold was made from the modified clay model, and then a silicon full-body model was made from that plaster mold. This silicon model is maintained as a master model.

Using this master model, silicon skin is made for the entire body. The thickness of the silicon skin in our current version is 5 mm. The mechanical parts, motors, and sensors are covered with polyurethane and the silicon skin. As shown in Fig. 2.3, the details are so finely represented that they cannot be distinguished from those of human beings in photographs.

Our current technology for replicating the human figure as an android has reached a fine degree of reality. It is, however, still not perfect. One issue is the detail of the wetness of the eyes. The eyes are the body part to which human observers are most sensitive. When confronted with a human face, a person first looks at the eyes. Although the android has eye-related mechanisms, such as blinking or making saccade movements, and the eyeballs are near-perfect copies of those of a human, we can still notice differences from those of a real human. Actually, making the wet surface of the eye and replicating the outer corners using silicone are difficult tasks, so further improvements are needed for this part.

Other issues are the flexibility and robustness of the skin material. The silicone used in the current manufacturing process is sufficient for representing the texture of the skin; however, it loses flexibility after one or two years, and its elasticity is insufficient for adapting to large joint movements.

## 2.3.2  Humanlike Movements

Very humanlike movement is another important factor in developing androids. Even if androids look indistinguishable from humans as static figures, without appropriate movements, they can be easily identified as artificial.

To achieve highly humanlike movement, we found that a child android was too small to embed the required number of actuators, which led to the development of an adult android. The right half of Fig. 2.3 shows our latest adult android.

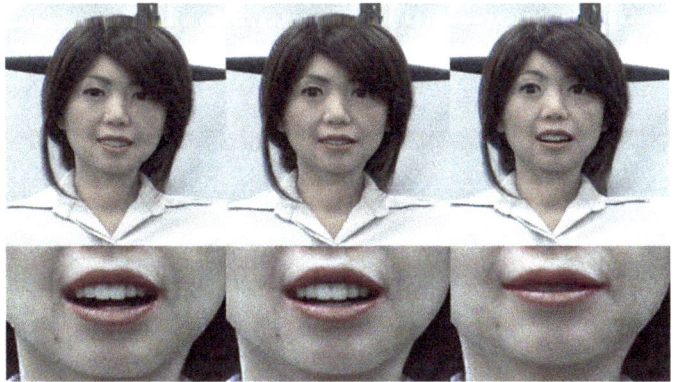

**Fig. 2.4**  Facial expressions generated by android Repliee Q2

This android, named Repliee Q2, contains 42 pneumatic (air) actuators in the upper torso. The positions of the actuators were determined by analyzing real human movements using a precise 3D motion tracker. With these actuators, both unconscious movements (such as breathing in the chest) and conscious large movements (such as head or arm movements) can be generated. Furthermore, the android is able to generate the facial expressions that are important for interacting with humans. Figure 2.4 shows some of the facial expressions generated by the android. To generate a smooth, humanlike expression, 13 of the 42 actuators are embedded in the head.

We decided to use pneumatic actuators for the androids instead of the DC motors used in most robots. The use of pneumatic actuators provides several benefits. First, they are very quiet, much closer to human-produced sound. DC servomotors require reduction gears, which generate non-humanlike noise that is very robotlike. Second, the reaction of the android to external force becomes very natural with pneumatic dampers. If we use DC servomotors with reduction gears, sophisticated compliance control is required to obtain the same effect. This is also important for ensuring safety in interactions with the android.

The disadvantage of pneumatic actuators is that they require a large and powerful air compressor. This requirement means that the current android cannot walk. For wider applicability, we need to develop new electric actuators that have similar specs to the pneumatic actuators.

The next issue is how to control the 42 air servo actuators used to achieve very humanlike movements. The simplest approach is to directly send angular information to each joint. However, as the number of actuators in the android is relatively large, this takes a long time. Another difficulty is that the skin movement does not simply correspond to the joint movement. For example, the android has more than five actuators around the shoulder for generating humanlike shoulder movements, with the skin moving and stretching according to the actuator motions. Already, we have developed methods such as using Perlin noise [11] to generate

**Fig. 2.5** Replicating human
motions with the android

smooth movements or employing a neural network to obtain the mapping between
the skin surface and actuator movements. There still remain some issues, such as
the limited speed of android movement due to the nature of the pneumatic dampers.
To achieve quicker and more humanlike behavior, speed and torque controls must
be designed in future studies.

After obtaining an efficient method for controlling the android, the next step is
the implementation of humanlike motions. A straightforward approach to this
challenge is to imitate real human motions in synchronization with the android's
master. By attaching 3D motion tracker markers on both the android and the master,
the android can automatically follow the motions of a human (Fig. 2.5).

This work is still in progress, but interesting issues have arisen with respect to
this kind of imitation learning. Imitation by the android means the representation of
complicated human shapes and motions in the parameter space of the actuators.
Although the android has a relatively large number of actuators compared to other
robots, this is still far fewer than in humans. Thus, the effect of data-size reduction
is significant. By carefully examining this parameter space and mapping, we may
find important properties of human body movements. More concretely, we expect
to develop a hierarchical representation of human body movements that consist of
two or more layers, such as small unconscious movements and large conscious
movements. With this hierarchical representation, we can expect to achieve more
flexibility in android behavior control.

### 2.3.3  Humanlike Perception

Androids require humanlike perceptual abilities in addition to humanlike appearance and movements. This problem has been tackled in the fields of computer
vision and pattern recognition under rather controlled environments. However, the
problem becomes extremely difficult when applied to robots in real-world situations, where vision and audition become unstable and noisy.

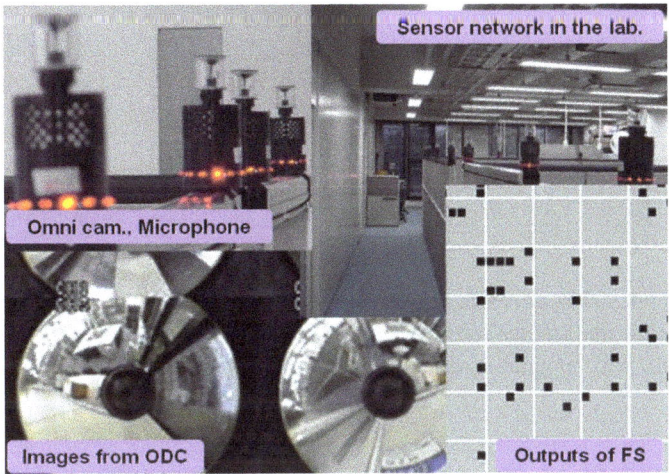

**Fig. 2.6** Distributed sensor system

Ubiquitous/distributed sensor systems can be used to solve this problem. The idea is to recognize the environment and human activities using many distributed cameras, microphones, infrared motion sensors, floor sensors, and ID tag readers in the environment (Fig. 2.6).

We have developed distributed vision systems [12] and distributed audition systems [13] in our previous work. To solve the present problem, these developments must be integrated and extended. Omnidirectional cameras observe humans from multiple viewing points and robustly recognize their behavior [14]. The microphones catch the human voice by forming virtual sound beams. The floor sensors, which cover the entire space, reliably detect human footprints.

The only sensors that should be installed on the robot are skin sensors. Soft and sensitive skin sensors are important, particularly for interactive robots. However, there has not been much work in this area in previous robotics research. We are now focusing on this important issue by developing original sensors. Our sensors are made by combining silicone skin and Piezo films (Fig. 2.7). This sensor detects pressure through the bending of the Piezo films. Furthermore, with increased sensitivity, it can detect the presence of humans very nearby from static electricity. That is, it can perceive that a human being is in the vicinity.

These technologies for very humanlike appearance, behavior, and perception enable us to develop feasible androids. These androids have undergone various cognitive tests, but this work is still limited. The bottleneck is long-term conversation during interactions with real humans. Unfortunately, current AI technology for developing humanlike brains has only limited ability, and thus, we cannot expect humanlike conversation with robots. When we meet humanoid robots, we usually expect to have a humanlike conversation. However, the technology is very far behind this expectation. Progress in AI takes time, and this is actually our final

**Fig. 2.7** Skin sensor

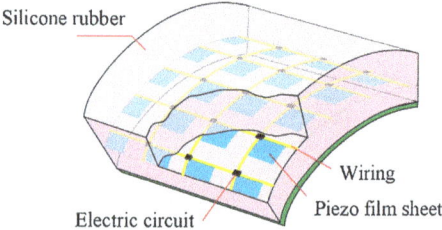

Silicone rubber

Wiring

Piezo film sheet

Electric circuit

goal in robotics. To arrive at this final goal, we need to use the technologies available today and, moreover, truly understand what a human is. Our solution to this problem is to integrate android and teleoperation technologies.

## 2.4 Geminoid

We have developed *Geminoid,* a new category of robot, to overcome the bottleneck issue. We coined the term "geminoid" from the Latin "geminus," meaning "twin" or "double," and added "oid," which indicates "similarity" or being a twin. As the name suggests, a geminoid is a robot that will work as a duplicate of an existing person. It appears and behaves as a person and is connected to that person by a computer network. Geminoids extend the applicable field of android science. Androids are designed for studying human nature in general. With geminoids, we can study such personal aspects as presence or personality traits, tracing their origins and implementation into robots. Figure 2.8 shows the robotic part of HI-1, the first geminoid prototype. Geminoids have the following capabilities:

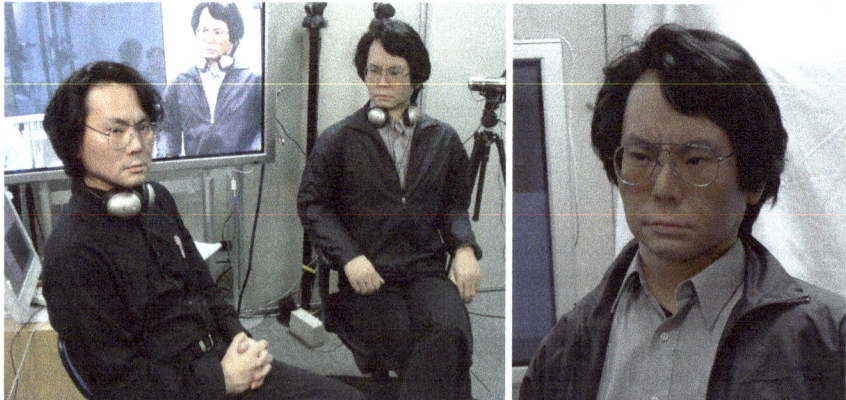

**Fig. 2.8** Geminoid HI-1

**Appearance and behavior highly similar to an existing person**

The appearance of a geminoid is based on an existing person and does not depend on the imagination of designers. Its movements can be made or evaluated simply by referring to the original person. The existence of a real person analogous to the robot enables easy comparison studies. Moreover, if a researcher is used as the original, we can expect that individual to offer meaningful insights into the experiments, which are especially important at the very first stage of a new field of study when beginning from established research methodologies.

**Teleoperation (remote control)**

Because geminoids are equipped with teleoperation functionality, they are not only driven by an autonomous program. By introducing manual control, the limitations in current AI technologies can be avoided, enabling long-term, intelligent conversational human–robot interaction experiments. This feature also enables various studies on human characteristics by separating "body" and "mind." In geminoids, the operator (mind) can be easily exchanged while the robot (body) remains the same. Additionally, the strength of connection, or what kind of information is transmitted between the body and mind, can be easily reconfigured. This is especially important when taking a top-down approach that adds/deletes elements from a person to discover the "critical" elements that comprise human characteristics. Before geminoids, this was impossible.

## 2.4.1  System Overview

The current geminoid prototype, HI-1, consists of three main elements: a robot, a central controlling server (geminoid server), and a teleoperation interface (Fig. 2.9).

**A robot that resembles a living person**

The robotic element has an essentially identical structure to that of previous androids [10]. However, there has been considerable effort to make a robot that appears to not only resemble a living person, but to be a copy of the original person.

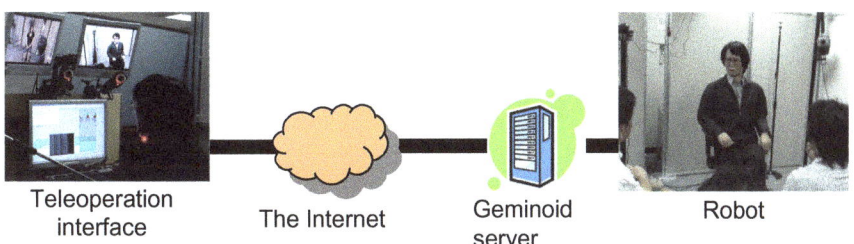

Teleoperation interface        The Internet        Geminoid server        Robot

**Fig. 2.9** Overview of geminoid system

Silicone skin was molded by a cast taken from the original person; shape adjustments and skin textures were painted manually based on MRI scans and photographs. Fifty pneumatic actuators drive the robot to generate smooth and quiet movements, which are important attributes when interacting with humans. The allocation of actuators was determined such that the resulting robot can effectively show the necessary movements for human interaction and simultaneously express the original person's personality traits. Among the 50 actuators, 13 are embedded in the face, 15 in the torso, and the remaining 22 move the arms and legs. The softness of the silicone skin and the compliant nature of the pneumatic actuators also provide safety when interacting with humans. As this prototype was intended for interaction experiments, it lacks the capability to walk around; it always remains seated. Figure 2.8 shows the resulting robot (right) alongside the original person, Dr. Ishiguro (author).

**Teleoperation interface**

Figure 2.10 shows the teleoperation interface prototype. Two monitors show the controlled robot and its surroundings, and microphones and headphones are used to capture and transmit utterances. The captured sounds are encoded and transmitted to the geminoid server by IP links from the interface to the robot and vice versa. The operator's lip corner positions are measured by an infrared motion capture system in real time, converted to motion commands, and sent to the geminoid server by the network. This enables the operator to implicitly generate suitable lip movement on the robot while speaking. However, compared to the large number of human facial muscles used for speech, the current robot only has a limited number of actuators in its face. In addition, the response speed is much slower than that of a human, partially due to the nature of the pneumatic actuators. Thus, simple transmission and playback of the operator's lip movement would not result in sufficient, natural robot motion. To overcome this issue, the measured lip movements are

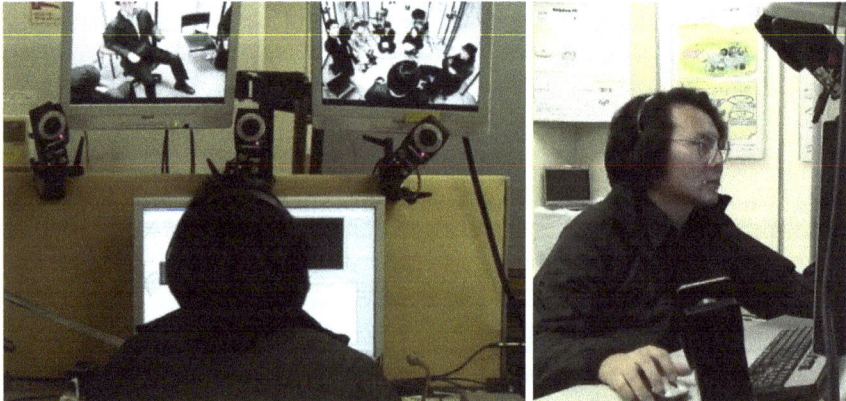

**Fig. 2.10** Teleoperation interface

currently transformed into control commands using heuristics obtained by observing the original person's actual lip movement.

The operator can also explicitly send commands to control the robot's behavior using a simple graphical user interface. Several selected movements such as nodding, opposing, or staring in a certain direction can be specified by a single mouse click. This relatively simple interface was prepared because the robot has 50 degrees of freedom, making it one of the world's most complex robots and basically impossible to manipulate manually in real time. A simple, intuitive interface is necessary so that the operator can concentrate on interaction and not on robot manipulation. Despite its simplicity, by cooperating with the geminoid server, this interface enables the operator to generate natural humanlike motions in the robot.

**Geminoid server**

The geminoid server receives robot control commands and sound data from the remote controlling interface, adjusts and merges the inputs, and sends and receives primitive controlling commands between the robot hardware. Figure 2.11 shows the data flow in the geminoid system. The geminoid server also maintains the state of human–robot interaction and generates *autonomous* or *unconscious* movements for the robot. As described above, as the features of the robot become more humanlike, its behavior should also become suitably sophisticated to retain a "natural" look [15]. One thing that can be seen in every human being, and that most robots lack, is the slight body movements caused by an autonomous system, such as breathing or blinking. To increase the robot's naturalness, the geminoid server emulates the human autonomous system and automatically generates these micro-movements depending on the interaction state at each point in time. When the robot is "speaking," it shows different micro-movements than when "listening." Such automatic robot motions, generated without the operator's explicit orders, are merged and adjusted with *conscious* operation commands from the teleoperation interface (Fig. 2.11). Simultaneously, the geminoid server applies a specific delay to the transmitted sounds, taking into account the transmission delay/jitter and the start-up delay of the pneumatic actuators. This adjustment serves to synchronize lip movements and speech, thus enhancing the naturalness of geminoid movement.

### 2.4.2   Experiences with the Geminoid Prototype

The first geminoid prototype, HI-1, was completed and presented to the press in July 2006. Since then, numerous operations have been held, including interactions with laboratory members and experiment subjects. Additionally, geminoid was demonstrated to a number of visitors and reporters. During these operations, we encountered several interesting phenomena. Here are some observations made by the geminoid operator:

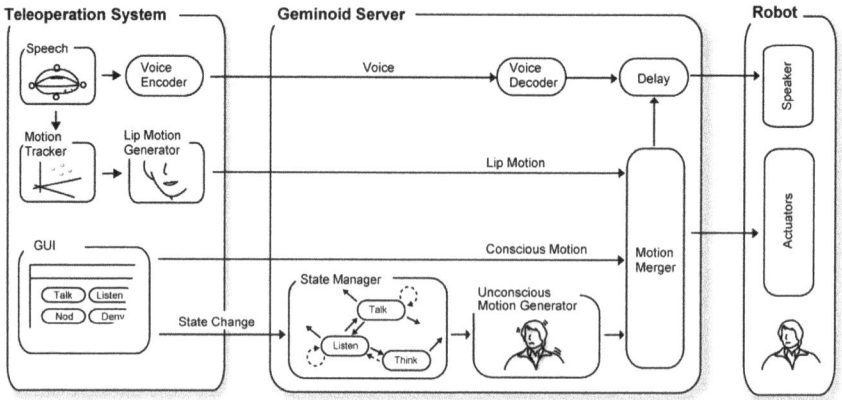

**Fig. 2.11** Data flow in the geminoid system

- When I (Dr. Ishiguro, the origin of the geminoid prototype) first saw HI-1 sitting still, it was like looking in a mirror. However, when it began moving, it looked like somebody else, and I couldn't recognize it as myself. This was strange, since we copied my movements to HI-1, and others who know me well say the robot accurately shows my characteristics. This means that we are not objectively recognizing our unconscious movements ourselves.
- While operating HI-1 with the operation interface, I find myself unconsciously adapting my movements to the geminoid movements. The current geminoid cannot move as freely as I can. I felt that not just the geminoid, but my own body is restricted to the movements that HI-1 can make.
- In less than five minutes, both the visitors and I can quickly adapt to conversation through the geminoid. The visitors recognize and accept the geminoid as me while talking to each other.
- When a visitor pokes HI-1, especially around its face, I get a strong feeling of being poked myself. This is strange, as the system currently provides no tactile feedback. Just by watching the monitors and interacting with visitors, I get this feeling.

We also asked the visitors how they felt when interacting through the geminoid. Most said that when they saw HI-1 for the very first time, they thought that somebody (or Dr. Ishiguro, if familiar with him) was waiting there. After taking a closer look, they soon realized that HI-1 was a robot and began to have some weird and nervous feelings. However, shortly after having a conversation through the geminoid, they found themselves concentrating on the interaction, and soon the strange feelings vanished. Most of the visitors were non-researchers unfamiliar with robots of any kind.

Does this mean that the geminoid has overcome the "uncanny valley" effect? Before talking through the geminoid, the initial response of the visitors seems to resemble the reactions seen with previous androids: Even though they could not immediately recognize the androids as artificial, they were nevertheless nervous about being with the androids. Is intelligence or long-term interaction a crucial factor in overcoming the valley and arriving at an area of natural humanness?

We certainly need objective means to measure how people feel about geminoids and other types of robots. In a previous android study, Minato et al. found that gaze fixation revealed criteria about the naturalness of robots [15]. Recent studies have shown different human responses and reactions to natural or artificial stimuli of the same nature. Perani et al. showed that different brain regions are activated while watching human or computer graphic arm movements [7]. Kilner et al. showed that body movement entrainment occurs when watching human motions, but not with robot motions [16]. By examining these findings with geminoids, we may be able to find some concrete measurements of human likeliness and approach the "appearance versus behavior" issue.

Perhaps HI-1 was recognized as a sort of communication device, similar to a telephone or a videophone. Recent studies have suggested a distinction in the brain process that discriminates between people appearing in videos and people appearing live [17]. While attending TV conferences or talking by cellular phones, however, we often experience the feeling that something is missing from a face-to-face meeting. What is missing here? Is there an objective means to measure and capture this element? Can we ever implement this in robots?

## 2.5 Summary and Further Issues

In developing the geminoid, our purpose was to study *Sonzai-Kan*, or human presence, by extending the framework of android science. The scientific aspect must answer questions about how humans recognize human existence/presence. The technological aspect must realize a teleoperated android that works on behalf of the person remotely accessing it. This will be one of the practical networked robots realized by integrating robots with the Internet.

The following summarizes our current challenges:

**Teleoperation technologies for complex humanlike robots**

Methods must be studied to teleoperate the geminoid so as to convey existence/presence, which is much more complex than traditional teleoperation for mobile and industrial robots. We are studying a method to autonomously control an android by transferring the motions of the operator measured by a motion capture system. We are also developing methods to autonomously control eye-gaze and humanlike small and large movements.

**Synchronization between speech utterances sent by the teleoperation system and body movements**

The most important technology for the teleoperation system is synchronization between speech utterances and lip movements. We are investigating how to produce natural behavior during speech utterances. This problem extends to other modalities, such as head and arm movements. Further, we are studying the effects of nonverbal communication by investigating not only the synchronization of speech and lip movements, but also facial expressions, head, and even whole body movements.

**Psychological test for human existence/presence**

We are studying the effect of transmitting Sonzai-Kan from remote places, such as meeting participation when the person themself cannot attend. Moreover, we are interested in studying existence/presence through cognitive and psychological experiments. For example, we are investigating whether the android can represent the authority of the person himself by comparing the person and the android.

**Application**

Although being developed as a research apparatus, the nature of geminoids may allow us to extend the use of robots in the real world. The teleoperated, semi-autonomous facility of geminoids allows them to be used as substitutes for clerks, for example, that can be controlled by human operators only when non-typical responses are required. In most cases, an autonomous AI response will be sufficient, so a few operators will be able to control hundreds of geminoids. Additionally, because their appearance and behavior closely resembles that of humans, geminoids could be the ultimate interface devices of the future.

**Acknowledgements** The establishment of android science as a new interdisciplinary framework has been supported by many researchers. Prof. Shoji Itakura, Kyoto University, is one of the closest collaborative researchers. He designed several android experiments from the perspective of cognitive science. Prof. Kazuo Hiraki, Tokyo University, is another key collaborator. He gave us many suggestions that originate in brain science and cognitive science. The authors appreciate their helpful support.

Dr. Takahiro Miyashita, ATR Intelligent Robotics and Communication Laboratories, initiated the android project with the authors and developed the skin sensors used for the androids and geminoid. Prof. Takashi Minato, Osaka University, developed software for controlling the androids and coordinated several cognitive experiments. Dr. Takayuki Kanda, ATR Intelligent Robotics and Communication Laboratories, collaborated with the authors in developing methods of evaluating human–robot interactions. Dr. Carlos Toshinori Ishii and Mr. Daisuke Sakamoto developed the geminoid server and helped us conduct studies with the geminoid.

Finally, the author appreciates the collaboration and support of Kokoro Co., Ltd. The adult androids of the Repliee Q series have been developed in a collaborative project with this company.

# References

1. Ishiguro, H., and S. Nishio. 2007. Building artificial humans to understand humans. *Journal of Artificial Organs* 10 (3): 133–142.
2. Pascalis, O., M. Haan, and C.A. Nelson. 2002. Is face processing species-specific during the first year of life? *Science* 296: 1321–1323.
3. Reeves, B., and C. Nass. 1996. *The media equation.* CSLI: Cambridge University Press.

4. Fong, T., I. Nourbakhsh, and K. Dautenhahn. 2003. A survey of socially interactive robots. *Robotics and Autonomous Systems* 42: 143–166.
5. Breazeal, C. 2004. Social interactions in HRI: The robot view. *IEEE Transactions on Man, Cybernetics and Systems: Part C* 34: 181–186.
6. Kanda, T., H. Ishiguro, M. Imai, and T. Ono. 2004. Development and evaluation of interactive humanoid robots. *Proceedings of the IEEE*, 1839–1850.
7. Perani, D., F. Fazio, N.A. Borghese, M. Tettamanti, S. Ferrari, J. Decety, and M.C. Gilardi. 2001. Different brain correlates for watching real and virtual hand actions. *NeuroImage* 14: 749–758.
8. Han, S., Y. Jiang, G.W. Humphreys, T. Zhou, and P. Cai. 2005. Distinct neural substrates for the perception of real and virtual visual worlds. *NeuroImage* 24: 928–935.
9. Blakemore, S.J., and U. Frith. 2004. How does the brain deal with the social world? *NeuroReport* 15: 119–128.
10. Ishiguro, H. 2005. Android science: Toward a new cross-disciplinary framework. In *Proceedings of toward social mechanisms of android science: A CogSci 2005 workshop*, 1–6.
11. Perlin, K. 1995. Real time responsive animation with personality. *IEEE Transactions on Visualization and Computer Graphics* 1: 5–15.
12. Ishiguro, H. 1997. Distributed vision system: A perceptual information infrastructure for robot navigation. In *Proceedings of international joint conference on artificial intelligence (IJCAI)*, 36–41.
13. Ikeda, T., T. Ishida., and H. Ishiguro. 2004. Framework of distributed audition. In *Proceedings of 13th IEEE international workshop on robot and human interactive communication (RO-MAN)*, 77–82.
14. Ishiguro, H. and T. Nishimura. 2001. VAMBAM: View and motion based aspect models for distributed omnidirectional vision systems. In *Proceedings of international joint conference on artificial intelligence (IJCAI)*, 1375–1380.
15. Minato, T., M. Shimada, S. Itakura, K. Lee, and H. Ishiguro. 2006. Evaluating the human likeness of an android by comparing gaze behaviors elicited by the android and a person. *Advanced Robotics* 20: 1147–1163.
16. Kilner, J.M., Y. Paulignan, and S.J. Blakemore. 2003. An interference effect of observed biological movement on action. *Current Biology* 13: 522–525.
17. Kuhl, P.K., F.M. Tsao, and H.M. Liu. 2003. Foreign-language experience in infancy: Effects of short-term exposure and social interaction on phonetic learning. *Proceedings of the National Academy of Sciences* 100: 9096–9101.

# Chapter 3
# Androids as a Telecommunication Medium with a Humanlike Presence

**Daisuke Sakamoto, Takayuki Kanda, Tetsuo Ono, Hiroshi Ishiguro and Norihiro Hagita**

**Abstract** In this study, we realize human telepresence by developing a remote-controlled android system called Geminoid HI-1. Experimental results confirm that participants feel a stronger presence of the operator when he talks through the android than when he appears on a video monitor in a video conference system. In addition, participants talk with the robot naturally and evaluate its humanlike-ness as equal to a man on a video monitor. We also discuss a remote-controlled system for telepresence that uses a humanlike android robot as a new telecommunication medium.

**Keywords** Android science · Humanoid robot · Telepresence Telecommunication

## 3.1 Introduction

Recently, many humanoid robots have been developed. For instance, Honda developed ASIMO, a famous humanoid robot that can walk on its biped legs. Breazeal et al. developed a face robot that can express facial emotions for

---

This chapter is a modified version of a previously published paper [1], edited to be comprehensive and fit with the context of this book.

---

D. Sakamoto (✉) · T. Kanda · T. Ono · H. Ishiguro · N. Hagita
ATR Intelligent Robotics Laboratories, 2-2-2 Hikaridai Seika-Cho Soraku-Gun,
Kyoto 619-0288, Japan
e-mail: d.sakamoto@acm.org

D. Sakamoto · T. Ono
Future University-Hakodate, 116-2 Kamedanakano-Cho, Hakodate, Hokkaido, Japan

H. Ishiguro
Osaka University, 1-1 Yamadaoka, Suita, Osaka 565-0871, Japan

© Springer Nature Singapore Pte Ltd. 2018
H. Ishiguro and F. Dalla Libera (eds.), *Geminoid Studies*,
https://doi.org/10.1007/978-981-10-8702-8_3

interacting with people [2], and a very humanlike robot, or android, has been developed [3]. Such humanlike body properties in robots can be used for natural human–robot interaction [4–7].

We believe that a humanoid robot, particularly an android, can also be used as a telecommunication medium for distant inter-human communication. Previous media, such as video conference systems, have problems with effective presence; people do not feel they are sharing a physical space [8], making it hard to identify eye gaze [9] and so forth.

In telepresence research, virtual reality techniques are often used; however, the advantages of robots as computer-graphic agents have been demonstrated. Kidd and Breazeal compared a robot and a computer-graphic agent and found that the robot was more suitable for communication regarding real-world objects [10]. Shinozawa et al. also argued that a robot is more appropriate for communication referring to real objects than a computer-graphic agent [11]. As we are interested in presence in the real world, we focus on an approach based on real robots.

Several telepresence studies are based on real robots. For instance, Sekiguchi et al. developed a telephone system in which two stuffed-animal-type robots were placed at either end and synchronized in motion so that people could exchange feelings of movement [12]. Tadakuma et al. used a humanoid robot whose operator was projected on its face [13]. However, this conveyed human presence is not yet as realistic as actual humans.

In this study, we utilize an android that resembles a human for distant inter-human communication. This paper reports an experiment to reveal how an android can convey human presence better than a speakerphone and a video.

## 3.2 Android Telecommunication System

We developed a remote-controlled android system that uses a very humanlike robot called Geminoid HI-1 (a copy of co-author Hiroshi Ishiguro).

As we focus on the realization of telepresence and communication using a telecommunication system, our initial system was developed as a teleoperation system. Previous media such as telephones and video conference systems are easy to use, and so telepresence and communication operations must also be easy. To achieve this, we will develop the android as a semi-automatic controlled system.

In this section, we introduce Geminoid HI-1 and the telecommunication system.

### 3.2.1  Geminoid III-1

Since Geminoid HI-1 was developed to resemble a human as closely as possible, its appearance is actually humanlike. This android is 140 cm tall when sitting in a chair (it cannot stand) and has 50 degrees of freedom (DOFs). Its face has 13 DOFs, which enable it to form natural facial expressions.

Figure 3.1 shows the Geminoid HI-1 and its real human counterpart.

**Natural and Humanlike Motions**

A very humanlike appearance involves natural and humanlike motion development. If we ignore this, then the android will immediately give a poor impression. We intend to employ this robot as a telecommunication medium, so we must avoid such negative impressions to realize natural telecommunication.

When preparing its motions to be natural, we tried to make them similar to the original motions of the human on which the android is based. Of course, the definition of "naturalness" is vague; however, as the android's appearance is very similar to the original man, we believe that this criterion works well.

The system continuously plays motion files to control the android's body movements. Motion builders (engineers who develop the robot's motions) define

**Fig. 3.1** Actual living man (left) and his copy named Geminoid HI-1

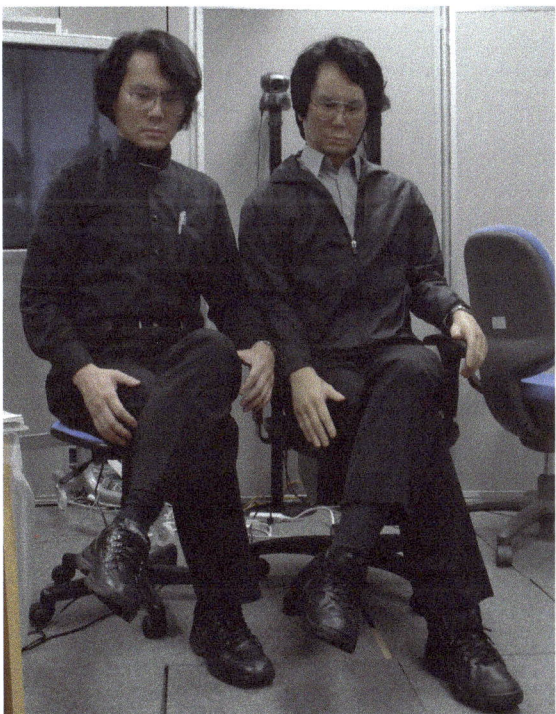

the motion files. A third person evaluates the developed motions to help the motion builders improve the naturalness of the motions.

## 3.2.2   Behavior Controller

There are two main difficulties in realizing natural behavior in the android during teleoperation: conscious and unconscious. For the conscious-level problem, the difficulty comes from the burden of excessive operations to control everything about the robot. Geminoid HI-1 has 50 actuators. It is hard for the operators to control all actuators directly. For this reason, we simplified the remote control by employing a semi-automatic system. The teleoperator simply switches the robot's behaviors.

Regarding unconscious-level difficulties, the problem concerns subtle expressions, which we generated by realizing unconscious behavior control. A humanlike robot such as Geminoid HI-1 can express delicate motions. However, if we do not design its motions expressively, humans will not react favorably to its appearance. This will obstruct the achievement of telepresence. The unconscious behavior controller automatically generates or embeds the robot's unconscious behaviors to realize natural motion that includes trepidation. A system overview is shown in Fig. 3.2.

**Fig. 3.2** System overview

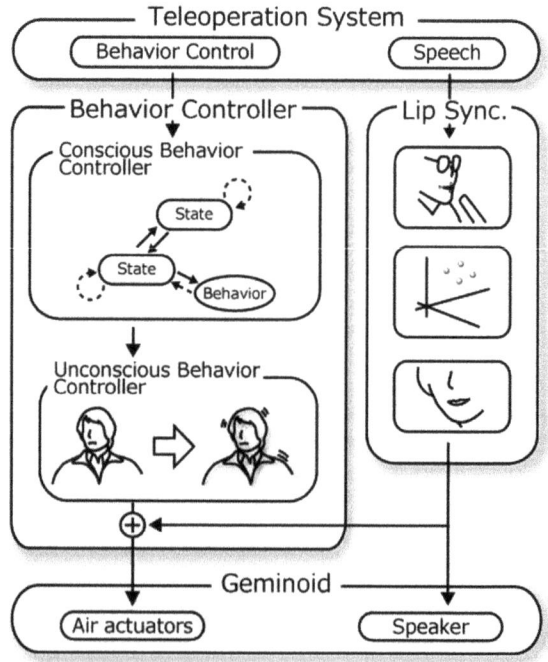

We present the details of the semi-automatic and remote controlled systems below.

### 3.2.2.1  Conscious Behavior Controller

This system has an internal "state" that can be automatically controlled. The state continues to randomly play individual motion files that define robot motion and realize automatic control. The present system has 20 unique motion files for each state. In addition, this system can play an individual motion file that is not in any state. In this case, the state returns to the last state when it has finished playing a file. The teleoperator controls the system that switches the states and plays individual behaviors.

**States**

The teleoperator can select from five defined states to match the situation:

1. Idle:
   The robot stares straight ahead, but slightly bends his head. He sometimes looks to the left or right.
2. Speaking:
   He stares straight ahead and sometime looks to the left or right. This behavior resembles a more active idle state.
3. Listening:
   This state is less active than the speaking state to project an image of listening.
4. Left-looking:
   This state looks to the left to make eye contact in that direction and resembles the speaking state.
5. Right-looking:
   This state looks to the right to make eye contact in that direction and resembles the speaking state.

**State Transition Example**

We now present a brief example of the state transition process. Figure 3.3 shows three states and one FILE-PLAYING state. The FILE-PLAYING state allows the android to speak particular sentences, such as greetings. When a command for a state transition is received from the teleoperator, the conscious behavior controller moves the state to that specified by the teleoperator. When the command to move to the FILE-PLAYING state with a motion filename is received, the conscious

**Fig. 3.3** Example of state
and behavior control

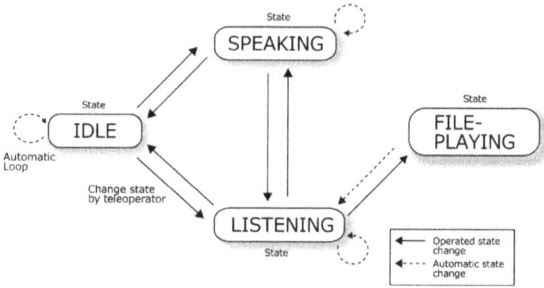

**Fig. 3.4** Behavior control
sequence

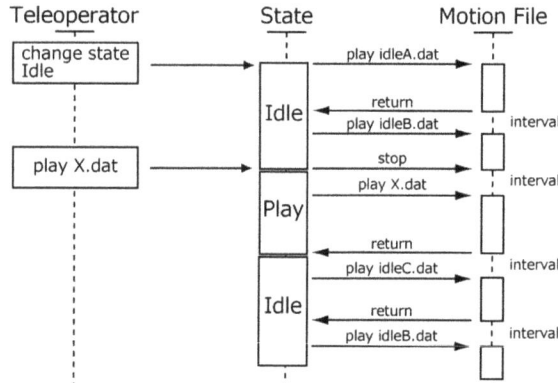

behavior controller plays the motion file specified by the teleoperator. When the motion has finished playing, the controller returns to the previous state.

Figure 3.4 shows the details of this flow, illustrating the sequence of state transitions and the treatment of motion files.

### 3.2.2.2 Unconscious Behavior Controller

Humans perform various unconscious behaviors such as breathing, blinking, and trembling. However, we do not notice most of them. Only when they are missing do we feel that something is wrong. Thus, unconscious behaviors must be expressed by androids for telepresence applications.

Our system treats this problem as unconscious behavior control. It adds subtle expressed motions to the original motions selected by a conscious behavior controller. In particular, when this system plays a motion file, the unconscious behavior controller adds breathing behavior to the original one. Currently, this controller has relatively few functions. However, we will improve it to realize humanlike motion in Geminoid HI-1.

### 3.2.3  Teleoperation System

The teleoperator remotely controls the android's behavior by choosing its state. We also prepared a function that synchronizes the lip motions between the teleoperator and the android. In this system, we employ Geminoid HI-1 as a telecommunication medium. It is important that the voice transfer is accompanied with mouth movement. If the speech and lip movements are not matched, people will not form a good impression of Geminoid HI-1's very humanlike mouth. Therefore, we control its mouth to mimic the teleoperator's mouth movements, which are measured by a motion capture system.

Figure 3.5 illustrates our remote-controlled system. Two monitors display the room condition, and one desktop computer is used by the teleoperator for remote control. In addition, we have five optical cameras for motion capture. In this way, the teleoperator remotely controls Geminoid HI-1 (Fig. 3.6).

The details of the remote-controlled system are explained below.

#### 3.2.3.1  Behavior Control

We developed a remote-controlled protocol to switch states and play individual motion files. In particular, we implemented a command system to quickly sort through different states and play motion files.

We briefly introduce the remote control commands.

**Change state <State name>**

This command is used to effect a state transition.

**Get state**

This command returns the current state of the robot.

**Fig. 3.5** Our remote-controlled system

**Fig. 3.6** View of remote
operation system

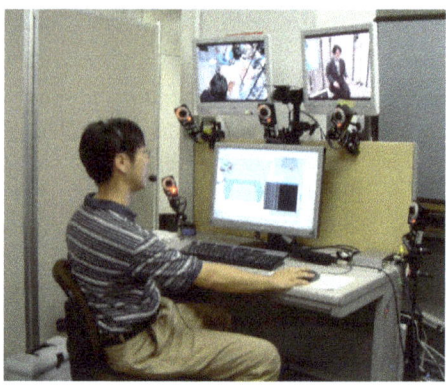

**Fig. 3.7** Geminoid controller
interface

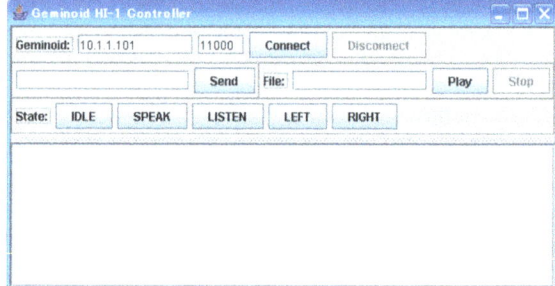

**Play motion <Filename>**

This command plays an individual motion file.

**Stop motion**

This command stops the current motion file.

Operators can easily realize complex robot behavior by combining these commands within an appropriate interface (Fig. 3.7).

### 3.2.3.2 Speech and Lip Synchronization

The remote-controlled system is not only controlled by commands. Because motion files are prepared in advance, it is impossible to synchronize the lip motion of the teleoperator with that of the robot only using motion files. Thus, we decided to use a motion capture system to synchronize the lip movements and speech output from the robot's speaker. The motion capture system, used to measure the teleoperator's lip movements, has five pairs of infrared cameras and markers that reflect infrared signals. These cameras were set around the desk. As shown in Fig. 3.8, we attached four markers around the teleoperator's mouth. The android expresses the same lip movements as those measured by the motion capture system for the teleoperator.

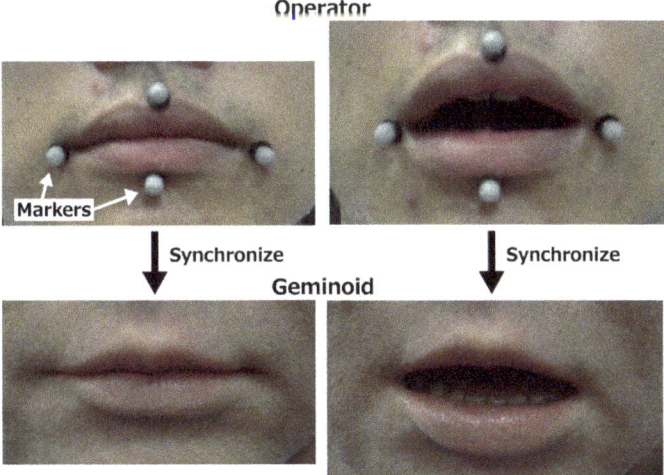

**Fig. 3.8** Operator lip synchronization with Geminoid HI-1

## 3.3 Experiments

### 3.3.1 Outline

Experiments were conducted to verify the usefulness of the android as a telecommunication medium. Our hypothesis argues that, in a sophisticated android robot, we can realize a humanlike presence that represents a remote person's presence. Thus, we are particularly interested in measuring presence, humanlike-ness, and naturalness. Presence is expected to be the android's "strong suit" as a telecommunication medium because of its physical existence. Moreover, we believe that it can preserve the humanlike-ness and naturalness of communication media such as telephones and teleconference systems.

In the experiments, an operator participated in a three-person conversation with the android. For comparison, they also conversed by video conference system and speakerphone. During the conversation session, the two participants mainly talked in the presence of the operator after he initiated the discussion. After the conversation, the participants evaluated the medium.

### 3.3.2  Method

#### 3.3.2.1  Participants

Thirty-four university students were recruited and paired to participate in experiments that featured a within-subject design, which means that each paired participant took part in three sessions for all conditions. The experiment was performed with a counterbalanced design. Participants were assigned completely at random.

#### 3.3.2.2  Conditions

There were three conditions:

**G condition**

The operator talked with participants through Geminoid HI-1 and controlled its motion to express nodding and looking behaviors (Fig. 3.9, top).

**V condition**

The operator talked with participants through a video conference system. Similar to the G condition, the operator only nodded and looked at the participants on his left and right without performing such behaviors as smiling, moving his hands, or

**Fig. 3.9** Eye contact with Geminoid HI-1 and teleoperator

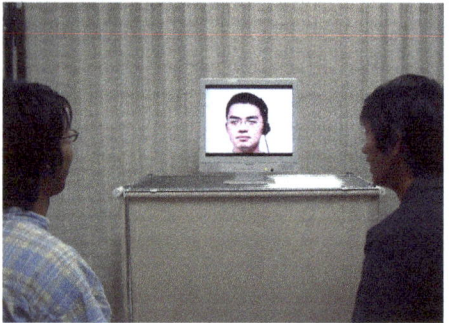

shaking his head. The direction of gaze was calibrated in advance so that partici
pants felt that the person was looking at them (Fig. 3.9, bottom).

**S condition**

The operator talked with participants through a speakerphone.

### 3.3.2.3  Environment

The experiment was conducted in a 3 m × 3 m room. Figure 3.10 shows the
setting of the experimental environments. The figure on the left shows the setting for
the G condition. Chairs for the participants and Geminoid HI-1 were placed in a 1-m
radius circle. In the V condition, we placed a monitor displaying the operator's face
at the same height as Geminoid HI-1's head (Fig. 3.10, center). In the S condition,
only the cameras and chairs were left in the room. In all conditions, cameras were
set up to allow the operator to look at the participants. In the V and S conditions, a
speaker was placed on the back wall. In the G condition, a speaker was placed on
the back of Geminoid HI-1. In all conditions, the participants could not see the
speakers. We coordinated the speaker volume to be the same in each condition.

### 3.3.2.4  Procedure

The experiment sessions proceeded as follows:

1.  An experimenter informed the paired participants that they must discuss a theme
    given remotely by a person who will participate in the conversation using some
    "equipment."
2.  They were led into the experimental room, where one of the three conditions
    was prepared (Fig. 3.10).
3.  After sitting down, the operator (not the experimenter) gave them the discussion
    theme, which they began to discuss.

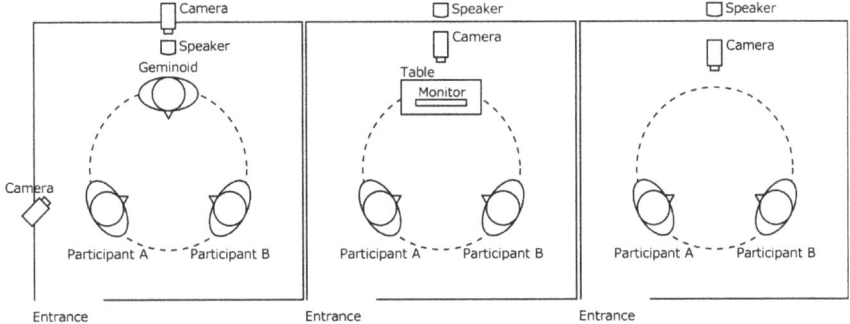

**Fig. 3.10** Experimental room setting (left: G condition, center: V condition, right: S condition)

4. One minute later, after the speaking participants had given their opinions, the operator gave them an additional discussion theme. This gave the operator an equal quality and quantity speaking chance in every experiment. Participants had difficulty in continuing the discussion longer than a minute, because Japanese students are not used to holding discussions.
5. After two minutes had elapsed, another theme was given, as in step 4.
6. When three minutes had elapsed and the speaking participants had again given their opinions, the operator concluded the experiment and the experimenter led the participants out of the experimental room.
7. Participants answered questionnaires.

The participants repeated this procedure three times for all conditions. We prepared three different themes (and six different additional themes) so that participants did not experience the same discussion theme twice. The discussion themes were counterbalanced within the conditions.

#### 3.3.2.5 Operators for Experiments

Two operators, who were not the Geminoid HI-1 model, appeared alternately. Regarding their utterances, the operator only gave the discussion theme to initiate conversation and responded to questions, but did not intrude into the discussions and avoided any vocal backchannel to control the experiment. We did not use a participant as the operator because we intended to control all three conditions similarly.

### 3.3.3 Evaluation

After each session, the participants answered a questionnaire that rated their impression of having a conversation with the third person on a 1–7 scale, with 7 being the most positive. The following items were scored:

**Presence**
Degree to which the participant felt the third person was present at the conversation.

**Humanlike**
Humanlike-ness of the third person's appearance, movements, and behavior.

**Naturalness**
Naturalness of the third person's appearance, movements, and behavior.

**Uncanny**

Uncanniness of the third person's appearance, movements, and behavior.

**Responsiveness**

Degree to which the third person responded to the participants' behavior and conversation.

**Eye contact**

Degree to which the third person's eye contact matched that of the participants.

### 3.3.4 Hypotheses and Expectations

The following are our hypotheses for the experiment:

**Hypothesis 1:**

With Geminoid HI-1, the operator conveys the strongest presence among these three media.

**Hypothesis 2:**

With Geminoid HI-1, the operator is perceived as humanlike and as natural as the other media.

**Other measures:**

We also measured uncanniness, responsiveness, and eye contact. Androids are generally considered uncanny, as explained by the uncanny valley theory [14]. Thus, we are interested in whether the android is deemed uncanny even when controlled by a human operator.

As previous papers have reported that one weakness of video conference systems is their eye contact capability [9], we investigated the strength of this component in the android.

## 3.4 Results

Figure 3.11 indicates the mean score and standard deviation of the questionnaire results. We conducted a within-subject design analysis of variance (ANOVA) to investigate the differences among conditions. The brackets above the boxes in Fig. 3.11 show the significant differences of multiple comparisons at a 5% level of significance.

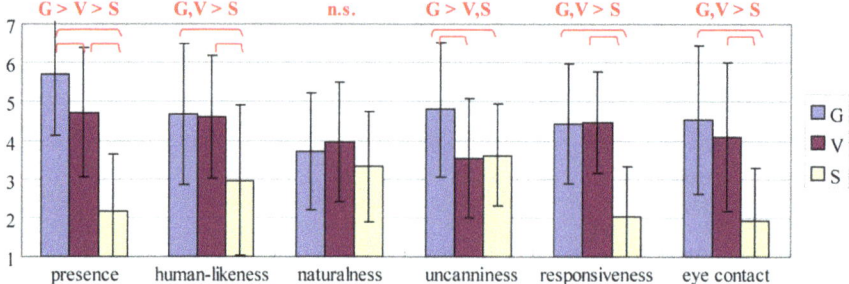

**Fig. 3.11** Participant impressions of three media: Geminoid android (G condition), video conference system (V condition), and speakerphone (S condition)

### 3.4.1 Hypothesis 1: Presence

A within-subject design ANOVA was conducted that showed significant differences in presence ($F(1, 33) = 50.762$, $p < 0.001$). The Bonferroni method was applied for multiple comparisons, and this proved that the rating of the G condition was significantly better than the V and S conditions and that the V condition was significantly better than the S condition (G > V, $p < 0.001$; G > S, $p < 0.001$; V > S, $p < 0.05$). This result indicates that, as a medium, the operator conveyed the strongest presence with Geminoid HI-1.

### 3.4.2 Hypothesis 2: Humanlike and Natural

A within-subject design ANOVA was conducted that showed significant differences in humanlike-ness ($F(1, 33) = 10.353$, $p < 0.001$). The Bonferroni method revealed that the G condition was significantly better than the S condition and that the V condition was also significantly better than the S condition (G > S, $p < 0.001$; V > S, $p = 0.001$). There were no significant differences in naturalness ($F(1, 33) = 1.777$, $p = 0.177$). These results indicate that, in the G condition, the operator gave similar humanlike and natural impressions to the V condition.

In the S condition, the operator was deemed less humanlike, although the same operator talked with the participants using the same quality of voice.

### 3.4.3 Analysis of Other Ratings

We also conducted a within-subject design ANOVA for the ratings of uncanniness, responsiveness, and eye contact. The results revealed significant differences among conditions ($F(1, 33) = 10.1$, $p < 0.001$; $F(1, 33) = 35.947$, $p < 0.001$; $F(1, 33) =$

20.143, p < 0.001). The Bonferroni method was applied for multiple comparisons The G condition was significantly better than the V and S conditions (G > V, p = 0.001, G > S, p < 0.001). Thus, the android condition was evaluated as being more uncanny than the video conference and speakerphone conditions.

Regarding responsiveness, the G and V conditions were significantly better than the S condition (G > S, p < 0.001; V > S, p < 0.001). The same trend was found in the eye contact rating (G > S, p < 0.001; V > S, p< 0.001), perhaps reflecting the calibration performed for the operators. In the V condition, the operator looked left and right to make eye contact. For this reason, the calibration for the operators may have affected the eye contact rating.

### 3.4.4  Summary of Results

The results indicate that, when the operator talked with participants through the Geminoid HI-1, they felt the strongest presence and a similar level of humanlike-ness and naturalness to a video conference system. Thus, we have confirmed that the Geminoid android is a good telecommunication medium that can convey humanlike presence by remotely representing human presence.

In the questionnaire, we asked about the participants' impressions of the operator. As there were differences in the ratings among conditions, even though the same operator appeared for the three conditions, participants actually evaluated the effect of the medium rather than the operator.

Despite being evaluated as humanlike and natural, the android was also described as uncanny.

## 3.5  Discussion and Conclusion

### 3.5.1  Summary

In this study, we developed a remote-controlled android system that uses a very humanlike robot called Geminoid HI-1. This system was developed to realize the telepresence of a teleoperator. To verify the presence of this system, we conducted experiments whose results confirmed that the presence conveyed through this system is stronger than that of a man appearing on a video monitor. In addition, the humanlike-ness of Geminoid HI-1 is equivalent to a man's image on a video monitor. However, participants felt that Geminoid HI-1 was uncanny. Based on its strong presence and these results, a humanlike robot could be used as a new telecommunication medium.

However, before using androids for telecommunication systems, one important problem must be solved: uncanniness. As the android gave a greater uncanny impression than the other media, and even though it was rated humanlike and as natural as a video conference system, its presence unfortunately increases feelings of uncanniness. Thus, more care is needed in the creation of androids to produce a stronger presence.

### 3.5.2 Contributions for Human–Robot Interaction

One major contribution of this research is that it has demonstrated a novel method of human–robot interaction: a "teleoperated communication robot" [15] in which an operator behind the robot interacts with people in front of the robot, particularly with spoken language, while the system autonomously controls such low-level robot behaviors as breathing and lip movements.

Moreover, we believe that studying such a teleoperated communication robot will offer insights for developing completely autonomous communication robots that currently lack verbal communication capabilities. The principal difficulty concerns speech recognition of colloquial utterances in noisy environments. Current technology is only capable of recognizing formal utterances in noiseless environments. Although research into robot audition is ongoing, the difficulties in daily environments are still beyond the grasp of current technology. When a robust speech recognition technique becomes available, the knowledge of teleoperated communication must be integrated to realize ideal communication robots.

### 3.5.3 Contributions to Android Science

We believe that this research also contributes to android science [16], which aims to reveal what is human by developing a humanlike robot that has humanlike appearance and behavior. Our developed system offers a strong platform for such studies. For now, the android is considered highly humanlike but uncanny, whereas a person in a video is evaluated as similarly humanlike but not uncanny. This might contradict the uncanny valley theory [14]. We believe that higher presence requires more humanlike-ness to avoid the uncanny valley. In addition, note that although the participants were told that the android was controlled by a human and were asked to rate the third person, they evaluated it as uncanny.

### 3.5.4 Applicability

Androids are still very expensive. One may be concerned with their applicability for telecommunication techniques. We suggest that in future, important people in industry and government such as presidents, politicians, and directors will use androids for telecommunication because they can afford them. For such people, "time is money," and sometimes physically meeting with people is more expensive in terms of time than in money.

At the same time, in part, the obtained techniques for controlling the android can also be applied to teleoperated communication robots.

### 3.5.5  Limitations

The generality of the findings is limited because this research only dealt with a particular robot, Geminoid HI-1, and two operators who were not the android model. The particular task only involved two people talking in the presence of the robot.

Regarding the generality of robots, the findings are limited to androids, and the effect depends on the android quality. If an android has equal or better humanlike-ness, we believe that the findings will be applicable to the android.

Regarding the generality of the operator, the same effect probably occurs when the operator is the model for the android; however, for people familiar with the modeled person, excessive similarity in appearance may cause a feeling of dissimilarity in motion and behavior. If the operator is very different, such as a woman who speaks in a shrill voice, the interacting person may feel less humanlike-ness because of the inconsistency between motion and speech.

As for the generality of tasks, the effects may change in relation to interaction complexity with the android. As this study's purpose was to verify the fundamental effects of humanlike presence, we focused on the role of the chairperson rather than those who actively joined the discussion. More complex interaction will require more sophisticated mechanisms, such as precise gaze direction control. In future work, we will investigate the humanlike presence of androids in more complex interactions.

**Acknowledgements**  We wish to thank Toshinori Carlos Ishii for developing the lip synchronizer and Kotaro Hayashi for operating the remote-controlled android system.

This research was supported by the Ministry of Internal Affairs and Communications of Japan.

## References

1. Sakamoto D., T. Kanda, T. Ono, H. Ishiguro, and N. Hagita. 2007. Android as a telecommunication medium with a human-like presence. In *2007 2nd ACM/IEEE international conference on human-robot interaction (HRI 2007)*, 193–200.
2. Breazeal, C., and B. Scassellati. 1999. A context-dependent attention system for a social robot. In *Proceedings of the 16th international joint conference* on *atificial intelligence*, 1146–1151.

3. Minato, T., M. Shimada., H. Ishiguro., and S. Itakura. 2004. Development of an android robot for studying human-robot interaction. In *Proceedings of the 17th international conference on industrial and engineering applications of artificial intelligence and expert systems (IEA/ AIE)*, 424–434.
4. Trafton, G., A. Schultz, D. Perznowski, M. Bugajska, W. Adams, N. Cassimatis, and D. Brock. 2006. Children and robots learning to play hide and seek. In *Proceedings of the 1st annual conference* on *human-robot interaction (HRI2006)*, 242–249.
5. Dautenhahn, K., M. Walters, S. Woods, K.L. Koay, C.L. Nehaniv, A. Sisbot, R. Alami, and T. Siméon. 2006. How may i serve you? a robot companion approaching a seated person in a helping context. In *Proceeding of 1st annual conference on human-robot interaction (HRI2006)*, 172–179.
6. Mutlu, B., S. Osman, J. Forlizzi, J. Hodgins, and S. Kiesler. 2006. Perceptions of ASIMO: An exploration on co-operation and competition with humans and humanoid robots. In *Proceedings of the 1st ACM SIGCHI/SIGART conference on Human-robot interaction*, 351–352.
7. Kanda, T., H. Ishiguro, M. Imai, and T. Ono. 2004. Development and evaluation of interactive humanoid robots. *Proceedings of the IEEE* 92 (11): 1839–1850.
8. Kraut, R.E., S.R. Fussell, and J. Siegel. 2003. Visual information as a conversational resource in collaborative physical tasks. *Human-Computer Interaction* 18: 13–49.
9. Morikawa, O., and T. Maesako. 1998. HyperMirror: Toward pleasant-to-use video mediated communication system. In *Proceedings of the 1998 ACM conference on computer supported cooperative work*, 149–158.
10. Kidd, C., and C. Breazeal. 2004. Effect of a robot on user perceptions. In *IEEE/RSJ international conference on intelligent robots and systems (IROS'04)*.
11. Shinozawa, K., F. Naya, J. Yamato, and K. Kogure. 2005. Differences in effect of robot and screen agent recommendations on human decision-making. *International Journal of Human-Computer Studies* 62: 267–279.
12. Sekiguchi, D., M. Inami, and S. Tachi. RobotPHONE: RUI for interpersonal communication. In *Extended abstract of CHI01*.
13. Tadakuma, R., Y. Asahara, H. Kajimoto, N. Kawakami, and S. Tachi. 2005. Development of anthropomorphic multi-D.O.F. master-slave arm for mutual telexistence. *IEEE Transactions on Visualization and Computer Graphics* 11 (6): 626–636.
14. Mori, M. 1970. "Bukimi no tani [The Uncanny Valley]", (in Japanese). *Energy* 7 (4): 33–35.
15. Koizumi, S., T. Kanda, S. Masahiro, H. Ishiguro, and N. Hagita. 2006. Preliminary field trial for teleoperated communication robots. In *IEEE international workshop on robot and human communication (ROMAN2006)*.
16. Hornyak T., T.B. Alert, S.E. Alert, and B.S.I. Alert. 2006. Android science. *Scientific American*.

# Chapter 4
# Generating Natural Motion in an Android by Mapping Human Motion

**Daisuke Matsui, Takashi Minato, Karl F. MacDorman and Hiroshi Ishiguro**

**Abstract** One of the main aims of humanoid robotics is to develop robots that are capable of interacting naturally with people. However, to understand the essence of human interaction, it is crucial to investigate the contribution of behavior *and* appearance. Our group's research explores these relationships by developing androids that closely resemble human beings in both aspects. If humanlike appearance causes us to evaluate an android's behavior from a human standard, we are more likely to be cognizant of deviations from human norms. Therefore, the android's motions must closely match human performance to avoid looking strange, including such autonomic responses as the shoulder movements involved in breathing. This paper proposes a method to implement motions that look human by mapping their three-dimensional appearance from a human performer to the android and then evaluating the verisimilitude of the visible motions using a motion capture system. Previous research has focused on copying and moving joint angles from a person to a robot. Our approach has several advantages: (1) in an android robot with many degrees of freedom and kinematics that differ from that of a human being, it is difficult to calculate which joint angles would make the robot's posture appear similar to the human performer; and (2) the motion that we perceive is at the robot's surface, not necessarily at its joints, which are often hidden from view.

**Keywords** Humanlike motion · Feedback error learning · Three-dimensional position mapping

This chapter is a modified version of a previously published paper [1], edited to be comprehensive and fit with the context of this book.

D. Matsui · T. Minato (✉) · H. Ishiguro
Department of Adaptive Machine Systems, Graduate School of Engineering,
Osaka University, Osaka, Japan
e-mail: minato@atr.Jp

K. F. MacDorman
School of Informatics, Indiana University, Bloomington, USA

## 4.1  Introduction

In recent years, there have been considerable efforts to develop mechanical-looking humanoid robots, such as Honda's Asimo and Sony's Qrio, with the goal of partnering them with people in daily situations. Just as an industrial robot's purpose determines its appearance, a partner robot's purpose will also determine its appearance. Partner robots generally adopt a roughly humanoid appearance to facilitate communication with people, because natural interaction is the only task that requires a humanlike appearance. In other words, humanoid robots mainly have significance insofar as they can interact naturally with people. Therefore, it is necessary to discover the principles underlying natural interaction to establish a methodology for designing interactive humanoid robots.

Kanda et al. [2] have tackled this problem by evaluating how the behavior of the humanoid robot "Robovie" affects human–robot interaction. However, Robovie's machine-like appearance distorts our interpretation of its behavior because of the way the complex relationship between appearance and behavior influences the interaction. Most research on interactive robots has neglected the effect of appearance (for exceptions, see [3, 4])—particularly in robots that closely resemble a person. Thus, it is not yet clear whether the most comfortable and effective human–robot communication would come from a robot that looks mechanical or human. However, we may infer that a humanlike appearance is important from the fact that human beings have developed neural centers specialized for the detection and interpretation of hands and faces [5–7]. A robot that closely resembles humans in both looks and behavior may prove to be the ultimate communication device insofar as it can interact with humans most naturally.[1] We refer to such a device as an android to distinguish it from mechanical-looking humanoid robots. Investigating the essence of how we recognize human beings as human will clarify how to produce natural interaction. Our study tackles the appearance and behavior problem with the objective of realizing an android and having it accepted as a human being [8].

Ideally, to generate humanlike movement, an android's kinematics should be functionally equivalent to the human musculoskeletal system. Some researchers have developed a joint system that simulates shoulder movement [9] and a muscle–tendon system to generate humanlike movement [10]. However, these systems are too bulky to be embedded in an android without compromising its humanlike appearance. Given current technology, we embed as many actuators as possible to provide many degrees of freedom insofar as this does *not* interfere with making the android look as human as possible [8]. Under these constraints, the main issue concerns how to move the android in a natural way so that its movement may be perceived as human.

A straightforward way to make a robot's movement more humanlike is to imitate human motion. Kashima and Isurugi [11] extracted essential properties of human arm trajectories and designed an evaluation function to generate corresponding robot arm trajectories. Another method is to copy human motion, as measured by a motion

---

[1]We use the term *natural* to denote communication that flows without seeming stilted, forced, bizarre, or inhuman.

capture system, to a humanoid robot. Riley et al. [12] and Nakaoka et al. [13] calculated a performer's joint trajectories from the measured positions of markers attached to the body and fed the results to the joints of a humanoid robot. In these studies, the authors assumed the kinematics of the robot to be similar to that of a human body. However, the more complex the robot's kinematics, the more difficult it is to calculate which joint angles will make the robot's posture similar to the performer's joint angles as calculated from the motion capture data. Therefore, it is possible that the assumption that the two joint systems are comparable will result in visibly different motion. This is a particular risk for androids, because their humanlike form makes us more sensitive to deviations from human ways of moving. Thus, slight differences could strongly influence whether the android's movement is perceived as natural or human. Furthermore, these studies did not evaluate the naturalness of robot motions.

Hale et al. [14] proposed several evaluation functions to generate a joint trajectory (e.g., minimization of jerk) and evaluated the naturalness of the generated humanoid robot movements according to how human subjects rated their naturalness. In the computer animation domain, researchers have tackled motion synthesis with motion capture data (e.g., [15]). However, we cannot apply their results directly; we must instead repeat their experiments with an android, because the results from an android testbed could be quite different from those of a humanoid testbed. For example, Mori described a phenomenon termed "uncanny valley" [16, 17], which relates to the relationship between a humanlikeness of a robot's appearance and a person's perception of familiarity. According to Mori, a robot's familiarity increases with its similarity until a certain point, at which time slight "nonhuman" imperfections cause the robot to appear repulsive (Fig. 4.1). This would be an issue if the similarity of androids fell into the trough. (Mori believes mechanical-looking humanoid robots lie on the left of the first peak.) This non-monotonic relationship can distort the evaluations proposed in existing studies. Therefore, it is necessary to develop a motion generation method in which the generated "android motion" is perceived as human.

This paper proposes a method to transfer human motion measured by a motion capture system to the android by copying changes in the positions of body surfaces. This method is necessary because the android's appearance demands movements that look human, but its kinematics are sufficiently different that copying the joint angle

**Fig. 4.1** Uncanny valley

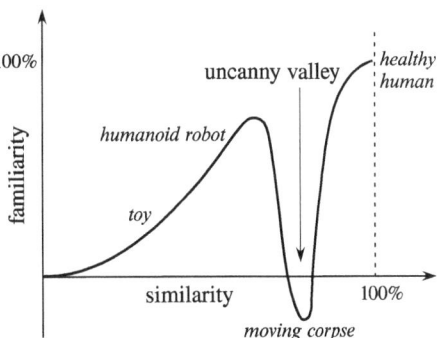

information would not yield good results. Comparing the similarity of the android's visible movement to that of a human being enables us to develop more natural movements for the android.

In the following sections, we describe the developed android and discuss the problem of motion transfer and our basic idea for a solution. We then describe the proposed method in detail and present experimental results from its application to the android.

## 4.2 The Android

Figure 4.2 shows the developed android, which is called *Repliee Q2*. The android has an Asian appearance because it is modeled on a Japanese woman. The standing height is about 160 cm. The skin is composed of a kind of silicone that has a humanlike feel and neutral temperature. The silicone skin covers the upper torso, neck, head, and forearms, with clothing covering other body parts. Unlike Repliee R1 [8, 18], silicone skin does not cover the entire body so as to facilitate flexibility and a maximal range of motion. The soft skin gives the android a human look and enables natural tactile interaction. To lend realism to the android's appearance, we took a cast of a person to mold the android's skin. Forty-two highly sensitive tactile sensors composed of piezo diaphragms are mounted under the android's skin and clothes throughout the body, except for the shins, calves, and feet. As the output

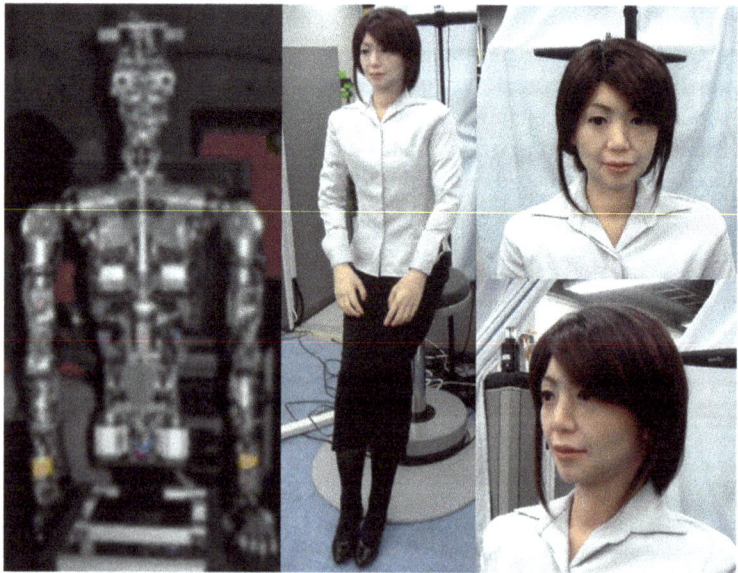

**Fig. 4.2** Developed android "Repliee Q2"

**Table 4.1**   DoF configuration of replie Q2

|  | Degree of freedom |
|---|---|
| Eyes | pan $\times$ 2 + tilt $\times$ 1 |
| Face | eyebrows $\times$ 1 + eyelids $\times$ 1 + cheeks $\times$ 1 |
| Mouth | 7 (including the upper and lower lips) |
| Neck | 3 |
| Shoulder | 5 $\times$ 2 |
| Elbow | 2 $\times$ 2 |
| Wrist | 2 $\times$ 2 |
| Fingers | 2 $\times$ 2 |
| Torso | 4 |

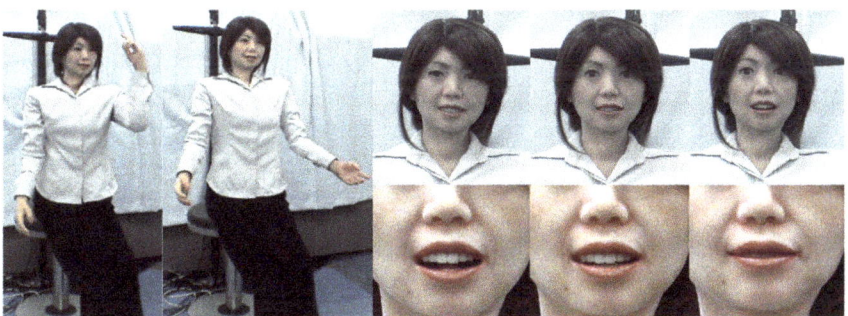

**Fig. 4.3**   Examples of motion and facial expressions

value of each sensor corresponds to its deformation rate, the sensors can distinguish different kinds of touch, ranging from stroking to hitting.

The android is driven by air actuators that give it 42 degrees of freedom (DoFs) from the waist up. (The legs and feet are not powered.) The configuration of the DoFs is shown in Table 4.1. The android can generate a wide range of motions and gestures, as well as various micro-motions such as the shoulder movements typically caused by human breathing. The DoFs of the shoulders enable them to move up and down and backwards and forwards. Furthermore, the android can make some facial expressions and mouth shapes, as shown in Fig. 4.3. The compliance of the air actuators makes for a safer interaction with movements that are generally smoother. Because the android has servo controllers, it can be controlled by a host computer that sends the desired joint positions. Parallel link mechanisms adopted in some parts complicate the kinematics of the android.

## 4.3   Transferring Human Motion

### 4.3.1   Basic Idea

One method of realizing humanlike motion in a humanoid robot is through imitation. Thus, we consider how to map human motion to the android. Most previous research assumes the kinematics of the human body are similar to that of the robot, albeit at a different scale. Thus, they aim to reproduce human motion by reproducing kinematic relations across time and, in particular, joint angles between links. For example, the three-dimensional locations of markers attached to the skin are measured by a motion capture system, and the angles of the body's joints are calculated from these positions, and these angles are transferred to the joints of the humanoid robot. It is assumed that, by using a joint angle space (which does not represent link lengths), morphological differences between the human subject and the humanoid robot can be ignored.

However, there is some potential for error in calculating joint angles from motion capture data. The joint positions are assumed to be the same between a humanoid robot and the human performer who serves as a model; however, the kinematics will actually differ. For example, the kinematics of Repliee Q2's shoulder differ significantly from those of human beings. Moreover, as human joints rotate, each joint's center of rotation changes, but joint-based approaches generally assume that this is not the case. These errors are perhaps more pronounced in Repliee Q2, because the android has many degrees of freedom and the shoulder has more complex kinematics than in existing humanoid robots. These errors are more problematic for an android than a mechanical-looking humanoid robot, because we expect natural human motion from something that looks human and are disturbed when the motion instead looks inhuman.

To create movement that appears human, we focus on reproducing positional changes at the body's surface rather than changes in the joint angles. We then measure the postures of a person and the android using a motion capture system and determine the control input to the android so that the postures of the person and the android become similar.

### 4.3.2   Method of Transferring Human Motion

We use a motion capture system to measure the postures of the human performer and the android. This system can measure the three-dimensional positions of markers attached to the surface of bodies in a global coordinate space. First, some markers are attached to the android so that all joint motions can be estimated. The reason for this will become clear later. The same number of markers are then attached to corresponding positions on the performer's body. We must assume the android's surface morphology is not too different from that of the performer.

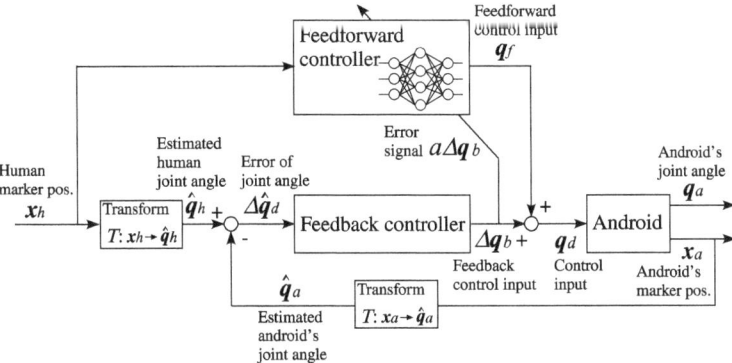

**Fig. 4.4** The android control system

We use a three-layer neural network to construct a mapping from the performer's posture to the android's control input, which is the desired joint angle. This network is required because it is difficult to obtain the mapping analytically. To train a neural network to map from $x_h$ to $q_a$ would require thousands of pairs of $x_h, q_a$ as training data, and the performer would need to assume the posture of the android for each pair. We avoid this prohibitively lengthy data collection task by adopting feedback error learning (FEL) to train the neural network. Kawato et al. [19] proposed FEL as a principle for learning motor control in the brain. FEL employs an approximate mapping of sensory errors to motor errors that can subsequently be used to train a neural network (or other method) by supervised learning. FEL neither prescribes the type of neural network employed in the control system nor the exact layout of the control circuitry. We use FEL to estimate the error between the postures of the performer and the android and feed the error back to the network.

Figure 4.4 shows a block diagram of the control system, where the network mapping is shown as the feedforward controller. The weights of the feedforward neural network are learned by means of a feedback controller. The method has a 2-DoF control architecture. The network tunes the feedforward controller to be the inverse model of the plant. Thus, the feedback error signal is employed as a teaching signal for learning the inverse model. If the inverse model is learned exactly, the output of the plant tracks the reference signal by feedforward control. The performer and android's marker positions are represented in their local coordinates $x_h, x_a \in \mathcal{R}^{3m}$; the android's joint angles $q_a \in \mathcal{R}^n$ can be observed by a motion capture system and a potentiometer, where $m$ is the number of markers and $n$ is the number of DoFs of the android.

The feedback controller is required to output the feedback control input $\Delta q_b$ so that the error in the marker's position $\Delta x_d = x_a - x_h$ converges to zero (Fig. 4.5a). However, it is difficult to obtain $\Delta q_b$ from $\Delta x_d$. To overcome this, we assume the performer has roughly the same kinematics as the android and obtain the estimated joint angle $\hat{q}_h$ by simply calculating the Euler angles (hereafter, the transformation

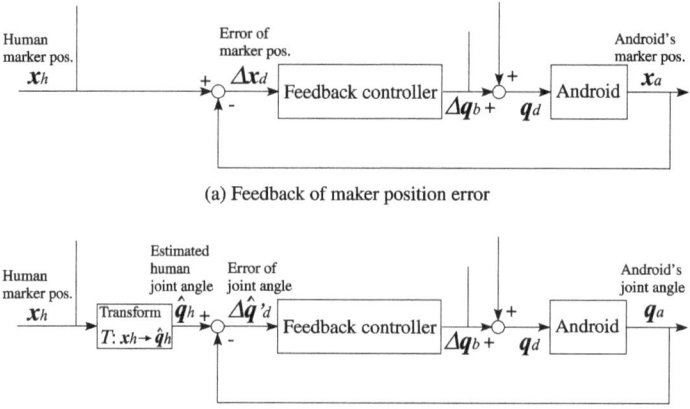

(a) Feedback of maker position error

(b) Error estimation with the android's joint angle measured by the potentiometer

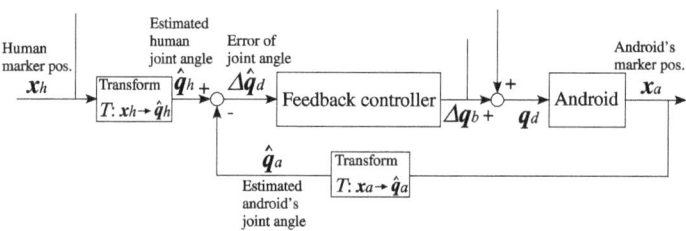

(c) Error estimation with the android's joint angle estimated from the android's marker position

**Fig. 4.5** Feedback controller with and without the estimation of the android's joint angle

from marker positions to joint angles is described as $T$).[2] Converging $\hat{q}_a$ to $q_h$ does not always produce identical postures, because $\hat{q}_h$ is an approximate joint angle that may include some transformation error (Fig. 4.5b). We obtain the estimated joint angle of the android $\hat{q}_a$ using the same transformation $T$ and the feedback control input to converge $\hat{q}_a$ to $\hat{q}_h$ (Fig. 4.5c). Using this technique, $x_a$ approaches $x_h$. The feedback control input approaches zero as learning progresses, while the neural network constructs the mapping from $x_h$ to the control input $q_d$. We can evaluate the apparent posture by measuring the android posture.

In this system, we could have made another neural network for the mapping from $x_a$ to $q_a$ using only the android. As long as the android's body surfaces are reasonably close to the performer's, we can use the mapping to generate the control input from $x_h$. Ideally, the mapping must learn every possible posture, but this is quite difficult. Therefore, it is still necessary for the system to evaluate the error in the apparent posture.

---

[2]Alternatives to using the Euler angles include angle decomposition [20], which has the advantage of providing a sequence-independent representation, and least squares to calculate the helical axis and rotational angle [21, 22]. The latter provides higher accuracy when many markers are used, but has an increased risk of marker crossover.

## 4.4 Experiment to Transfer Human Motion

### 4.4.1 Experimental Setting

To verify the proposed method, we conducted an experiment to transfer human motion to the android Repliee Q2. We used 21 of the android's 42 DoFs by excluding the 13 DoFs of the face, the four of the wrists, and the four of the fingers ($n = 21$). We used a Hawk Digital System,[3] which can track more than 50 markers in real time. The system is highly accurate, with a measurement error of less than 1 mm. Twenty markers were attached to the performer and another 20 to the android, as shown in Fig. 4.6 ($m = 20$). Because the android's waist is fixed, the markers on the waist set the frame of reference for an android-centered coordinate space. To facilitate learning, we introduced a representation of the marker position $x_h, x_a$, as shown in Fig. 4.7. The effect of waist motions is removed with respect to the markers on the head. To avoid accumulating position errors at the end of the arms, vectors connecting neighboring pairs of markers represent the positions of these markers. We used arc tangents for the transformation $T$, whereby the joint angle is the angle between two neighboring links (a link consists of a straight line between two markers).

The feedback controller outputs $\Delta q_b = K \Delta \hat{q}_d$, where the gain $K$ consists of a diagonal matrix. There are 60 nodes in the input layer (20 markers × $x, y, z$), 300 in the hidden layer, and 21 in the output layer (for the 21 DoFs). Using 300 units in the hidden layer provides a good balance between computational efficiency and accuracy. Using significantly fewer units resulted in too much error, whereas using

**Fig. 4.6** Marker positions corresponding to each other

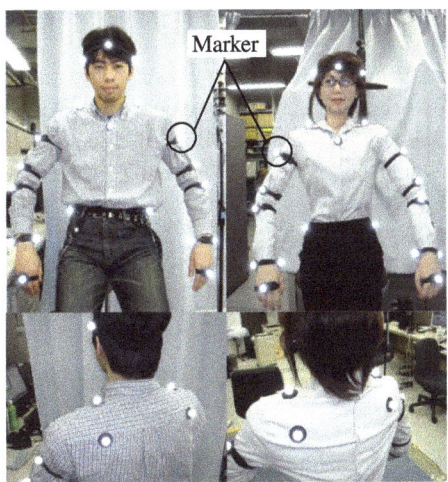

Performer        Android

---

[3] Motion Analysis Corporation, Santa Rosa, California. http://www.motionanalysis.com/.

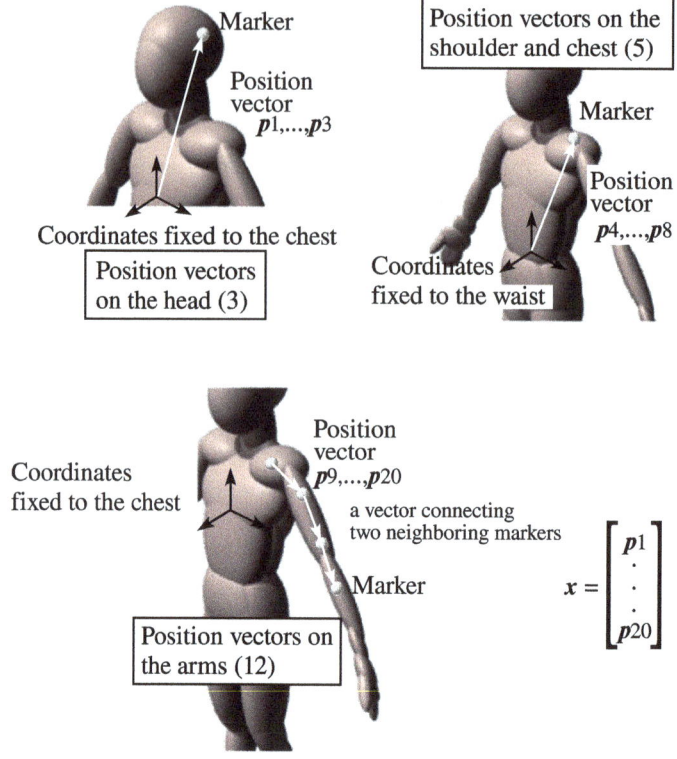

**Fig. 4.7** Representation of the marker positions. A marker's diameter is about 18 mm

significantly more units provided only marginally higher accuracy but at the cost of slower convergence. The error signal to the network is $t = \alpha \Delta q_b$, where the gain $\alpha$ is a small number. The sampling time for capturing the marker positions and controlling the android is 60 ms. Another neural network with the same structure previously learned the mapping from $x_a$ to $q_a$ to set the initial values of the weights. We obtained 50,000 samples of training data ($x_a$ and $q_a$) by moving the android randomly. The learned network was used to set the initial weights of the feedforward network.

## 4.4.2  *Experimental Results and Analysis*

### 4.4.2.1  Surface Similarity Between the Android and Performer

The proposed method assumes a surface similarity between the android and the performer. However, the male performer whom the android imitates in the experiments was 15 cm taller than the women after whom the android was modeled. To check the

similarity, we measured the average distance between corresponding pairs of mark
ers when the android and performer assumed each of the given postures; the value
was 31 mm (see Fig. 4.6). This gap is small compared to the size of their bodies, but
it is not small enough.

#### 4.4.2.2   Learning the Feedforward Network

To show the effect of the feedforward controller, Fig. 4.8 illustrates the feedback
control input averaged among the joints while learning from the initial weights. The
abscissa denotes the time step (the sampling time is 60 ms). Although the value of the
ordinate does not have a direct physical interpretation, it corresponds to a particular
joint angle. The performer exhibited various fixed postures. When the performer
started to make the posture at step 0, the error increased rapidly because network
learning had not yet converged. The control input decreases as learning progresses.
This shows that the feedforward controller learns, pushing the feedback control input
to zero.

Figure 4.9 shows the average position error of a pair of corresponding markers.
The performer also assumed an arbitrary fixed posture. The position errors and feedback control input both decrease as the feedforward network learning converges.
This result shows that the feedforward network learns the mapping from the performer's posture to the android control input, which allows the android to adopt the
same posture. The android's posture cannot match the performer's posture when the
weights of the feedforward network are at their initial values. This is because the
initial network does not know every possible posture in the pre-learning phase. This
demonstrates the effectiveness of the method to evaluate the apparent posture.

**Fig. 4.8** Change in the
feedback control input with
network learning

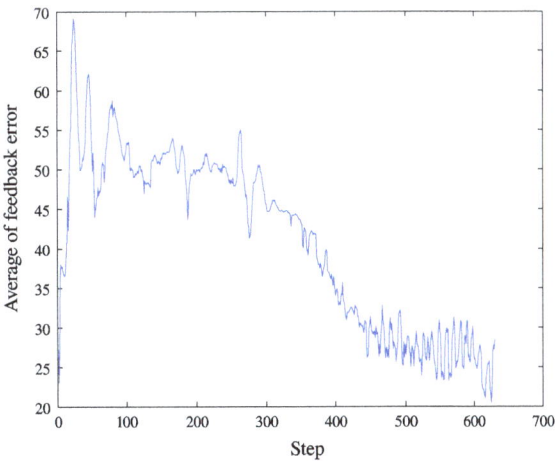

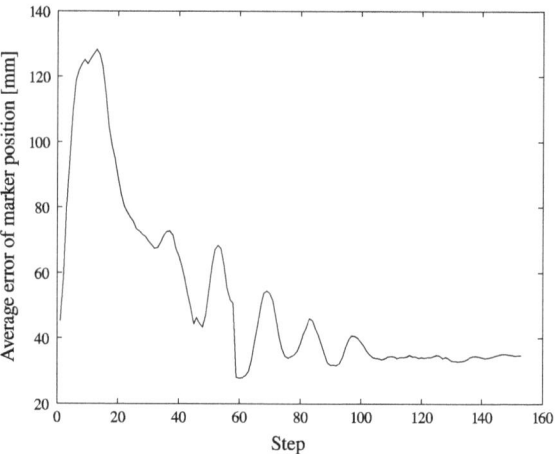

**Fig. 4.9** Change in the position error with network learning

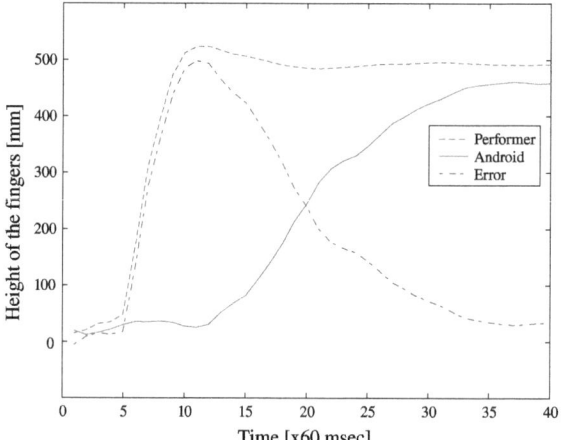

**Fig. 4.10** Step response of the android

### 4.4.2.3 Performance of the System in Following Fast Movements

To investigate the performance of the system, we obtained a step response using the feedforward network after sufficient learning. The performer puts his right hand on his knee and quickly raised the hand above his head. Figure 4.10 shows the height of the fingers of the performer and the android. The performer started to move at step 5 and reached the final position at step 9, approximately 0.24 s later. In this case, the delay is 26 steps or 1.56 s. The arm moved at roughly the maximum speed permitted by the hardware. The androids arm cannot quite reach the performer's position, because the performer's position is outside the android's range of motion. Clearly, the speed of the performer's movement exceeds the android's capabilities.

**Fig. 4.11** Generated android motion compared to the performer's motion. The numbers represent steps

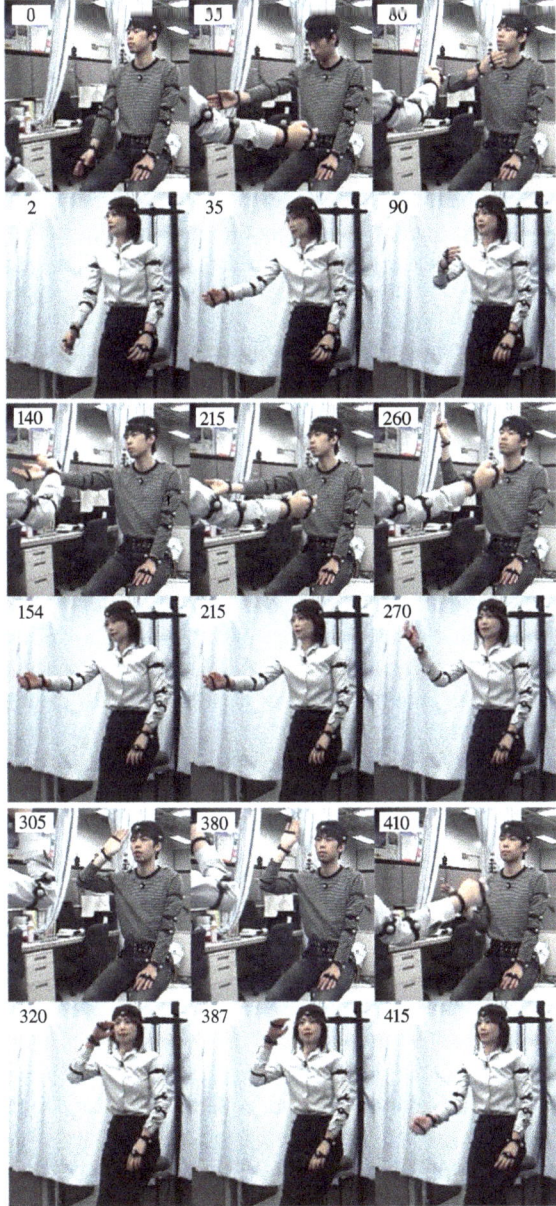

This experiment represents an extreme case. For less extreme gestures, the delay will be much shorter. For example, for the sequence in Fig. 4.11, the delay averages seven steps or 0.42 s.

#### 4.4.2.4  Generated Android Motion

Figure 4.11 shows the performer's postures during a movement and the corresponding postures of the android. The values denote the time step. The android followed the performer's movement with some delay (the maximum is 15 steps, that is, 0.9 s). The trajectories of the android's markers are considered to be similar to those of the performer, but some errors remain that cannot be ignored. Although we recognize that the android is making the same gesture as the performer, the quality of the movement is not the same. There are a three major causes of this:

- The kinematics of the android are too complicated to be represented with an ordinary neural network. To avoid this limitation, it is possible to introduce the constraint of the body's branching in the network connections. Another idea is to introduce a hierarchical representation of the mapping. Human motions can be decomposed into a dominant motion, which is at least partly conscious, and secondary motions, which are mainly nonconscious (e.g., contingent movements to maintain balance, autonomous responses such as breathing). We are trying to construct a hierarchical representation of motion, not only to reduce the computational complexity of learning, but also to make the movement appear more natural.
- The method handles motion as a sequence of postures; it does not precisely reproduce higher-order properties of motion such as velocity and acceleration, because varying delays can occur between the performer's movement and the android's imitation. If the performer moves very quickly, the apparent motion of the android differs. Moreover, the lack of higher-order properties prevents the system from adequately compensating for the dynamic characteristics of the android and the delay of the feedforward network.
- The proposed method is limited by the speed of motion. It is necessary to consider various properties to overcome this restriction, although the android has absolute physical limitations such as a fixed compliance and a maximum speed that is less than that of a typical human being.

Although physical limitations cannot be overcome by any control method, there are ways of finessing them to ensure that the movements still look natural. For example, although the android lacks the opponent musculature of human beings, which affords a variable compliance of the joints, the wobbly appearance of movements such as rapid waving, which are both high speed and high frequency, can be overcome by slowing the movement and removing repeated closed curves in the joint angle space to eliminate the lag caused by the slowed movement. If the goal is humanlike movement, one approach may be to query a database of movements that are known to be humanlike to find the one that is most similar to the movement made by the performer, although this raises the question of where those movements came from in the first place. Another method is to establish criteria for evaluating the naturalness of a movement [11]. This is an area for future study.

### *4.4.3 Required Improvements and Future Work*

In this paper, we have focused on reproducing positional changes at the body's surface rather than changes in the joint angles to generate the android's movement. Figure 4.5a illustrates a straightforward method to implement this idea. This paper has described the transformation $T$ from marker positions to estimated joint angles. It is difficult to derive a feedback controller that produces the control input $\Delta q_b$ analytically using only the error in the marker's positional error $\Delta x_d$. We do not actually know which joints should be moved to remove positional errors at the body's surface. This relation must be learned; however, the transformation $T$ could disturb the learning. Hence, it is not generally guaranteed that a feedback controller which converges the estimated joint angle $\hat{q}_a$ to $\hat{q}_h$ will enable the marker's position $x_a$ to approach $x_h$. The assumption that the android's body surfaces are reasonably close to those of the performer could avoid this problem, but the feedback controller shown in Fig. 4.5a is essentially necessary for mapping the apparent motion. It is possible to determine how the joint changes relate to the movements of body surfaces by analyzing the weights of the neural network of the feedforward controller. A feedback controller could be designed to output the control input based on the error in the marker's position using the analyzed relation. Concerning the design of the feedback controller, Oyama et al. [23–25] proposed several methods for learning both feedback and feedforward controllers using neural networks. This is one potential method to obtain the feedback controller shown in Fig. 4.5a. Assessing and compensating for the deformation and displacement of the human skin, which cause marker movement with respect to the underlying bone [26], are also useful in designing the feedback controller.

We have not dealt with the android's gaze and facial expressions in the experiments reported here; however, if gaze and facial expressions are unrelated to hand gestures and body movements, the appearance is often unnatural, as we have found in our experiments. Therefore, to make the android's movement appear more natural, we must consider a method to implement the android's eye movements and facial expressions.

## 4.5 Conclusion

This paper has proposed a method of implementing humanlike motions by mapping their three-dimensional appearance to an android using a motion capture system. By measuring the android's posture and comparing it to the posture of a human performer, we propose a new method to evaluate motion sequences along bodily surfaces. Unlike other approaches that focus on reducing joint angle errors, we consider how to evaluate differences in the android's apparent motion, that is, motion at its visible surfaces. The experimental results show the effectiveness of the evaluation: the method can transfer human motion. However, the method is restricted by the speed of the motion. It is therefore necessary to introduce a method to deal with the dynamic characteristics and physical limitations of the android. We should also evaluate the method with different performers. It is expected that the most natural and accurate movements will be generated using a female performer who is approximately the same height as the woman on which the android is based. Moreover, it is important to evaluate the humanlikeness of the visible motions in terms of the subjective impressions the android gives experimental subjects and the responses it elicits, such as eye contact [27, 28], autonomic responses. Research in these areas is in progress.

**Acknowledgements** We developed the android in collaboration with Kokoro Company, Ltd.

# References

1. Matsui, Daisuke, Takashi Minato, Karl F. MacDorman, and Hiroshi Ishiguro. 2005. Generating natural motion in an android by mapping human motion. In *2005 IEEE/RSJ international conference on intelligent robots and systems, 2005. (IROS 2005)*, 3301–3308. IEEE.
2. Kanda, T., H. Ishiguro, T. Ono, M. Imai, and K. Mase. 2002. Development and evaluation of an interactive robot "Robovie". In *Proceedings of the IEEE international conference on robotics and automation*, 1848–1855.
3. Goetz, J., S. Kiesler, and A. Powers. 2003. Matching robot appearance and behavior to tasks to improve human-robot cooperation. In *Proceedings of the workshop on robot and human interactive communication*, 55–60.
4. DiSalvo, C.F., F. Gemperle, J. Forlizzi, and S. Kiesler. 2002. All robots are not created equal: The design and perception of humanoid robot heads. In *Proceedings of the symposium on designing interactive systems*, 321–326.
5. Grill-Spector, K., N. Knouf, and N. Kanwisher. 2004. The fusiform face area subserves face perception, not generic within-category identification. *Nature Neuroscience* 7 (5): 555–562.
6. Farah, M.J., C. Rabinowitz, G.E. Quinn, and G.T. Liu. 2000. Early commitment of neural substrates for face recognition. *Cognitive Neuropsychology* 17: 117–123.
7. Carmel, D., and S. Bentin. 2002. Domain specificity versus expertise: Factors influencing distinct processing of faces. *Cognition* 83: 1–29.
8. Minato, T., K.F. MacDorman, M. Shimada, S. Itakura, K. Lee, and H. Ishiguro. 2004. Evaluating humanlikeness by comparing responses elicited by an android and a person. In *Proceedings of the 2nd international workshop on man-machine symbiotic systems*, 373–383.
9. Okada, M., S. Ban, and Y. Nakamura. 2002. Skill of compliance with controlled charging/discharging of kinetic energy. In *Proceeding of the IEEE international conference on robotics and automation*, 2455–2460.
10. Yoshikai, T., I. Mizuuchi, D. Sato, S. Yoshida, M. Inaba, and H. Inoue. 2003. Behavior system design and implementation in spined musle-tendon humanoid "Kenta". *Journal of Robotics and Mechatronics* 15 (2): 143–152.
11. Kashima, T., and Y. Isurugi. 1998. Trajectory formation based on physiological characteristics of skeletal muscles. *Biological Cybernetics* 78 (6): 413–422.
12. Riley, M., A. Ude, and C.G. Atkeson. 2000. Methods for motion generation and interaction with a humanoid robot: Case studies of dancing and catching. In *Proceedings of AAAI and CMU workshop on interactive robotics and entertainment*.
13. Nakaoka, S., A. Nakazawa, K. Yokoi, H. Hirukawa, and K. Ikeuchi. 2003. Generating whole body motions for a biped humanoid robot from captured human dances. In *Proceedings of the 2003 IEEE international conference on robotics and automation*.

14. Hale, J.G., F.E. Polliek, and M. Tzonova. 2003. The visual categorization of humanoid move ment as natural. In *Proceedings of the third IEEE international conference on humanoid robotics*.
15. Gleicher, Michael. Retargestting motion to new characters. 1998. In *Proceedings of the 25th annual conference on computer graphics and interactive techniques, SIGGRAPH '98, New York, NY, USA*, 33–42. ACM.
16. Mori, M. 1970. Bukimi no tani [the uncanny valley] (in Japanese). *Energy* 7 (4): 33–35.
17. Fong, T., I. Nourbakhsh, and K. Dautenhahn. 2003. A survey of socially interactive robots. *Robotics and Autonomous Systems* 42: 143–166.
18. Minato, T., M. Shimada, H. Ishiguro, and S. Itakura. 2004. Development of an android robot for studying human-robot interaction. In *Proceedings of the 17th international conference on industrial & engineering applications of artificial intelligence & expert systems*, 424–434.
19. Kawato, M., K. Furukawa, and R. Suzuki. 1987. A hierarchical neural network model for control and learning of voluntary movement. *Biological Cybernetics* 57: 169–185.
20. Grood, E.S., and W.J. Suntay. 1983. A joint coordinate system for the clinical description of three-dimensional motions: application to the knee. *Journal of Biomechanical Engineering* 105: 136–144.
21. Challis, J.H. 1995. A procedure for determining rigid body transformation parameters. *Journal of Biomechanics* 28: 733–737.
22. Veldpaus, F.E., H.J. Woltring, and L.J.M.G. Dortmans. 1988. A least squares algorithm for the equiform transformation from spatial marker co-ordinates. *Journal of Biomechanics* 21: 45–54.
23. Oyama, E., N.Y. Chong, A. Agah, T. Maeda, S. Tachi, and K.F. MacDorman. 2001. Learning a coordinate transformation for a human visual feedback controller based on disturbance noise and the feedback error signal. In *Proceedings of the IEEE international conference on robotics and automation*.
24. Oyama, E., K.F. MacDorman, A. Agah, T. Maeda, and S. Tachi. 2001. Coordinate transformation learning of a hand position feedback controller with time delay. *Neurocomputing*, 38–40(1–4).
25. Oyama, E., A. Agah, K.F. MacDorman, T. Maeda, and S. Tachi. 2001. A modular neural network architecture for inverse kinematics model learning. *Neurocomputing* 38–40 (1–4): 797–805.
26. Leardini, Alberto, Lorenzo Chiari, Ugo Della Croce, and Aurelio Cappozzo. 2005. Human movement analysis using stereophotogrammetry: Part 3. soft tissue artifact assessment and compensation. *Gait & Posture* 21 (2): 212–225.
27. Minato, Takashi, Michihiro Shimada, Shoji Itakura, Kang Lee, and Hiroshi Ishiguro. 2005. Does gaze reveal the human likeness of an android? In *Proceedings of the 4th International Conference on Development and Learning*, 106–111. IEEE.
28. Macdorman, Karl F., Takashi Minato, and Michihiro Shimada. 2005. Assessing human likeness by eye contact in an android testbed. In *Proceedings of the XXVII annual meeting of the cognitive science society*.

# Chapter 5
# Formant-Based Lip Motion Generation and Evaluation in Humanoid Robots

**Carlos T. Ishi, Chaoran Liu, Hiroshi Ishiguro and Norihiro Hagita**

**Abstract** Generating natural motion in robots is important for improving human–robot interaction. We have developed a teleoperation system in which the lip motion of a remote humanoid robot is automatically controlled by the operator's voice. In the present work, we introduce an improved version of our proposed speech-driven lip motion generation method, where lip height and width degrees are estimated based on vowel formant information. The method requires the calibration of only one parameter for speaker normalization. Lip height control is evaluated in two types of humanoid robots (Telenoid-R2 and Geminoid-F). Subjective evaluations indicate that the proposed audio-based method can generate lip motion with superior naturalness to vision-based and motion capture-based approaches. Partial lip width control is shown to improve lip motion naturalness in Geminoid-F, which also has an actuator for stretching the lip corners. Issues regarding online real-time processing are also discussed.

**Keywords** Lip motion generation · Formant · Teleoperation
Human–robot interaction

## 5.1 Introduction

We have been developing teleoperation systems for transmitting human telepresence through humanoid robots like androids. Previous teleoperation systems have used motion capture or vision-based lip tracking techniques. However, the performance of vision-based approaches is known to be dependent on the speaker, as well as on other factors such as good lighting conditions and image resolution.

---

This chapter is a modified version of a previously published paper [1], edited to be comprehensive and fit with the context of this book.

---

C. T. Ishi (✉) · C. Liu · H. Ishiguro · N. Hagita
ATR Intelligent Robotics and Communication Labs, Kyoto 619-0288, Japan
e-mail: carlos@atr.jp

© Springer Nature Singapore Pte Ltd. 2018
H. Ishiguro and F. Dalla Libera (eds.), *Geminoid Studies*,
https://doi.org/10.1007/978-981-10-8702-8_5

Motion capture systems are more robust to these factors, but remain somewhat expensive. Thus, we developed a teleoperation system whereby the lip motion of a remote humanoid robot is automatically controlled by the operator's voice.

Several approaches have been proposed for converting speech or text to lip motion. When text or phonetic transcriptions are available, methods like concatenation, trajectory generation, or dominance functions (which are linear combinations of trajectories selected according to a phonetic transcription) can be applied [2, 3]. Lip motion generation methods based on audio alone can be categorized as phone-based methods or direct audiovisual conversion. Phone-based methods model the audiovisual data using different phone models, mainly artificial neural networks (ANNs) and hidden Markov models (HMMs) [4–7].

Direct audiovisual conversion, without using phone models, has been shown to be more effective than HMM-based phone models [8, 9]. For example, in [10], these two approaches are directly compared, and the ANN-based method is judged to be significantly better than the HMM method. An explanation for this is that ANN-based approaches typically work on a frame-wise basis and can offer closer, more direct synchronization with the acoustic signal than HMM, in which the mapping is mediated through longer phone-sized units.

Another class of direct audiovisual conversion methods uses Gaussian mixture models (GMMs) [10]. In [10], a maximum likelihood estimation of the visual parameter trajectories is adopted using an audiovisual joint GMM, and a minimum converted trajectory error approach is proposed for refining the converted visual parameters.

Finally, another direct audiovisual conversion approach is based on formants (resonances of the vocal tract) [11]. Although most approaches use Mel-frequency cepstral coefficients (MFCC) as acoustic parameters, the interpretation of their values with regard to phonetic content is not a straightforward means of determining the formant frequencies. Further, all MFCC-based methods require dedicated models to be constructed prior to their use. Thus, in the present work, we adopt a formant-based approach to generate lip motion.

Vowels can be represented in a two-dimensional space formed by the first and second formants (F1, F2). It is well known that there is relationship between the formant space and vocal tract area functions (including lips). For example, F1 is related to the jaw lowering, whereas F2 is related to the front–back position of the tongue. However, lip opening and closing may occur without jaw lowering, so the relationship between F1 and lip opening is not straightforward. In our previous work [12], we proposed a method of estimating the degree of lip opening from the first and second formants.

In the present work, we improve the parameter extraction and evaluate its performance in two types of humanoid robots compared with vision- and motion capture-based approaches. The effects of partial lip width control are also evaluated in the android robot. The term "partial" is used because the android cannot round its lips, but has an actuator for stretching the lip corners.

The remainder of this paper is organized as follows. Section 5.2 describes the proposed method for lip motion generation from speech signals. In Sect. 5.3, the audio-based lip motion (lip height and width) generation method is evaluated. In Sect. 5.4, the constraints of the present method are discussed. Finally, Sect. 5.5 concludes the paper.

## 5.2  Proposed Method

The proposed method is divided into two parts. The first is the operator side, and the second is the robot side. Figure 5.1 shows a block diagram for the lip motion generation procedure in the operator side.

Firstly, formant extraction is conducted on the input speech signal. Then, a transformation of the formant space by the first and second formants is realized according to some speaker-dependent parameter to obtain a normalized vowel space. This normalization is necessary because the vowel space differs depending on speaker-specific features such as gender, age, and height. The lip shape is then estimated from the normalized vowel space, and actuator commands are generated from the lip shapes. Audio packets and actuator commands are sent to a remote robot at 10-ms intervals.

Figure 5.2 shows a block diagram of the lip motion generation on the robot side. The audio packets and lip motion actuator commands are received, actuator commands are sent to the robot for moving the lip actuators, and the received audio packets are delayed so as to synchronize the two streams.

The following sections describe each block in detail.

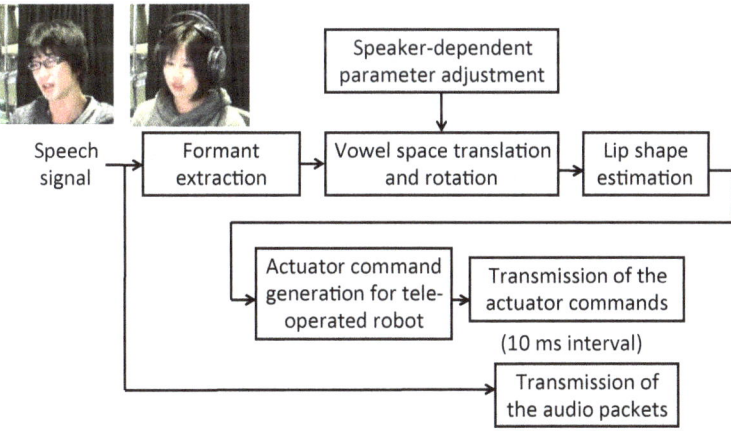

**Fig. 5.1**  Block diagram of the lip motion generation in the operator side

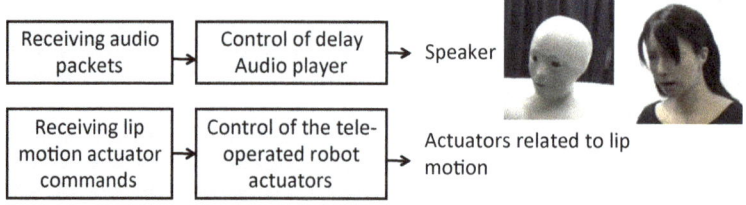

**Fig. 5.2** Block diagram of the lip motion generation in the robot side

## 5.2.1 Formant Extraction

The formant extraction implemented in the present work is a conventional method of picking the peaks of the Linear Predictive Coding (LPC)-smoothed spectrum [13], as shown in the block diagram of Fig. 5.3. Nonetheless, any other formant extraction method with better performance could be used instead.

The input signal is captured at 16 kHz/16-bit resolution, pre-emphasized by $1 - 0.97z^{-1}$ to reduce the effects of the glottal waveform and the lip radiation, and framed by a 32-ms Hamming window at 10-ms intervals. The 19th-order LPC coefficients $a_k$ ($k = 0 - 18$, with $a_0 = 1$) are then extracted for each frame. This allows the LPC-smoothed spectrum to be obtained according to the following expression.

$$\left|H(e^{jw})\right|^2 = \frac{1}{\left|1 + \sum_{k=1}^{P} a_k e^{-jw}\right|^2} \tag{5.1}$$

In practice, the smoothed spectrum is obtained by taking a 512-point fast Fourier transform of the LPC coefficients $a_k$ with zero-padding.

Finally, the first and second peaks are picked from the LPC-smoothed spectrum, searching from low to high frequencies, for the first and second formants (F1 and F2).

In our previous work, the LPC analysis order was fixed to 19. However, it is known that the order should be selected according to the actual number of formants in the analyzed bandwidth. Thus, the LPC order should be lower for female voices and higher for male voices. Thus, in the present work, the LPC order varies according to the speaker normalization factor, as explained in the following subsection.

**Fig. 5.3** Block diagram of the formant extraction process

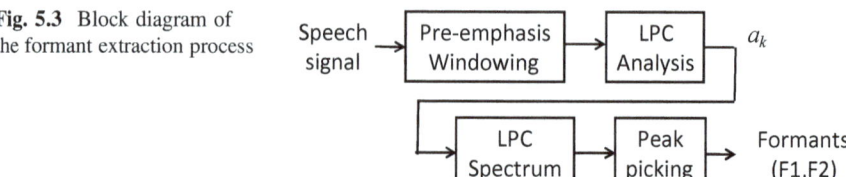

## 5.2.2  Calibration and Speaker Normalization

As a first step for speaker normalization, the origin of the coordinates in the formant space (centerF1; centerF2) is adjusted according to the speaker, because the vowel space changes with gender, age, and height. Specifically, the new origin is moved to the center of the speaker's vowel space (corresponding to the schwa vowel in English) in the logF1 versus logF2 space (i.e., the space given by the logarithms of the first and second formants).

The center point of the speaker's vowel space is adjusted through a graphical user interface (GUI). After uttering isolated vowels, the user (operator) visually identifies the approximate position of the center of his/her vowel system. In practice, only centerF1 is adjustable, in the range 400–800 Hz in steps of 10 Hz. The centerF2 can be automatically estimated from centerF1 using the following expression:

$$centerF2 = 2.9 \times centerF1 \tag{5.2}$$

Theoretically, an open–closed straight tube (which is a quarter-wave resonator) would have F2 equal to three times F1 [14]. However, the neutral vocal tract configuration is not quite a straight tube, and expression (5.2) was found to give a better fit in our preliminary analysis.

The next step of speaker normalization is to scale the coordinates. This step is equivalent to vocal tract length normalization, where the logF1 and logF2 coordinates are stretched or enlarged. Our preliminary analysis indicated that scaling factors of around 2 (male) and 1.8 (female) are good approximations. In the present work, the scaling factor is automatically estimated from centerF1, so that values of 450–500 Hz (average for male speakers) produce scaling factors of around 2 and values of 540–600 produce scaling factors of around 1.8.

The LPC order described in the previous subsection was also adjusted according to the centerF1 value, giving values of 15–19 for female voices and 19–23 for male voices. This improves the correspondence between the formants and the peaks in the LPC spectrum.

## 5.2.3  Vowel Space Rotation and Lip Shape Estimation

After moving the origin of the formant space to the center vowel position, the axes are rotated counterclockwise by about 25°. After rotation, the new coordinate axes are represented by logF1' and logF2'. This rotation process is motivated by the observations that the logF1' axis (after rotation) has a good correspondence with the (vertical) aperture of the lips. Figure 5.4 shows examples of the distribution of the formant maps for isolated vowels uttered by two speakers (one male and one female), superimposed with the average vowel spaces for Japanese male and female

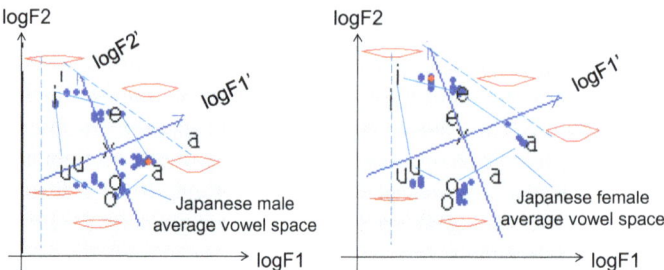

**Fig. 5.4** Examples of the distributions of single vowels uttered by a male speaker and a female speaker

speakers. The new coordinates after translation to the center of the vowel space and rotation are also shown. Note that the logF1' values are ordered as $/a/ > /e/ > 0 > /i/ \cong /o/ > /u/$, which correspond to the relative lip height variations between the different vowels. The center (schwa) vowel has logF1' = 0. Note also that /i/a nd /o/ have different lip widths (lip spreading in /i/ and lip rounding in /o/), but have approximately the same lip height.

Normalized lip height values are estimated from the formants according to the following expression:

$$lip\_height = 0.5 + height\_scale \times logF1',  \qquad (5.3)$$

where lip_height = 0 corresponds to a closed mouth, lip_height = 1 corresponds to a maximally opened mouth, the factor 0.5 corresponds to the aperture for the center (schwa) vowel, and height_scale is the scaling factor described in Sect. 5.2.2.

For the lip width, we use the F2 value before rotation. Although F2 is known to be more closely related to the front–back tongue position, there is also a relationship between lip spreading and rounding in most languages. F2 (or logF2) values are ordered as $/i/ > /e/ > /a/ > /u/ > /o/$, which correspond to the degree of lip spreading in /i/to lip rounding in /o/. Lip width values can be estimated from the formants according to the following expression:

$$delta\_lip\_width = width\_scale \times (logF2 - logcenterF2)  \qquad (5.4)$$

For delta_lip_width, positive values are obtained when F2 is higher than centerF2, as in /i/ and /e/, where the lips are spread, whereas negative values are obtained when F2 is lower than centerF2, as in /o/ and /u/, where the lips are rounded. The scaling factor width_scale determines the degree of lip spreading/rounding relative to lip height. Values around 0.5 produce humanlike lip shapes. Nonetheless, it is important to clarify that the relationship between F2 and lip width does not hold for languages with labialized vowels (as in French) and is used here as a first approximation.

Representative lip shapes generated for each vowel are shown in Fig. 5.4. In the figure, we can observe that similar lip shapes are generated for the vowels of different speakers, a result of the normalization processing.

The mapping between formants and lip shapes can be constructed in vowel or semivowel intervals. In consonants, where there is a constriction in the vocal tract, the formants are more difficult to estimate, and their relationship with lip shape is less straightforward.

In this case, the formant range and power constraints are used to discriminate consonants and fix a lip height of 0.35, corresponding to the average aperture in consonants. In the present work, the constraints for accepting the detected formants for lip motion generation were improved by establishing an upper limit for the vowel formant space and detecting fricative and affricative consonants such as /s/, /sh/, /ts/, and /ch/. The constraints are as follows:

$$F1 > centerF1 \times 0.5, \tag{5.5}$$

$$\log F2 < f(\log F1), \tag{5.6}$$

$$sonorant\_power > power\_threshold, \tag{5.7}$$

$$fricative\_power < sonorant\_power, \tag{5.8}$$

where $f(.)$ is a straight-line function defining the upper limit of vowel formants in the logF1 versus logF2 space, sonorant_power is the power value computed in the frequency band of 100–3000 Hz, where the power of vowels is concentrated, and fricative_power is the power in the frequency band of 3000–8000 Hz, where the power of fricative and affricative consonants is concentrated. The coefficients of the straight line $f(.)$ are obtained from two points: (log centerF1; log centerF2 + 0.55/height_scale) and (log centerF1 + 0.7/height_scale; log centerF2). These points were determined from observations of the distribution of the vowel formants in the logF1 versus logF2 space. They are shown by dashed lines in Fig. 5.4.

If the low-power interval exceeds a threshold of 200 ms, it is judged to be a non-speech interval, and the mouth is gradually closed by a multiplying factor of 0.95, so that the mouth is totally closed after 200–400 ms.

Finally, a moving average smoothing filter with nine taps (four past and four future points, with intervals of 10 ms between two points) is passed through the generated lip height and width sequences.

## 5.2.4  Actuator Command Generation and Data Transmission

The actuator command generation can be realized either on the operator side or the robot side. The normalized lip motion generated by the method described in Sect. 5.2.3 is mapped to the actuator commands by a linear function.

Two types of humanoid robots are evaluated in the present work: a humanoid robot with a neutral face (Telenoid-R2) and a female android (Geminoid-F). The external appearances of these two robots are shown in Fig. 5.2.

The lip motion in both robots is basically controlled by the jaw actuator, which linearly controls the degree of mouth opening (or, equivalently, the lip height). The actuator command values range from 0 to 255, where 0 corresponds to a closed mouth and 255 corresponds to the maximum opening. For Geminoid-F, the maximum actuator command value was limited to 200 to prevent excessive opening of the mouth during speech.

None of the robots have dedicated actuators for rounding or spreading the lips, so the lip width control cannot be evaluated. However, in Geminoid-F, the lip width can be partially controlled by a smiling actuator, which stretches but also raises the lip corners in the outer and upward directions.

Finally, the generated actuator commands, obtained in 10-ms frame intervals, and the audio packets are sent by TCP/IP to the robot side.

## 5.2.5  Speech and Motion Synchronization (Robot Side)

To synchronize the lip motion with the speech utterances, a suitable delay must be applied to the speech signal to account for:

(1) audio capture delay on the operator side,
(2) the processing time required to generate the lip motion parameter,
(3) network transmission delay in the transmission channel, and
(4) mechanical delay on the android side.

The audio capture delay depends on the audio capture device, being around 50–100 ms. For the processing time to generate the motion parameters, it takes a frame size (32 ms) plus four future frames for smoothing, resulting in about 70–80 ms. The network transmission in our experiment was a local network system with delays of tens of milliseconds. The mechanical delay, i.e., the time taken from the instant an actuator command is sent to the control box until the instant the robot achieves the target position, is around 100–200 ms for an air actuator (as in the android Geminoid-F) and around 50–100 ms for a servomotor (as in the robot Telenoid-R2). Considering the above delays, we set a fixed delay of 400 ms in our experiments.

Further, the transmitted and received audio packets may deviate because of:

(1)  missing packets during transmission, or
(2)  slight differences in the clocks of the capture and player devices.

These deviations can be compensated by removing or replaying audio packets in such a way as to keep a constant delay between the capture and the player.


## 5.3  Evaluation of the Proposed Method

### 5.3.1  Experimental Setup for Evaluation of Lip Height Control

Simultaneous recordings of audio, vision-based face parameter data, and motion-captured data were conducted for seven speakers (four males and three females) talking in several languages. Table 5.1 summarizes the languages spoken by the different speakers (M1–M4 are male and F1–F3 are female speakers). The speakers are researchers, intern students, or research staff in our laboratory who can speak more than one language.

To capture motion, the Hawk system from Motion Analysis was used, while for vision-based face parameter extraction, a Logitech web camera and the FaceAPI software from Seeing Machines were used. A headband condenser microphone (DPA4066) was used to capture audio. For the motion capture, four markers were attached (one in the center of the upper lips, one in the center of the lower lips, and two in the left and right lip corners), so that variations in lip height and lip width could be directly measured. Lip height and width values were also obtained directly from the face parameters extracted by FaceAPI.

Motion was generated in two humanoid robots: Telenoid-R2 (humanoid robot with neutral face) and Geminoid-F (female android). The female voices were evaluated using Geminoid-F, whereas both male and female voices were evaluated using Telenoid-R2.

**Table 5.1** List of the speakers' origins and spoken languages

| Speaker ID | Origin | Spoken languages |
|---|---|---|
| M1 | Canadian | English, Japanese |
| M2 | Chinese | Chinese, Korean, Japanese |
| M3 | Mexican | Spanish, English, Japanese |
| M4 | French | French, English, Japanese |
| F1 | Japanese | Japanese, English |
| F2 | Iranian | Persian, Japanese, English |
| F3 | Turkish | Turkish, English |

The lip height values estimated from both motion-captured data and vision-based face parameter data were subtracted from the lip height value when the lips are closed, and then normalized to the actuator command range.

For each of the speakers, video clips were recorded for each of the three motion types, audio-based ("audio"), vision-based ("vision"), and motion capture-based ("mocap"). Segments of 10–20 s were selected from the utterances spoken by each speaker and each language (as presented in Table 5.1), resulting in eleven video stimuli for male speakers and seven for female speakers for each of the three motion types.

In the evaluation experiment, the stimuli (video clips) for the three different motion types were played in random order for the motion types. Subjects were asked to grade the naturalness scores on a 7-point scale, where 1 is the most unnatural, 7 is the most natural, and 4 is "difficult to decide." We also asked the subjects to give their reasons for scoring a motion as "unnatural."

Twenty subjects (ranging from teenagers to those in their 40 s) participated in the evaluation test; the subjects are not involved in robotics research.

### 5.3.2   Evaluation Results and Analysis for Lip Height Control

Figure 5.5 shows the subjective naturalness scores for each motion type ("audio," "video," "mocap") in both robots. For Telenoid-R2, both male and female voices were evaluated, whereas for Geminoid-F, only female voices were evaluated. The Telenoid-R2 results for male and female voices were separated (denoted by (M) and (F), respectively) to allow for a direct comparison between the female voices in Geminoid-F. The naturalness of individual motion types was obtained by normalizing the scores to a scale from 0 (most unnatural) to 100 (most natural). A repeated measure analysis of variance (ANOVA) test was conducted to determine the significance of the scores.

**Fig. 5.5** Subjective naturalness scores for lip motion generated by different motion types in different robot types

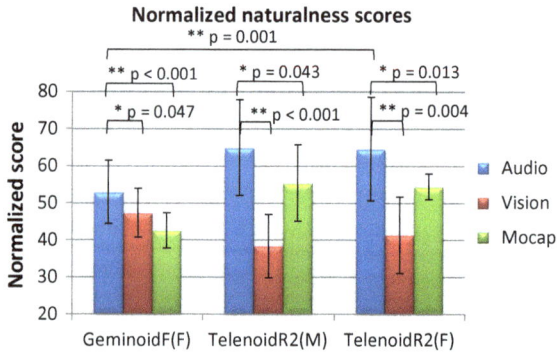

The results in Fig. 5.5 show that the proposed audio based method was judged to be more natural than both vision- and motion capture-based approaches in both robot types.

Comparing the scores for Geminoid-F(F) and Telenoid-R2(F), the latter scores higher in the proposed audio-based and motion capture-based methods.

We then analyzed why subjects judged a motion as being unnatural. Different reasons were attributed for different motion and robot types.

The reasons for the worst performance of the "vision" approach were that the motion was "not synchronized" (subjects felt there was a mismatch between lip motion and speech) in female voices and because there was "little motion" (subjects felt that the lips should move more) in male voices. An explanation for the results in male voices is that lip tracking by FaceAPI often resulted in errors in two of the male speakers (one of whom had a mustache and beard), resulting in fewer motions.

For the "mocap" approach, the main reasons for low naturalness scores in Geminoid-F(F) were that motion was "not synchronized" and there was "excessive openness" (subjects felt that the mouth remained open most of time). Possible explanations for these results are that speakers do not always close their mouth completely or move their lips in non-speech intervals. Although such motion "naturally" appears in humans, it might be undesirable in robots. Note that in "mocap," the mean subjective naturalness was around 50 for Telenoid-R2 and around 40 for Geminoid-F (on a scale of 0–100). This implies that reproducing the actual lip motion of the operator does not necessarily imply an impression of natural motion in the robot.

The main reasons for unnaturalness in the "audio" case for Geminoid-F(F) were that there was "little motion" and "insufficient openness" (subjects felt that the mouth should open more). Better control of scaling may improve the naturalness of Geminoid-F. Another possible explanation for lower scores for Geminoid-F compared to Telenoid-R2 is the appearance of the two robot types, with subjects applying a harsher evaluation for robots whose appearance is closer to that of a human. These are topics for future investigation.

### 5.3.3  Evaluation of Partial Lip Width Control

The lip width control cannot be thoroughly evaluated because neither Geminoid-F nor Telenoid-R2 has actuators for rounding the lips. However, Geminoid-F has one actuator for stretching the lip corners in the outer and upward directions to realize smiling behavior. Although this lip corner stretching is not strictly in the same direction as the lip corner spreading in /i/ and /e/ (because of the presence of a simultaneous upward direction), we considered that lip width control could be partially evaluated in Geminoid-F.

We then mapped the lip width control parameters to this lip corner actuator and conducted an experiment to verify whether or not the use of this actuator improves the perceived naturalness.

**Fig. 5.6** Subjective naturalness scores for lip motion generated by jaw actuator only ("jaw only") and by both jaw and lip corner actuators ("jaw + lip corner") for the audio-based method in Geminoid-F(F)

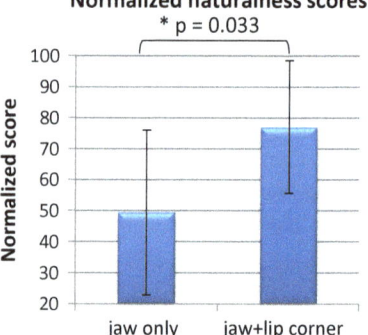

The lip corner actuator commands were generated by a linear mapping between the lip width parameters to the lip corner actuator according to the following expression:

$$LipCornerActuator = delta\_lip\_width \times 64 + 64 \qquad (5.9)$$

For this actuator, the maximum value was limited to 127 to prevent the appearance of a strong smiling expression.

The motions for the female voices (in Table 5.1) were generated in Geminoid-F using the lip corner actuators. Ten subjects evaluated the naturalness of the generated motions in the same way as in the previous experiment.

Figure 5.6 shows the subjective naturalness scores for lip motion generated by jaw actuator only ("jaw only") and by both jaw and lip corner actuators ("jaw + lip corner").

The subjective scores in Fig. 5.6 indicate that the use of the lip corner actuator increases the overall naturalness of the lip motion. However, some of the subjects preferred "jaw only" to "jaw + lip corner." The main reason given for unnaturalness was that the lip corner motion was a little bit "jerky" in some of the stimuli. This jerky motion was mainly caused by formant extraction errors in /o/ and /u/, where F2 was misdetected as the true F3, and in consonant portions where high F2 values were detected. These eventual misdetections in F2 caused a sudden change in lip corner actuation, resulting in unnatural motion. However, the results show that, despite problems in formant extraction, the lip corner actuator can be used to generate more natural lip motion.

## 5.4 Discussion

Regarding the constraints of the present method, it is desirable to have reasonably good formant extraction. We have observed that the simple method of peak picking in the LPC-smoothed spectrum often fails when the first two formants are close (such as in /o/ and /u/), A relatively natural motion could be generated by the lip height control of the present method, even under the current formant extraction limitations. However, we also observed that errors in formant extraction cause more severe problems in lip width control. Further improvements in the formant extraction method are an issue for future work.

Another remaining issue is lip shape estimation for consonants. With the current approach, the lip closure can be correctly detected in the bilabials /m/ and /b/ when the transitional parts to/from the neighbor vowels exhibit F1 and F2 lowering curves, but these curves are seldom observed in /p/ (where the lips should close). There is also a trade-off in smoothing the generated actuator commands, as this prevents the occurrence of jerky motion but also prevents the lips from closing completely in bilabials.

However, even though the proposed method cannot generate perfect lip motion, the evaluation results show that the method is more effective than vision-based approaches for the purpose of generating natural lip motion, so that the user feels that the robot is speaking.

Finally, no clear differences were found among different languages, so the audio-based lip motion was consistently judged as being more natural than other motion types, regardless of language. However, the absolute naturalness judgment could be different for subjects with different origins. This is another topic for future investigation.

## 5.5  Conclusion

With the aim of teleoperating the lip motion of a remote humanoid robot in synchrony with the operator's voice, we developed and evaluated a formant-based lip motion (height and width) generation method.

An evaluation of lip height control indicates that the proposed method can generate lip motion with superior subjective naturalness to motion generated by both motion-captured data and vision-based lip shape extraction methods, which try to reproduce the actual lip motion of the operator.

By evaluating the partial lip width control, we found that additional control using the lip corner actuators improves the perceived naturalness.

However, an analysis of the low naturalness scores revealed some remaining issues concerning formant extraction and the detection of bilabial consonants. These will be the targets of future work.

**Acknowledgements**  This work was supported by JSPS Grant-in-Aid for Scientific Research No. 25220004 and by JST CREST Studies on cell phone-type teleoperated androids transmitting human presence.

# References

1. Ishi, C., C. Liu, H. Ishiguro, and Hagita, N. 2012. Evaluation of formant-based lip motion generation in tele-operated humanoid robots. In *2012 IEEE/RSJ international conference on intelligent robots and systems (IROS 2012)*, 2377–2382.
2. Cohen, M., and D. Massaro. 1993. Modeling coarticulation in synthetic visual speech. In *Models and techniques in computer animation*.
3. Tamura, M., S. Kondo, T. Masuko, and T. Kobayashi. 1998. Text-to-visual speech synthesis based on parameter generation from HMM. In *Proceedings of ICASSP98*, 3745–3748.
4. Hong, P., Z. Wen, and T. Huang. 2002. Real-time speech-driven face animation with expressions using neural networks. *IEEE Transactions on Neural Networks* 13 (4): 916–927.
5. Beskow, J., and M. Nordenberg, 2005. Data-driven synthesis of expressive visual speech using an MPEG-4 talking head. In *Proceedings of interspeech 2005*, 793–796.
6. Hofer, G., J. Yamagishi, and H. Shimodaira. 2008. Speech-driven lip motion generation with a trajectory HMM. In *Proceedings of the interspeech 2008*, 2314–2317.
7. Salvi, G. 2006. Dynamic behaviour of connectionist speech recognition with strong latency constraints. *Speech Communication* 48 (7): 802–818.
8. Takacs, G. 2009. Direct, modular and hybrid audio to visual speech conversion methods—a comparative study. In *Proceedings of the Interspeech09*, 2267–2270.
9. Hofer, G., and K. Richmond. 2010. Comparison of HMM and TMDN methods for lip synchronization. In *Proceedings of the Interspeech 2010*, 454–457.
10. Zhuang, X., et al. 2010. A minimum converted trajectory error (MCTE) approach to high quality speech-to-lips conversion. In *Proceedings of interspeech 2010*, 1726–1739.
11. Wu, J., et al. 2008. Statistical correlation analysis between lip contour parameters and formant parameters for Mandarin monophthongs. In *Proceedings of the AVSP2008*, 121–126.
12. Ishi, C., C. Liu, H. Ishiguro, and N. Hagita. 2011. Speech-driven lip motion generation for tele-operated humanoid robots. In *Proceedings of the Auditory-Visual Speech Processing, 2011 (AVSP2011)*, 131–135.
13. Markel, J.D., and A.H. Gray. 1976. *Linear prediction of speech*. Berlin, Heidelberg, New York: Springer.
14. Titze, I.R. 1994. *Principles of voice production*, 136–168. NJ: Prentice Hall.

# Chapter 6
# Analysis of Head Motions and Speech, and Head Motion Control in an Android Robot

Carlos Toshinori Ishi, Hiroshi Ishiguro and Norihiro Hagita

**Abstract** With the aim of automatically generating head motions during speech utterances, analyses are conducted to verify the relations between head motions and linguistic and paralinguistic information carried by speech utterances. Motion-captured data are recorded during natural dialogue, and the rotation angles are estimated from the head marker data. Analysis results show that nods frequently occur during speech utterances, not only for expressing specific dialogue acts such as agreement and affirmation, but also to indicate syntactic or semantic units, which appear at the last syllable of the phrases, in strong phrase boundaries. The dependence on linguistic, prosodic and voice quality information of other head motions, including shakes and tilts, is also analyzed, and the potential for using this to automatically generate head motions is discussed. Intra-speaker variability and inter-speaker variability on the relations between head motion and dialogue acts are also analyzed. Finally, a method for controlling the head actuators of an android based on the rotation angles is proposed, and the mapping from human head motions is evaluated.

**Keywords** Head motion · Nonverbal communication · Paralinguistic information · Prosody · Android science

## 6.1 Introduction

Head motion occurs naturally during speech utterances. Sometimes, this motion is intentional and carries a clear meaning in communication; for example, nods are frequently used to express agreement, whereas head shakes are used to express

---

This chapter is a modified version of previously published papers, [1, 2], edited to be comprehensive and fit with the context of this book.

---

C. T. Ishi (✉) · H. Ishiguro · N. Hagita
ATR Intelligent Robotics and Communication Labs, Kyoto, Japan
e-mail: carlos@atr.jp

© Springer Nature Singapore Pte Ltd. 2018
H. Ishiguro and F. Dalla Libera (eds.), *Geminoid Studies*,
https://doi.org/10.1007/978-981-10-8702-8_6

disagreement. Most head motions are produced unconsciously. However, they are somehow synchronized with the speech utterances. One of our motivations for the present work is to obtain a method for generating head motion from speech signals in order to automatically control the head motion of a humanoid robot (such as an android).

Head motion analyses generally focus on two problems: one is how to recognize the user's head motion and interpret its role in communication (e.g., [3, 4]); the other is how to generate the robot's head motions and synchronize them with the robot's speech. The analysis in the present work focuses on the latter problem of generating natural head motion while the robot is speaking.

Many studies have attempted to find a correspondence between head motions and prosodic features such as the fundamental frequency (F0) contours (which represent pitch movements) and energy contours in several languages [5–10].

For example, in [5], head motions were associated with speech over F0. Experiments using speech utterances read by one American English speaker (ES) and one Japanese speaker (JS) showed a mean accuracy of 73% for JS and 88% for ES in terms of estimating F0 from head motions. However, the inverse prediction (from F0 to head motion) showed only 25% mean accuracy for JS and 50% for ES. In addition, the correlation among F0 and the 6 degrees of freedom (DOF) (rotation and translation) of the head motion ranged from 0.39 to 0.52 for ES and 0.22–0.30 for JS. This shows that using F0 alone is not sufficient to generate head motions.

Variations in speech according to head movement have been observed for English sentences [6]. Emphasis on a particular word is often accompanied by head nodding, and a rise of the head often corresponds with a rise in the voice. These movements are known as "visual prosody." In [7], facial parameters (including head motions) were analyzed for short Swedish utterances in which the focal accent was systematically varied in a variety of expressive modes including certainty, confirmation, questioning, uncertainty, happiness, angriness, and neutrality. The results indicate that, in all expressive modes, words with a focal accent are accompanied by a greater variation in facial parameters than words in non-focal positions.

Additionally, most previous studies have analyzed read speech or acted emotional speech data. In [11], relations between head movements and the semantics of utterances are analyzed in Japanese spoken dialogue by considering speaking turn and speech functions. Regarding head motion generation, a system that uses corpus-based selection strategies to specify the head and eyebrow motion of an animated talking head has been reported [12]. The system considers syntactic, prosodic and pragmatic context for generating the motions. However, only data for one speaker were analyzed.

We consider the relationship between prosodic features and head motions to be language dependent, as the function of the prosodic features may differ if, for example, the language is tonal (such as Chinese and Thai), a pitch-accent language (such as Japanese), or non-tonal (such as English and other European languages).

In the present work, we consider the use of local prosodic features (e.g., phrase-final tones) instead of the global utterance features, voice quality features (especially when F0 cannot be measured, as for vocal fry and whisper) besides the commonly used prosodic features and linguistic information linked to dialogue act information. Further, although most research on head motion analysis use read or acted speech, we use spontaneous speech data for analysis. We also analyze the relationship between head motion and speech for several speakers and discuss the intra-speaker variability and inter-speaker variability.

Finally, we consider the problem of generating natural head motions in humanoid robots during speech utterances as a three-step process. The first step is to investigate the relationship between speech and head motions. The second is to verify how natural head motions can be realized in a humanoid robot. The third step, which is the final goal of the present work, is to generate head motions from speech signals. In the present work, we focus on the first two steps.

This chapter is organized as follows. Section 6.2 presents analysis of relationship between head motions and several types of linguistic/paralinguistic information carried by the speech utterances. The potential to predict head motions from speech is discussed. In Sect. 6.3, inter-speaker variability and intra-speaker variability analysis on the relationship between head motions and dialogue acts are presented. Section 6.4 treats the problem of reproducing human head motions in a specific android, and Sect. 6.5 concludes the chapter. The present chapter is based on the previous works [13, 14].

## 6.2  Analysis of Head Motion and Speech: The Roles of Linguistic, Prosodic, and Dialogue Act Information

This section presents analysis of head motion and several types of linguistic and paralinguistic information conveyed by speech.

### 6.2.1  Data Collection and Annotation

#### 6.2.1.1  Data

About 30 min of free dialogue between two Japanese graduate students who are familiar with one another was recorded. The target is a female speaker, and the interlocutor is a male speaker.

Audio, video and motion capture data were simultaneously recorded. The Hawk system from nac Image Technology Inc. was used for the motion capture. Four infrared cameras were arranged in an arc in front of the speaker. Thirty-eight hemispherical passive reflective markers were applied to the speaker's face, head, and shoulders, as shown in Fig. 6.1. Most of the markers placed on the face were used to capture lip motions and eye blinks and were disregarded in the present analysis. Only

**Fig. 6.1** Markers and angles
used to describe head motions

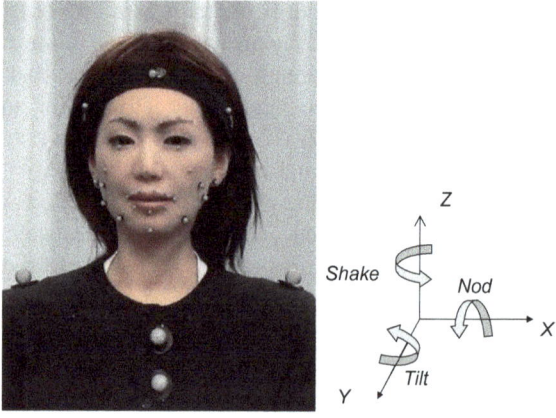

six markers (placed on the head, nose, and earlobes), which provide something of a
static reference frame for the head, were used to characterize the head motions.

The three rotation angles shown in Fig. 6.1 were used to describe the head
motions. We use the terms "nod," "shake," and "tilt" in correspondence with the
terms "pitch," "yaw," and "roll" used in aerodynamics. The head rotation angles
were estimated from the markers based on singular value decomposition [15]
according to the following expression:

$$\left[\mathbf{U}, \mathbf{D}, \mathbf{V}^{\mathrm{T}}\right] = svd(reference \times target), \tag{6.1}$$

where reference and target are the 3D marker set of the neutral and current positions, respectively, translated to new coordinates with their centroids as the origin.
The neutral positions were obtained in intervals where the subjects were looking
directly ahead. The rotation matrix was obtained by

$$\mathbf{R} = \mathbf{V}\mathbf{U}^{\mathrm{T}}. \tag{6.2}$$

The rotation angles were then obtained from the elements of the rotation matrix
**R** according to the following expressions:

$$\text{tilt angle} = atan2(\mathrm{R}(2,1), \mathrm{R}(1,1)), \tag{6.3}$$

$$\text{nod angle} = atan2(-\mathrm{R}(3,1), sqrt(\mathrm{R}(3,2)^{\wedge}2 + \mathrm{R}(3,3)^{\wedge}2)), \tag{6.4}$$

$$\text{shake angle} = atan2(\mathrm{R}(3,2), \mathrm{R}(3,3)), \tag{6.5}$$

where *sqrt* is the square root function, ^ is the power function and *atan2* is the
arctangent function of MATLAB.

#### 6.2.1.2   Head Motion Tags

The following tag set was used to annotate the head motions. These are thought to cover the meaningful events in speech communication.

- **no**: no head motion.
- **nd** (nod): single nod (down-up motion).
- **mnd** (multiple nods): multiple nods occur during the phrase.
- **fd** (face down): the face moves down.
- **ud** (up-down): single up-down motion.
- **fu** (face up): the face moves up.
- **ti** (tilt): head tilts occur within the phrase.
- **sh** (shake): shakes (left-right motions) occur during the phrase.

Nods were not necessarily realized by perfect vertical head motions, and may be accompanied by a slight head tilt. In this case, we considered the motion with the strongest magnitude.

#### 6.2.1.3   Linguistic Information

A preliminary observation of the head motions in the data indicated that nods frequently occurred in particles (such as "ne," "de," "kara"). Interjections (such as "un," "ee," "hee") were also often accompanied by a head motion. To verify the relations between such morphemes and the head motions, we segmented the utterances in phrase units ("bunsetsu" in Japanese), because such morphemes usually appear in the boundary of the phrases.

The speech utterances were manually segmented in phrase units, and the last morpheme of each phrase was transcribed by a native Japanese speaker.

The annotation process resulted in the segmentation of 535 phrases. Special labels were also annotated for laughing and breathing.

#### 6.2.1.4   Dialogue Act Tags

Dialogue act tags were annotated for each phrase in the dataset according to the following set, which is based on the tags proposed in [16]. The tags consider dialogue acts such as affirmative or negative reactions, expression of emotions such as surprise, and turn-taking functions.

- **k** (keep): the speaker is keeping the turn; a short pause or a clear pitch reset occurs at strong phrase boundaries.
- **k2** (keep): weak phrase boundaries in the middle of an utterance (when no pause exists between phrases).
- **k3** (keep): the speaker lengthens the end of the phrase, usually when thinking, but keeping the turn (may or may not be followed by a pause).

- **f** (filler): the speaker is thinking or preparing the next utterance, e.g., "uuun," "eeee," "eettoo," "anoo" ("uhmmm").
- **f2** (conjunctions): can be considered as non-lengthened fillers, e.g., "dakara," "jaa," "dee" ("I mean," "so").
- **g** (give): the speaker has finished talking and is giving the turn to the interlocutor.
- **q** (question): the speaker is asking a question or seeking confirmation from the interlocutor.
- **bc** (backchannels): the speaker is producing backchannels (agreeable responses) to the interlocutor, e.g. "un" usually accompanied by a fall pitch movement, "hai" ("uh-huh," "yes").
- **su** (admiration/surprise/unexpectedness): the speaker is producing an expressive reaction (admiration, surprise) to the interlocutor's utterances, e.g., "heee," "uso!," "ah!" ("wow," "really?").
- **dn** (denial, negation): For example, "iie" and "uun" accompanied by a fall-rise pitch movement ("no," "uh-uh").

We are aware that the above set of dialogue act categories is not complete, but we consider it a basic set for applications to human–robot interaction.

### 6.2.1.5  Prosodic and Voice Quality Features

Although automatic procedures could be conducted to extract the phrase-final tones [17, 18], in the present work, we hand-annotated the prosodic and voice quality features for analysis purposes according to the following tags.

- **rs (rise)**: rising tone.
- **fa (fall)**: falling tone (includes reset-fall tones).
- **fr (fall-rise)**: fall pitch movement followed by a rise movement.
- **hi (high)**: high pitch.
- **mi (mid)**: middle-height pitch.
- **lo (low)**: low pitch.
- **cr (creaky)**: creaky voice or vocal fry (a voice quality characterized by very large intervals between glottal excitation pulses), when F0 is lowered and cannot be reliably measured.
- **wh (whisper)**: whisper (absence of F0).

For each phrase, the tone tags were annotated by one subject with experience in prosody and voice quality annotation.

## 6.2.2   Relationship Between Head Motions and Speech

### 6.2.2.1   Head Motions and Morphemes

Regarding the relations between morphemes and head motions, the analysis results firstly indicated that nods frequently occur at the boundary of phrases, regardless of the morpheme at the phrase boundaries. This frequently occurs with particles, which usually appear at the boundary of phrases, but was not restricted to the particles. Rather, nods seem to be more closely related with the dialogue act functions carried by the morphemes at the phrase boundaries, as will be shown in the next subsection.

Although only a small number of shakes were observed in the data, they occurred in utterances expressing negation or rejection. In the current dataset, shakes appeared in the morpheme "uun" (accompanied by a fall-rise intonation) and in utterances ending with "…nai," which expresses negation.

### 6.2.2.2   Head Motions and Dialogue Acts

Table 6.1 summarizes the relationship between the head motions and the dialogue act functions.

From Table 6.1, it is clear that the nod (**nd**) motion occurred most frequently. Firstly, an expected result was that nods were present in almost all backchannels (**bc**). Nods were also frequently observed at the strong phrase boundaries (**k, g, q**), regardless of the presence or the type of particle at the phrase final.

A surprising result is that nods were more frequent than up-down or face-up motions, even in questions (**q**), where phrase finals are usually accompanied by a rising intonation. This is one of the factors that reduce the correlation between pitch and head motions.

**Table 6.1**  Distribution of dialogue acts (rows) and head motions (columns)

|      | Total | nd  | fd  | ud  | fu  | ti  | sh  | mnd | no  |
| ---- | ----- | --- | --- | --- | --- | --- | --- | --- | --- |
|      |       | *142* | 24 | 28 | 20 | 33 | 4 | 3 | 189 |
| k    | 61    | *35* | *5* | 1 | 0 | 3 | 0 | 0 | 17 |
| k2   | 137   | 2 | 2 | 6 | 6 | 11 | 0 | 0 | *106* |
| k3   | 28    | 4 | 1 | 2 | 1 | 4 | 0 | 0 | *16* |
| f1   | 15    | 3 | 1 | 0 | 1 | 2 | 0 | 0 | *8* |
| f2   | 22    | 2 | 0 | 0 | 3 | 5 | 0 | 0 | *12* |
| g    | 79    | *29* | *11* | *12* | *2* | 5 | 2 | 1 | 14 |
| q    | 25    | *9* | 3 | 3 | 2 | 0 | 0 | 0 | 7 |
| bc   | 71    | *58* | 1 | 1 | 1 | 0 | 0 | 2 | 7 |
| su   | 12    | 0 | 0 | *2* | *4* | 3 | 0 | 0 | 1 |
| dn   | 4     | 0 | 0 | 1 | 0 | 0 | *2* | 0 | 1 |

A particular result was observed in pre-pause phrases when the subject is keeping turn (**k**): nods (**nd**) frequently occurred, while upward motions (**ud, fu**) never occurred.

Nods occur with less frequency at "weak" phrase boundaries in the middle of an utterance (**k2**), and at phrase boundaries where the speaker is thinking or indicates that he/she has not finished their utterance (**k3, f, f2**). In these dialogue act categories, the absence of head motions (**no**) is dominant (Table 6.1).

Phrases expressing surprise/admiration/unexpectedness (**su**) are usually accompanied by upward (**fu, ud**) or tilt motions (**ti**). As the number of **su** phrases is small in the present dataset, more detailed analysis with a larger database is necessary to verify these trends in the expression of different emotions.

Nods sometimes occurred at the beginning of the phrase (10 phrases removed from Table 6.1). This is thought to be a kind of signal to the interlocutor that the speaker will take the turn and start to utter. Face-up motions sometimes occur at the beginning of the phrases with the same purpose.

Regarding the motion shapes, we observed that nods are often accompanied by a small upward motion before the usual down-up motion.

A sequence of multiple nods (**mnd**) occurs along the whole utterance when the speaker is expressing agreement. When multiple nods occur in a sequence of backchannels, such as in "*un un un un,*" the first nod is usually larger than the others.

Only four shakes (**sh**) were observed in the data. They occurred in utterances expressing negation or rejection and were more dependent on the linguistic content, as discussed in the previous subsection.

Finally, head tilts (**ti**) occurred in almost all dialogue acts, excluding backchannels (**bc**), questions (**q**), and denial (**dn**). However, tilts were observed to occur over longer durations than nods and shakes, including multiple phrases. These tilts often occurred during or right after a filler or a disfluency. Further, although the direction of tilts (right or left) was also annotated, no significant differences were observed between their distributions.

### 6.2.2.3 Head Motions and Prosodic and Voice Quality Features

A relationship between pitch and head motions exists, but the correlation is not high, as pointed out in [5]. Japanese is a pitch-accent language, so many pitch movements occur within utterances because of lexical accents. As described in previous sections, our analyses indicated that head motions do not occur at every pitch accent nucleus, but, rather, occur more frequently at the phrase boundaries.

However, even for the phrase boundary tones, a straight relation between pitch movements (tones) and head motions could not be observed. Table 6.2 presents the distribution of head motions and phrase-final tones, and Table 6.3 roughly, we can say that falling tones (**fa**) and a creaky voice (**cr**) are usually accompanied by nods

**Table 6.2** Distribution of head motions (rows) and phrase-final tones (columns)

| Total | rs | fa | fr | hi | mi | lo | cr | wh |
|---|---|---|---|---|---|---|---|---|
| | 25 | 108 | 3 | 26 | 205 | 14 | 58 | 13 |
| nd | 8 | *83* | 1 | 0 | 26 | 4 | *20* | 4 |
| fd | 2 | *4* | 0 | 1 | 5 | 2 | *7* | 2 |
| ud | 5 | 1 | 0 | 1 | 11 | 3 | 6 | 1 |
| fu | 3 | 2 | 0 | 4 | 11 | 0 | 1 | 1 |
| ti | 1 | 4 | 0 | 2 | 20 | 1 | 4 | 1 |
| sh | 0 | 0 | 2 | 0 | 1 | 0 | 1 | 0 |
| nm | 0 | 0 | 0 | 0 | 3 | 0 | 0 | 0 |
| no | 6 | 14 | 0 | *18* | *126* | 3 | 18 | 4 |

**Table 6.3** Distribution of dialogue acts (rows) and phrase-final tones (columns)

| | rs | fa | fr | hi | mi | lo | cr | wh |
|---|---|---|---|---|---|---|---|---|
| k | 1 | *45* | 0 | 1 | 12 | 0 | 2 | 0 |
| k2 | 0 | 1 | 0 | *12* | *110* | 2 | 11 | 1 |
| k3 | 0 | 10 | 0 | 4 | 12 | 0 | 2 | 0 |
| f1 | 0 | 3 | 0 | 0 | 11 | 0 | 1 | 0 |
| f2 | 0 | 2 | 0 | 2 | 14 | 0 | 4 | 0 |
| g | 2 | 1 | 0 | 1 | 20 | *9* | *36* | *10* |
| q | *19* | 1 | 1 | 1 | 1 | 0 | 1 | 0 |
| bc | 0 | *45* | 0 | 0 | 19 | 3 | 1 | 2 |
| su | 2 | 0 | 0 | 5 | 5 | 0 | 0 | 0 |
| dn | 1 | 0 | 2 | 0 | 1 | 0 | 0 | 0 |

(**nd**), whereas high and middle-height pitch tones (**hi**, **mi**) are not usually accompanied by any head motion (**no**).

Although a clear correspondence could not be found between tones and head motions, Table 6.3 suggests a better correspondence between tones and dialogue acts. Rising tones (**rs**) basically appear in questions (**q**), falling tones (**fa**) appear frequently in turn-keeping functions (**k**) and backchannels (**bc**), high and middle-height tones are frequent in weak phrase boundaries (**k2**), and low pitch, creaky, and whisper tones are frequent in turn-giving functions (**g**). The use of morpheme information along with this tone information would produce a better correspondence with the dialogue act functions [14, 16].

Finally, regarding the relation between head motions and other voice qualities (not included in the tables above), possible correspondences were mainly observed in nods and tilts. In confident utterances, accompanied by a normal or more pressed voice quality, nods tend to be more frequent, whereas in non-confident utterances, usually accompanied by a more lax or breathy voice quality, nods tend to be of smaller magnitude, and tilts or no head motions become more frequent. A deeper study would be necessary to verify such trends.

### 6.2.3 Some Rules for Head Motion Generation

The analysis results in the present section showed that there is some relationship between prosodic features and head motion, but among the information conveyed by speech, dialogue act functions are closely related to head motions. On the other hand, dialogue act functions can be predicted from linguistic and prosodic features. Based on the analysis results, some rules for generating head motions from speech can be summarized as follows.

- backchannels ("*un*," "*hai*," …) + {falling or mid-height tones}: high percentage of nods occurring.
- strong phrase boundaries + {low pitch, falling tones, creaky, whisper}: high percentage of nods occurring.
- weak phrase boundaries + falling tone: lower percentage of nods occurring.
- denial/rejection words ("*uun*" + fall-rise tone, "*iie*," …): high percentage of shakes occurring.
- questions (usually accompanied by a rising tone): select nods or face-up motions.
- fillers and disfluencies: high percentage of tilts occurring.

## 6.3 Analysis of Head Motion and Speech: Intra-Speaker Variability and Inter-Speaker Variability

The analysis for a single speaker in the previous section showed that head motion is closely related to the dialogue acts expressed by speech. In the present section, intra-speaker variability and inter-speaker variability are analyzed for multiple speakers.

### 6.3.1 Data Collection and Annotation Procedure

#### 6.3.1.1 Data

Data for seven speakers (four males and three females) were used for analysis. Table 6.4 lists the speakers, their respective interlocutors, and the relationships between them. The relationship between the dialogue partners will be useful for interpreting the intra-speaker variability.

Several sessions of 10–15 min of free dialogue conversation between dialogue partners were recorded. A total of 19 sessions were recorded (FMH–FKH (two sessions), FKN–FKH (two), FMH–MHI (one), FKN–MHI (one), FMH–MSN (one), FKN–MSN (one), FMH–MIT (three), FKN–MIT (three), FMH–MSR (five)).

**Table 6.4** List of speakers, interlocutors and their relationships

| Speaker (age) | Interlocutor (relationship with speaker) |
|---|---|
| FMH (30) | FKH (mother), MHI (boss of the boss), MSN (colleague of the boss), MSR (boyfriend) |
| FKN (30) | FKH (mother of the colleague), MHI (boss of the boss), MSN (colleague of the boss) |
| FKH (50 s) | FMH (daughter), FKN (daughter's friend) |
| MHI (40 s) | FMH (subordinate of subordinate), FKN (subordinate of subordinate) |
| MSN (38) | FMH (subordinate of colleague), FKN (subordinate of colleague) |
| MIT (30) | FMH (friend of friend), FKN (friend) |
| MSR (29) | FMH (girlfriend) |

The speakers were instructed to talk freely about any topics. The resulting dialogues were mostly everyday conversations on topics such as past events, future plans for trips, self-introductions, topics about a common known person, topics regarding family and work and past experiences.

Simultaneous recordings of audio, video and motion data were conducted for both dialogue partners. The distance between the subjects was set as large as possible while allowing the motion of both subjects to be captured, resulting in about 1 m of separation. Directional microphones (Sanken CS-1) were positioned pointing toward each subject.

The Hawk motion capture system from Motion Analysis was used. Ten infrared cameras were arranged in a rectangle around a room to capture the motions of both speakers. Seven hemispherical passive reflective markers were applied to the speaker's head, nose, and chin, as shown in Fig. 6.2. The number of markers was reduced in comparison to the analysis in the previous section, to reduce the efforts for post-process capture errors. The markers on the head and nose provided a static reference frame for the (rigid) head, with the marker on the chin (relative to the nose marker) was used to align the motion data with the speech audio data, as systematic errors sometimes occurred in the synchronization process.

**Fig. 6.2** Markers and angles used to describe head motions

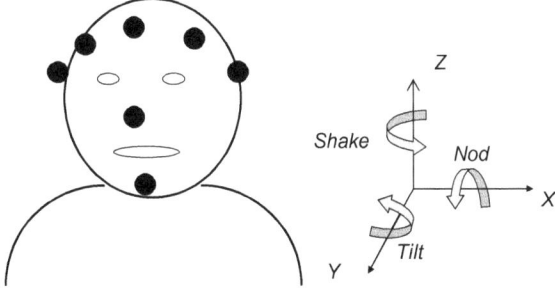

Note that although the situation with motion capture markers is unnatural, all speakers agreed that the feeling of unnaturalness only occurred at the beginning of the first session. After a while (1–2 min), they simply forgot about the markers and could talk naturally.

The speech utterances were manually segmented in phrase units and transcribed by a native Japanese speaker. The segmentation resulted in a total of 16920 phrases. Head rotation angles were computed based on the same approach described in Sect. 6.2.1.1.

### 6.3.1.2  Head Motion Tags

The head motion tags, defined in Sect. 6.2.1.2, were used to annotate head motion in the present section. Head motion tags were annotated by one subject based on the three measured angles and the video information, and then checked/corrected by a second subject (both research assistants). The second subject modified 5.3% of the labels.

Table 6.5 presents the distribution of the annotations for each head motion tag. Note that nods (**nd** plus **mnd**) are the head motions that occur most frequently. The column "others" includes phrases where subjects were unsure about choosing a specific head motion tag.

### 6.3.1.3  Dialogue Act Tags

Dialogue act tags were annotated using the tag list described in Sect. 6.2.1.4. For each phrase, dialogue act tags were annotated by one subject and then checked/ corrected by a second subject (same subjects as for head motion tag annotation in the previous subsection). The second subject modified 5.9% of the labels.

Table 6.6 shows the distribution of annotations for each dialogue act tag. The column "others" includes phrases where subjects were unsure about choosing a specific dialogue act tag, as well as sub-categories of **g** (give the turn) such as greetings and interjections other than **bc** (simple backchannels) and **su** (surprise, admiration). A detailed analysis of these sub-categories is left for future work.

**Table 6.5** Distribution of the annotations for each head motion tag. The second line shows the number of phrases, and the third line shows the percentage

| no | nd | mnd | fd | ud | fu | ti | sh | Others |
|------|------|------|-----|-----|------|------|-----|--------|
| 5198 | 2843 | 1401 | 556 | 644 | 1364 | 1553 | 146 | 3215 |
| 30.7 | 16.8 | 8.3 | 3.3 | 3.8 | 8.1 | 9.2 | 0.9 | 19.0 |

**Table 6.6** Distribution of the annotations for each dialogue act tag. The second line shows the number of phrases, and the third line shows the percentage

| g | q | bc | k | k2 | k3 | f | f2 | dn | su | Others |
|------|-----|------|------|------|-----|-----|------|----|-----|--------|
| 2589 | 969 | 2425 | 1607 | 2435 | 344 | 253 | 1944 | 54 | 226 | 4074 |
| 15.3 | 5.7 | 14.3 | 9.5 | 14.4 | 2.0 | 1.5 | 11.5 | 0.3 | 1.3 | 24.1 |

## 6.3.2 Relation Between Head Motion and Speech

### 6.3.2.1 Head Motion and Dialogue Acts

The panels in Fig. 6.3 show the distribution of head motions for each dialogue act function, arranged for each speaker and interlocutor group (indicated in the x-axis of the bottom panel). The y-axis represents the percentage of occurrence for each head motion tag.

The general trends of head motion for each dialogue act are first analyzed by observing the overall distribution of each graph in Fig. 6.3, disregarding the intra-speaker variability and inter-speaker variability.

It can be noted that nods (**nd**) and multiple nods (**mnd**) occur with high frequency during backchannels (**bc**). Nods were also frequently observed at strong phrase boundaries (**k, g, q**). Even in questions (**q**), where the end of the phrase is usually accompanied by a rising intonation, nods were more frequent than up-down or face-up motions. This is one factor that contributes to the reduced correlation between pitch (F0 contours) and head motions.

Nods occur with less frequency at weak phrase boundaries in the middle of an utterance (**k2**) and at phrase boundaries where the speaker is thinking or indicating that he/she has not completed the utterance (**k3, f, f2**). In these dialogue act categories, the predominance of no head motions (**no**) can be observed.

A more detailed analysis on the sequence of multiple nods (**mnd**) indicates that they tend to occur during the whole utterance when the speaker is expressing strong agreement, deep understanding, or interest in the interlocutor.

Phrases expressing surprise/admiration/unexpectedness (**su**) occurred with less frequency in the database. No head motion (**no**), face-up motions (**fu**), and tilt motions (**ti**) were predominant.

These results are all in agreement with the results for one female speaker presented in the previous section.

### 6.3.2.2 Intra-Speaker Variability and Inter-Speaker Variability

The analysis results for inter-speaker variability indicate that the frequency of head motions varies according to the speaker.

For example, two of the four male speakers (MSN and MHI) exhibited much less head motion than the others. One hypothesis for this fact is that their social

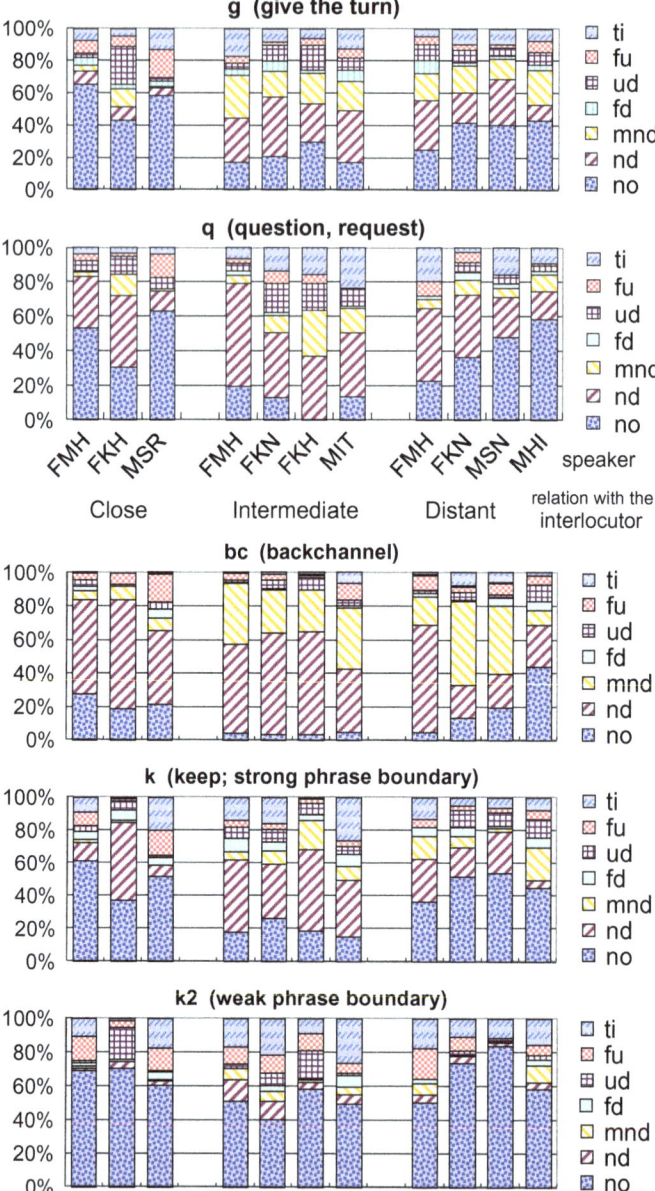

**Fig. 6.3** Distribution of head motions for each dialogue act, arranged by the relationship between speakers and interlocutors

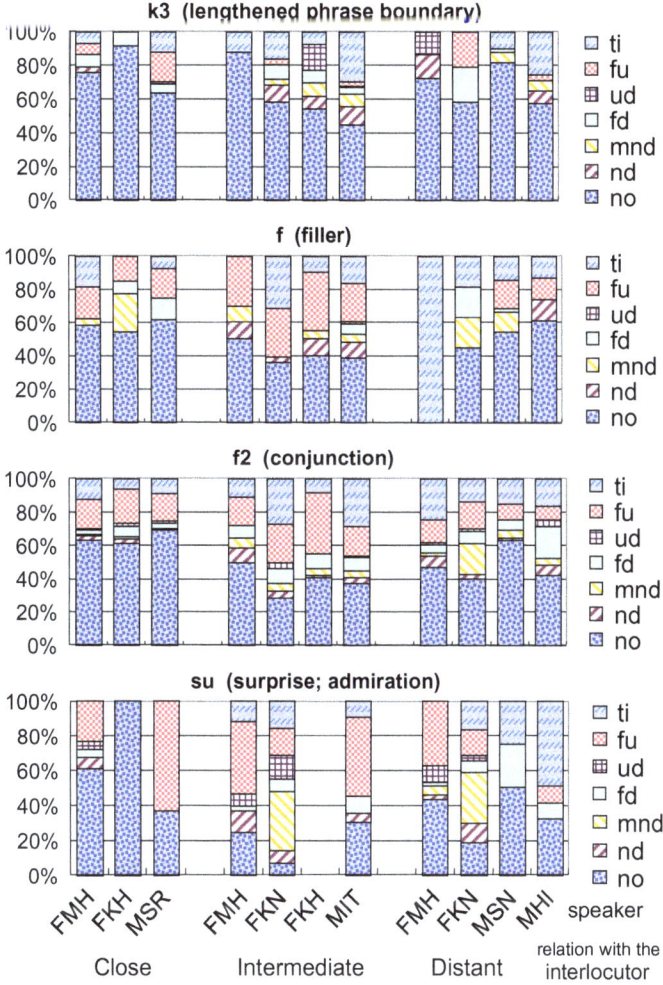

**Fig. 6.3** (continued)

status is higher than that of their dialogue partners (research assistants). Another hypothesis concerns the difference in age between the speakers and the interlocutors. Further analysis is necessary to verify these hypotheses.

The analysis of intra-speaker variability indicates that the frequency of head motion differs depending on the inter-personal relationship with the interlocutor. In Fig. 6.3, the relationship between speakers and interlocutors is arranged according to three levels: "Close" relationship (family members, boyfriend), "Distant" relationship (different generations and different social status), and "Intermediate" relationship (friend, friend of friend, relatives of friend). To distinguish between "distant" and "intermediate," besides the difference in status, we took into account the actual inter-personal relationship, so that an additional factor was whether or not the subjects were meeting for the first time.

It was observed that head motion had significantly lower frequency when the speaker had a close relationship with the interlocutor. For example, in speaker FMH, it can be noted that no head motion (**no**) occurs with higher frequency and nods (**nd**, **mnd**) occur with lower frequency in **g**, **q**, **bc** and **k** when FMH talks with her mother or with her boyfriend (FMH—Close). Additionally, most head motions in backchannels (**bc**) are single nods (**nd**) in FMH—Close, whereas the frequency of multiple nods (**mnd**) increases when FMH is talking with dialogue partners whom she is meeting for the first time (FMH—Distant). One explanation for this is that head motions (or, more specifically, nods) are used to express attitudes such as showing interest in the interlocutor's dialogue. For family members, careless behavior tends to occur, so that the head motions become less frequent. Regarding the turn-keeping utterances (**k**), nods are frequent in FMH—Inter, but less frequent in FMH—Distant. An explanation for this is that FMH did not assert herself when talking with MSN and MHI ("Distant" dialogue partners), whereas she spoke more confidently (making strong assertions) when talking with MIT ("Intermediate" dialogue partner), who is a friend of a friend and is close in age to FMH.

Regarding FKN, it can be observed in **bc** that multiple nods (**mnd**) occurred more frequently than single nods (**nd**) when talking with FKH, MHI and MSN, who are older than her and whom she is meeting for the first time, whereas single nods are predominant when talking to her friend MIT. In **k** and **g**, it is observed that no head motion is predominant when talking with FKH, MHI, and MSN. The explanation for this is the same as for FMH; i.e., FKN did not assert herself when talking with older people she is meeting for the first time (FKH, MHI, and MSN), whereas she spoke more confidently with MHI.

For speaker FKH, the frequency of no head motion (**no**) is relatively higher when talking with her daughter (FMH) than when talking with the daughter's friend (FKN). It can also be noted that multiple nods (**mnd**) occur more frequently when talking with FKN.

## 6.3.3  Summary of the Analysis Results

Firstly, the overall distributions of head motions for different dialogue act categories by multiple speakers confirmed the trends presented in the analysis results in Sect. 6.2 for a single speaker.

Inter-speaker variability analysis indicated that the frequency of head motion may vary according to the speaker's age or status, whereas intra-speaker variability indicated that the frequency of head motion may differ depending on the inter-personal relationship with the interlocutor. The following factors were found to increase the frequency of nods during dialogue speech.

- the speaker is talking with confidence;
- the speaker is expressing interest in the interlocutor's speech;
- the speaker is talking cheerfully or with enthusiasm. Such attitudes mainly appeared when there is a difference in the social status of the dialogue partners.

More details on the analysis between head motions and speech can be found in [19].

## 6.4   Mapping Head Motions to an Android

In the present section, we evaluate a method for mapping the head motions from a human to an android, which is a minimum requirement (from a hardware view-point) for guaranteeing that natural head motions can be realized in the android.

### 6.4.1   Android Head Actuators

We use an android called Repliee Q2 [20], which was cast from a Japanese woman, as a test bed for evaluating the head motion control system. The android has three actuators for controlling its head motions, as shown in Fig. 6.4. Actuators 14 and 15 move the head diagonally from lower-left to upper-right and from lower-right to upper-left, respectively. Up to down (vertical) motions can be realized by changing both actuators 14 and 15 simultaneously with the same command value. Actuator 16 moves the head in a shaking pathway from left to right.

The actuation values range from 0 to 255 for actuators 14 and 15, and from 50 to 205 for actuator 16. All three actuators have 127 as their central (neutral) position.

The head motions of the android are, unfortunately, limited when compared with the human head-movement range. Table 6.7 compares the range of head rotation angles measured in a human subject and in Repliee Q2, relative to the respective neutral positions. *Angle_X, _Y,* and *_Z* represent the rotation angles around the X, Y, and Z axes (coordinates shown in Fig. 6.2).

**Fig. 6.4**  Android actuators
(14, 15, 16 are head actuators)

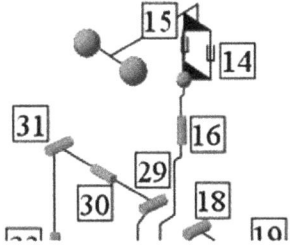

**Table 6.7** Range of the head rotation angles (in degrees) measured in the human subject and in the android relative to the neutral positions

|  | Human | Android |
|---|---|---|
| Minimum *Angle_X* (downward direction) | −50 | −12 |
| Maximum *Angle_X* (upward direction) | 30 | 12 |
| Minimum *Angle_Y* (slantwise left direction) | −30 | −12 |
| Maximum *Angle_Y* (slantwise right direction) | 30 | 12 |
| Minimum *Angle_Z* (left direction) | −50 | −20 |
| Maximum *Angle_Z* (right direction) | 40 | 20 |

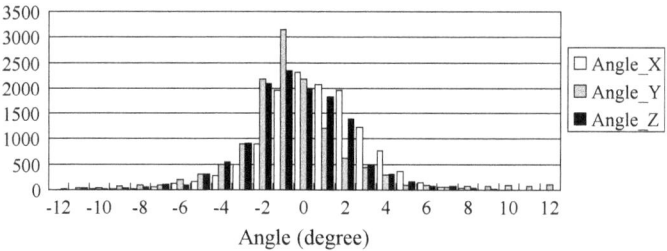

**Fig. 6.5** Histogram of the measured head rotation angles for the subject during natural conversation

Although humans have a large maximum range of head rotation angles, as shown in Table 6.7, during natural dialog, the head motions are limited to a smaller range. Figure 6.5 presents histograms of the head rotation angles measured in the subject during natural dialogue conversation over a period of 4 min.

The histograms in Fig. 6.5 show that the rotation angles range from −12° to 12°, which indicates that the actuation range of the android is sufficient to realize the human head motions during face-to-face dialogue interactions.

### 6.4.2 Mapping Rotation Angles to Android Actuator Commands

For the angle computation, we required hand-selected human head positions in addition to the neutral positions. We took the position of the head in the neutral position as a reference for calculating the angle for every frame during the speech.

The actuation values for each actuator are given by:

$$
\begin{aligned}
act[14] = act\_neutral + (Angle\_X/MaxAngle\_X*127) \\
+ (Angle\_Y/MaxAngle\_Y*127),
\end{aligned}
\tag{6.6}
$$

$$act[15] = act\_neutral + (Angle\_X/MaxAngle\_X*127) \\ - (Angle\_Y/MaxAngle\_Y*127), \tag{6.7}$$

$$act[16] = act\_neutral + (Angle\_Z/MaxAngle\_Z*78), \tag{6.8}$$

where *act_neutral* is the activation value for the neutral position (127 for all actuators), *Angle_X, _Y, _Z* are the target rotation angles, and *maxAngle_X, _Y, _Z* are the maximum angles measured in the android.

The actuation values obtained above are clipped to the limit ranges for each actuator.

To evaluate the proposed method, we mapped the human motions to the android motions. The motion-captured data for the human head motions were used as input,

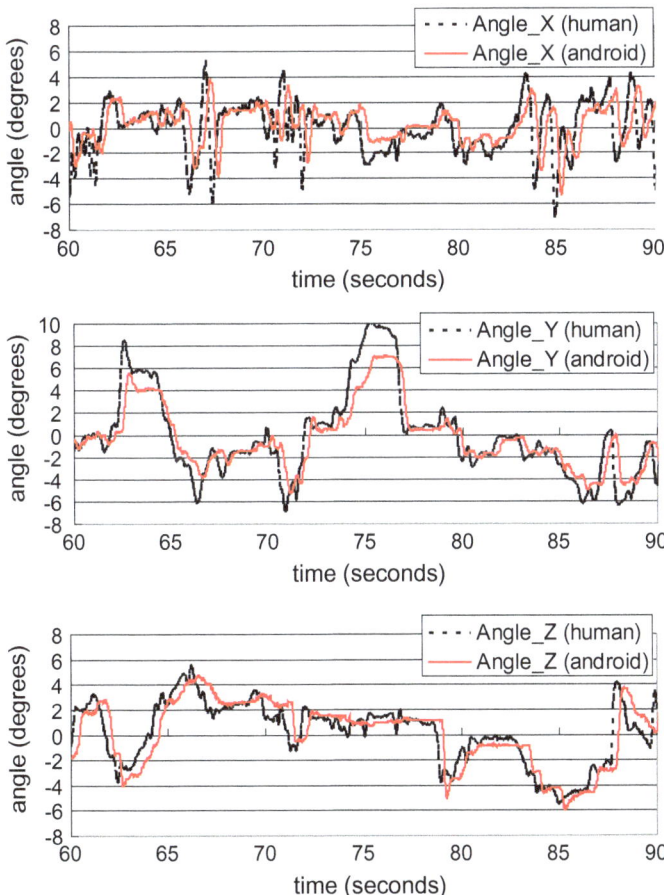

**Fig. 6.6** Head rotation angles measured in the subject (human) during natural conversation and in the android after mapping from the subject motions

with the head rotation angles estimated by the procedure described in Sect. 6.2.1.1. The android head actuation values were estimated by Eqs. (6.6)–(6.8). Figure 6.6 illustrates the correspondence between the head rotation angles measured in the subject (human) during natural conversation and those in the android after mapping from the subject motions. The target angles could not be achieved in some intervals of *Angle_X* and *Angle_Y*, probably because of the angle limitations resulting from their codependence on actuators 14 and 15. Correlation coefficients of 0.74, 0.94, and 0.91 were achieved for rotation angles *Angle_X*, *_Y*, and *_Z,* respectively.

A preliminary subjective evaluation also indicated a high degree of naturalness and good visual correspondence between the human and android motions.

## 6.5  Conclusion

With the aim of generating head motions from speech signal, analyses were conducted to verify the relations between head motions and dialogue acts, prosodic features and linguistic information. Among the several head motions, nods were the most frequent, appearing not only when expressing dialogue acts such as agreement or affirmation, but also to indicate syntactic or semantic units when appearing at the last syllable of the phrases in strong phrase boundaries. Tilts were found to occur most often in utterances where the speaker is thinking or is not confident and were frequently observed alongside fillers or disfluencies. They may possibly be accompanied by a lax or breathy voice quality. Shakes are used to express negation/denial and are more dependent on the content of the speech utterance.

Inter-speaker variability analysis indicated that the frequency of head motion may vary according to the speaker's age or status, whereas intra-speaker variability indicated that the frequency of head motion may differ depending on the inter-personal relationship with the interlocutor. The frequency of nods during dialogue speech was found to increase when the speaker is talking with confidence, expressing interest in the interlocutor's speech, or is talking cheerfully or with enthusiasm. Such attitudes mainly appeared when there is a difference in the social status of the dialogue partners.

We also presented a method for controlling the head actuators of an android based on the three head rotation angles and verified the naturalness and good correspondence of the mapping between the human and android head motions. To achieve a complete system of head motion generation from speech in a tele-operated android system, the automatic extraction of both linguistic and paralinguistic information (such as phrase boundaries and disfluencies) would be necessary to apply the results of the present analysis.

**Acknowledgements**  This research is supported by the Ministry of Internal Affairs and Communications. We thank Eri Takano and Judith Haas for helping with the motion data collection, and Freerk Wilbers, Kyoko Nakanishi, and Maiko Hirano for helping with the data analysis.

# References

1. Ishi, C.T., J. Haas, F. P. Wilbers, H. Ishiguro, N. Hagita. 2007. Analysis of head motions and speech, and head motion control in an android. In *Proceedings of the 2007 IEEE/RSJ International Conference on Intelligent Robots and Systems (IROS 2007)*, 548–553.
2. Ishi, C.T., C. Liu, H. Ishiguro, N. Hagita. 2010. Head motions during dialogue speech and nod timing control in humanoid robots. In *Proceedings of the 5th ACM/IEEE International Conference on Human-robot interaction (HRI 2010)*, 293–300.
3. Sidner, C., C. Lee, L.-P. Morency, C. Forlines. 2006. The effect of head-nod recognition in human-robot conversation. In Proceedings of the HRI 2006, 290–296.
4. Morency, L.-P., C. Sidner, C. Lee, and T. Darrell. 2007. Head gestures for perceptual interfaces: The role of context in improving recognition. *Artificial Intelligence* 171 (8–9): 568–585.
5. Yehia, H.C., T. Kuratate, and E. Vatikiotis-Bateson. 2002. Linking facial animation, head motion and speech acoustics. *Journal of Phonetics* 30: 555–568.
6. Sargin, M.E., O. Aran, A. Karpov, F. Ofli, Y. Yasinnik, S. Wilson, E. Erzin, Y. Yemez, and A.M. Tekalp. 2006. Combined gesture-speech analysis and speech driven gesture synthesis. In *Proceedings of the IEEE international conference on multimedia*.
7. Munhall, K.G., J.A. Jones, D.E. Callan, T. Kuratate, and E. Vatikiotis-Bateson. 2004. Visual prosody and speech intelligibility—Head movement improves auditory speech perception. *Psychological Science* 15 (2): 133–137.
8. Graf, H.P., E. Cosatto, V. Strom, and F.J. Huang. 2002. Visual prosody: Facial movements accompanying speech. In *Proceedings of the IEEE international conference on automatic face and gesture recognition (FGR'02)*.
9. Beskow, J., B. Granstrom, and D. House. 2006. Visual correlates to prominence in several expressive modes. In *Proceedings of the interspeech 2006—ICSLP*, 1272–1275.
10. Busso, C., Z. Deng, M. Grimm, U. Neumann, and S. Narayanan. 2007. Rigid head motion in expressive speech animation: analysis and synthesis. *IEEE Transactions on Audio, Speech and Language Processing*.
11. Iwano, Y., S. Kageyama, E. Morikawa, S. Nakazato, and K. Shirai. 1996. Analysis of head movements and its role in spoken dialogue. In *Proceedings of the ICSLP'96*, 2167–2170.
12. Foster, M.E., and J. Oberlander. 2007. Corpus-based generation of head and eyebrow motion for an embodied conversational agent. *Language Resources and Evaluation* 41 (3–4): 305–323.
13. Ishi, C.T., J. Haas, F.P. Wilbers, H. Ishiguro, and N. Hagita. 2007. Analysis of head motions and speech, and head motion control in an android. In Proceedings of the IROS 2007, 548–553.
14. Ishi, C.T., C. Liu, H. Ishiguro, and N. Hagita. Head motion during dialogue speech and nod timing control in humanoid robots. In *Proceedings of 5th ACM/IEEE International Conference on Human-Robot Interaction (HRI 2010)*, 293–300, 2010.
15. Stegmann, M.B., D.D. Gomez. 2002. *A brief introduction to statistical shape analysis*, Published online.
16. Ishi, C.T., H. Ishiguro, N. Hagita. 2006. Analysis of prosodic and linguistic cues of phrase finals for turn-taking and dialog acts. In *Proceedings of the interspeech'2006—ICSLP*, 2006–2009.
17. Ishi, C.T. 2005. Perceptually-related F0 parameters for automatic classification of phrase final tones. *Transactions on Information and Systems* E88-D(3), 481–488.
18. Ishi, C.T., H. Ishiguro, and N. Hagita. 2008. Automatic extraction of paralinguistic information using prosodic features related to F0, duration and voice quality. *Speech Communication* 50 (6): 531–543.

19. Ishi, C.T., H. Ishiguro, and N. Hagita. 2013. Analysis of relationship between head motion events and speech in dialogue conversations. *Speech Communication* 57 (2014): 233–243.
20. Minato, T., M. Shimada, H. Ishiguro, and S. Itakura. 2004. *Development of an android robot for studying human-robot interaction*, 424–434. Springer Verlag: Innovations in applied artificial intelligence.

# Chapter 7
# Generation of Head Motion During Dialogue Speech, and Evaluation in Humanoid Robots

Carlos T. Ishi, Chaoran Liu and Hiroshi Ishiguro

**Abstract** Head motion occurs naturally and in synchrony with speech during human dialogue communication and may carry paralinguistic information such as intentions, attitudes, and emotions. Therefore, natural-looking head motion by a robot is important for smooth human–robot interaction. Based on rules inferred from analyses of the relationship between head motion and dialogue acts, we proposed a model for generating nodding and head tilting and evaluated for different types of humanoid robot. Analysis of subjective scores showed that the proposed model including head tilting and nodding can generate head motion with increased naturalness compared to nodding only or directly mapping people's original motions without gaze information. We also found that an upward motion of the face can be used by robots that do not have a mouth in order to provide the appearance that an utterance is taking place. Finally, we conducted an experiment in which participants act as visitors to an information desk attended by robots. Evaluation results indicated that our model is equally effective as directly mapping people's original motions with gaze information in terms of perceived naturalness.

**Keywords** Head motion · Dialogue acts · Eye gazing · Motion generation

## 7.1 Introduction

To allow smooth dialogue communication between humans and robots, both verbal (linguistic) information and nonverbal information are important. Nonverbal information includes head motions that express paralinguistic information such as intentions, attitudes, and emotions and increase the robot's perceived lifelikeness.

This chapter is a modified version of previously published papers, [1, 2] edited to be comprehensive and fit with the context of this book.

C. T. Ishi (✉) · C. Liu · H. Ishiguro
ATR Intelligent Robotics and Communication Labs, Kyoto, Japan
e-mail: carlos@atr.jp

Head motion naturally occurs during speech utterances and can be either intentional or unconscious. In the former case, the motion may carry clear meanings in communication. For example, nods are frequently used to express agreement, whereas headshakes are used to express disagreement. Most of the time, however, head motion is produced unconsciously. Regardless of this difference, both types of motions are somehow synchronized with the speech utterances and transmit non-verbal information.

One of our motivations for the present work is to obtain a method for generating head motion from speech signals in order to automatically control the head motion of a teleoperated humanoid robot (such as an android). In this way, the lifelikeness of a robot can be enhanced by imitating a human's natural head motion, and smooth human–robot communication can be expected.

In our previous work [3, 4], we analyzed several free dialogue conversations between Japanese speakers and found a strong relationship between head motion and dialogue acts (including turn-taking functions). Nods occurred frequently during dialogue speech, not only to express dialogue acts such as agreement and affirmation, but also to indicate syntactic or semantic units, appearing at the last syllable of phrases with strong phrase boundaries. At weak phrase boundaries where the speaker is thinking or indicates that he/she has not finished his/her speech, head tilts were frequently observed.

In the present work, we propose head motion generation models based on rules inferred from human head motion analysis results. We then evaluate the proposed models in different types of humanoid robots: a typical humanoid robot with fewer degrees of freedom ("Robovie" and "Telenoid") and female android robots ("Repliee Q2" and "Geminoid F") [4, 5]. We start evaluation from nod motion generation, then extend to head tilt generation, and finally introduce eye gaze control. The effects of an additional "face-up" motion during utterances are also evaluated, with the goal of reducing perceived unnaturalness in robots that do not have a mouth (i.e., movable lips).

The remainder of this chapter is organized as follows. Section 7.2 describes the previous related work. In Sects. 7.3 and 7.4, rule-based nodding and head tilting generation models are described and evaluated in two types of humanoid robots. In Sect. 7.5, a motion that suggests speech in robots that do not have movable lips is described and evaluated using a humanoid robot. Section 7.6 investigates the effects of gaze in our model and evaluates human–robot face-to-face interactions. Section 7.7 concludes the chapter.

## 7.2 Related Work

Head motion analyses can focus on either of two problems: how to recognize a user's head motion and interpret its role in communication (e.g., [6, 7]), or how to generate a robot's head motions in synchrony with the robot's speech. The present work focuses on the latter problem of generating natural head motion while the robot is speaking.

Many studies have attempted to find a correspondence between head motions and prosodic features such as the fundamental frequency (F0) contours (which represent pitch movements) and energy contours in several languages [8–13]. For example, emphasis of a word is often accompanied with head nodding, and a rise of the head can correspond with a rise in voice in English [11]. In Swedish, words with a focal accent are accompanied by a greater variation of the facial parameters (including head motions) than words in non-focal positions in all expressive modes [12]. However, it has been reported that the correlation among F0 and the six degrees of freedom (rotation and translation) of the head motions ranged from 0.39–0.52 for English speakers and 0.22–0.30 for Japanese speakers [8]. This shows that the use of F0 alone is not sufficient to generate head motions and that the correspondence between F0 and head motions is language-dependent. In addition, most of these studies analyzed read speech or acted emotional speech data. In [14], the relations between head movements and the semantics of utterances are analyzed in Japanese speech by considering speaking turn and speech functions. Regarding head motion generation, a system that uses corpus-based selection strategies to specify the head and eyebrow motion of an animated talking head is described in [15]. The system considers syntactic, prosodic, and pragmatic context for generating the motions. However, this study only analyzed data from one speaker.

In [16], strong relationships between head motion and dialogue acts (including turn-taking functions) and between dialogue acts and prosodic features were reported. In the present work, we focus on the relationship between dialogue acts and head motion.

Dialogue act tags have been annotated for each phrase in a database of dialogue between several pairs of speakers. The annotation used the following tag set, which is based on the tags proposed in [17], and took into account dialogue acts such as affirmative or negative reaction, the expression of emotions such as surprise, and turn-taking functions.

- **k** (keep): The speaker is keeping the turn; a short pause or a clear pitch reset is accompanied at strong phrase boundaries.
- **k2** (keep): Weak phrase boundaries in the middle of an utterance (when no pause exists between phrases).
- **k3** (keep): The speaker lengthens the end of the phrase, usually when thinking, but keeps the turn (may or may not be followed by a pause).
- **f** (filler): The speaker is thinking or preparing the next utterance, e.g., "uuun," "eeee," "eettoo," "anoo" ("uhmmm").
- **f2** (conjunctions): It can be considered as non-lengthened fillers, e.g., "dakara," "jaa," "dee" ("I mean," "so").
- **g** (give): The speaker has finished talking and is giving the turn to the interlocutor.
- **q** (question): The speaker is asking a question or seeking confirmation from the interlocutor.

- **bc** (backchannels): The speaker is producing backchannels (agreeable responses) to the interlocutor, e.g., "un," "hai" ("uh-huh," "yes").
- **su** (admiration/surprise/unexpectedness): The speaker is producing an expressive reaction (admiration, surprise) to the interlocutor's utterances, e.g., "heee," "uso!," "ah!" ("wow," "really?").
- **dn** (denial, negation): For example, "iie" and "uun" accompanied by a fall–rise pitch movement ("no," "uh-uh").

Among the analysis results in [3] for the relationship between head motion and dialogue acts based on dialogue between several pairs of speakers, it was shown that nodding occurs most frequently during dialogue speech and that this motion appears most often in backchannels (bc) and at the last syllable of strong phrase boundaries (k, g, q). At the weak phrase boundaries where the speaker is thinking, embarrassed, or indicates his/her speech utterance has not been concluded (f, k3), head tilting was frequently observed.

In the present work, we exploit these analysis results to create a rule-based head motion generation model.

## 7.3   Evaluation of Head Nodding Generation

From the analysis results in [3], it was shown that the nodding head motion occurs most frequently during dialogue. In this section, we propose a rule-based nodding generation model and evaluate the effects of nodding control in two types of humanoid robot.

### 7.3.1   Rule-Based Nodding Generation

We first propose a very simple model "**NOD**" that controls only the timing of the nods. In this model, nods are generated in the center of the last syllable of utterances with strong phrase boundaries (**k, g, q**) and backchannels (**bc**). The utterance segmentation and the respective dialogue act tags provide the timing of the nods.

Figure 7.1 shows examples of nod shapes extracted from human–human dialogue database and an average nod shape used for nodding generation. Note the presence of a slight upward motion, which often occurs before the characteristic down-up motion of nods. In this experiment, single nods with similar shapes were used (i.e., the intensity or the duration of the nods remained the same), so that the specific effects of timing could be evaluated.

In the second model "**NOD+**", the nod timing generation of "NOD" was superimposed onto a slight face-up motion (3 degrees) during the speech utterance intervals. This upward motion during speech is often observed in natural speech and

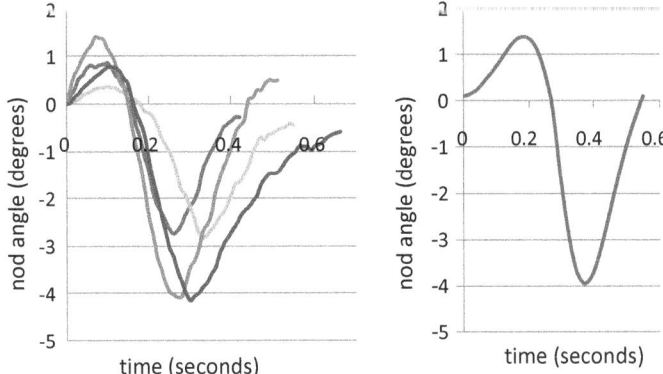

**Fig. 7.1** Examples of observed nod shapes extracted from the database (left) and the nod shape used in the nodding generation model

is expected to improve the naturalness of motion with regard to using only single nods, as in "NOD".

### 7.3.2   Experimental Setup

Six conversation passages including relatively frequent nods were randomly selected from our database, and the rotation angles (nod, shake, and tilt angles) were extracted for each utterance. The duration of the conversation passages was limited to 10–20 s, because the subjects have to compare a pair of motions for the same speech utterances, and this becomes difficult if each video is too long. In addition, as a dialogue act is attributed for each phrase unit, utterances of 10–20 s usually contain more than 10 phrases, i.e., some context information is still present in the dialogue passage.

For each conversation passage, two types of motion were generated, one for each of the nodding generation models described in the previous subsection "NOD", "NOD+".

For comparison, we prepared two more motion types. One is a reproduction of the head rotations extracted from the original motion capture data "**ORIGINAL**". The second is the original head motions shifted backward by 1 s "**SHIFTED**". This shifted type was evaluated to confirm that appropriate timing control of head motions is important. Preliminary experiments were conducted for the original head motions shifted forward by 1 s, and similar effects were found in comparison with shifting backward. Thus, in the present paper, the results for shifting backward are presented.

The motions were generated in a small-sized humanoid robot (Robovie Mini-R2) and a female android (Repliee Q2 [18]), as shown in Fig. 7.2.

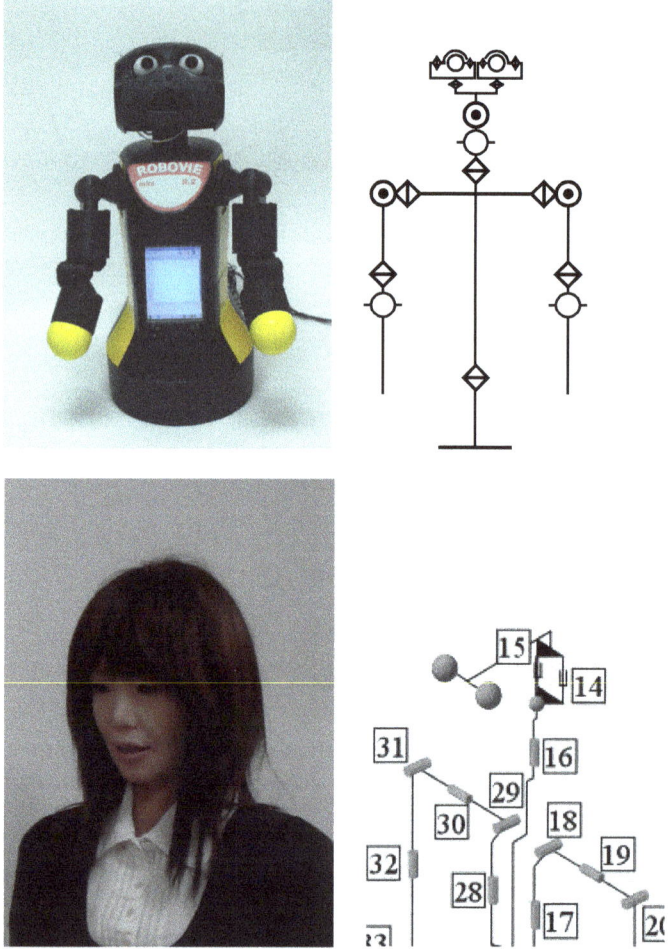

**Fig. 7.2** Robovie Mini-R2, the android Repliee Q2, and their respective actuators

Robovie Mini-R2 has three degrees of freedom for its head, so the rotation angles were directly mapped to the actuator commands by a linear mapping. Repliee Q2 also has three actuators in its head (as shown in Fig. 7.2), but their correspondence to the three rotation angles is not direct. Thus, the mapping function between the rotation angles and the android actuator commands proposed in our previous work was applied [16]. The commands were sent every 20 ms for Repliee Q2 and every 100 ms for Robovie Mini-R2 according to constraints in each robot's hardware, but these rates are thought to be sufficient for head motion control purposes. Further, for the android, the lip motions (jaw lowering motions) were reproduced from the distances between the nose and chin markers in the original motions.

Video clips were recorded for each motion type, resulting in 24 videos (six conversation passages and four motion types) for each robot type. For each trial, subjects watched a sequence of two videos with different head motion control types and the same conversation passage, enabling us to compare the effect of changing the head motion control.

Pairs of video were presented to subjects in the following order:

- NOD versus NOD+
- SHIFTED versus NOD+
- ORIGINAL versus NOD+

Subjects were asked to rate the naturalness of the motion for each video and assign preference scores for each pair of video according to the following questionnaire. (Only one item could be chosen for each question.)

- Comparing the two videos, which one is more natural?

The first is clearly more natural (−2) | the first is slightly more natural (−1) | difficult to decide (0) | the second is slightly more natural (1) | the second is clearly more natural (2)

- Is the motion of the robot natural?

Unnatural (−2) | slightly unnatural (−1) | difficult to decide (0) | slightly natural (1) | natural (2)

Ten subjects (aged in their 20 and 30 s) participated in the experiment. None of the subjects are involved in robotics research.

### 7.3.3  Experimental Results

Figure 7.3 shows the subjective preference scores between pairs of motion types for each robot. Subjective scores are quantified using a scale from −2 to 2. The y-axis represents the preference scores between pairs of motion. Positive scores indicate that "NOD+" is preferred compared to other motion types.

First, "NOD+" (single nods with appropriate timing plus face-up during utterances) was judged to be slightly more natural than "NOD" (only single nods with appropriate timings), showing the effectiveness of the face-up motions in addition to the single nods. Next, "NOD+" was judged to be clearly more natural than "SHIFTED" (the original shifted by 1 s), showing the importance of an appropriate timing control. Finally, surprisingly, there was confusion between "NOD+" and "ORIGINAL"; some of the preference scores indicated that "ORIGINAL" was (slightly) more natural, whereas most of the preference scores indicated that "NOD+" was (slightly) more natural. Possible reasons for this result are discussed in the next section.

**Fig. 7.3** Preference between pairs of motion types for Robovie Mini-R2 (top) and Repliee Q2 (bottom)

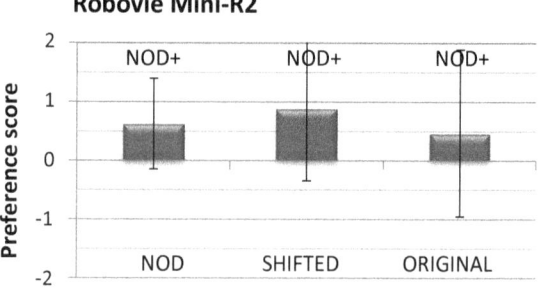

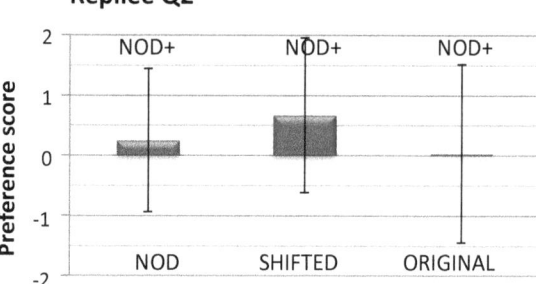

Figure 7.4 shows the subjective naturalness for individual motion types for each robot.

Regarding the naturalness of individual motions, "NOD+" was judged to be "slightly natural" in both robot types. A clear difference between the robot types was found in the "SHIFTED" motion type, which was judged unnatural in Robovie Mini-R2, but slightly unnatural to slightly natural for Repliee Q2. One reason could be that the android also moves its lips, whereas Robovie only moves its head, so that the unsynchronized head motion is more apparent in Robovie. This would indicate that head motion control is more important in robots whose lips do not move.

Finally, better control of the other rotation angles and the intensity of the nod may improve naturalness. However, the present results show that "slightly natural" motion can be achieved using a very simple nodding model.

## 7.3.4 Discussions

The subjective preference results in the present section showed that some subjects consider the proposed single-nod motion types (slightly) more natural than the original head motions. One possible reason could be that the axis for the tilt rotations is in an upper position, in the case of Robovie Mini-R2 (see head actuators in Fig. 7.2). Thus, the measured (original) head motion cannot be perfectly

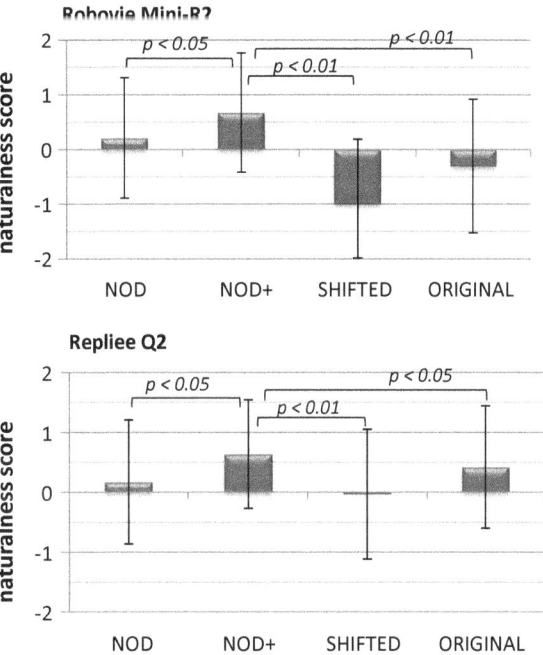

**Fig. 7.4** Subjective naturalness for each motion type for Robovie Mini-R2 (top) and Repliee Q2 (bottom)

reproduced, so that controlling only nod angles looks more natural. However, similar results were also obtained for the android Repliee Q2, which can better reproduce the original head motion. A careful observation of the videos leads us to think that this is because of the lack of any control over eye gaze—the unchanged gaze during head motion in the shake axis leads to a very unnatural impression. As the proposed model only generates movements around the nod axis, the negative effects of unchanged gaze would be less apparent compared to the original motion, where the head moves around all rotation angles. The same explanation could be given for Robovie Mini-R2. In Sect. 7.6, we investigate the effects of gaze control along with head motion control.

## 7.4 Evaluation of Head Tilt Generation

In the previous section, nod generation was evaluated for utterances without thinking behaviors. In this section, our head motion generation model for nodding is extended to include head tilting for improving naturalness of utterances expressing thinking behaviors. The effects of head tilting control are evaluated in two types of humanoid robot.

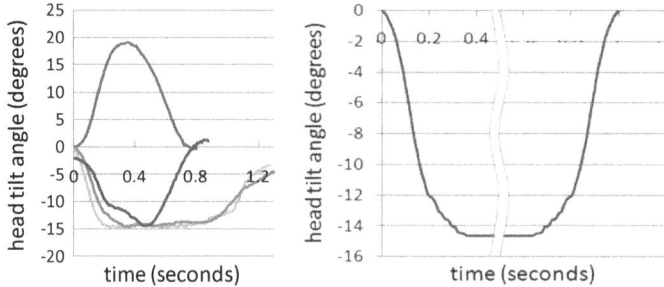

**Fig. 7.5** Examples of observed head tilt shapes extracted from the database (left) and the head tilt shape used in the head motion generation model

## 7.4.1 Proposed Method for Head Motion Generation

In this proposed model, nods are generated in the center of the last syllable of utterances with strong phrase boundaries (k, g, q) and backchannels (bc), as in our previously proposed nod generation model [3], whereas head tilts are generated in the weak phrase boundaries where the speaker lengthens the end of the phrase or pauses in the middle of a sentence because of disfluencies (f, k3). The utterance segmentation and respective dialogue act tags were used to determine the timing of the nods.

The left panel of Fig. 7.5 shows examples of head tilt shapes found in the database. The duration of the nods varied from 0.4 to 0.7 s. Note the presence of a slight upward motion, which often occurs before the characteristic down-up motion of nods. The duration of the head tilt samples varied in length from 0.8–1.5 s and was more dependent on the phrase length. The head tilt shape used in the motion generation model is shown in the right panel of Fig. 7.5. The nodding generation model is the same as that used in our previous work [4], which is described in the previous section.

Fixed shapes for single nods and head tilts were used (i.e., the intensity and the duration of the nods and the intensity of head tilts were kept the same), allowing the effects of timing to be evaluated. Regarding head tilt, the tilt angle (15 degrees) was maintained until the phrase had finished, so that the length of the motion was determined according to the inter-phrase interval lengths.

## 7.4.2 Experimental Setup

Eleven conversation passages with durations of 10–20 s, including fillers and turn-keeping functions (**f** and **k3**), were randomly selected from our database, and the rotation angles (nod, shake, and tilt) were extracted for each utterance. The duration of the conversation passages was limited to 10–20 s because subjects have to compare a pair of motions for the same speech utterances, and this would be difficult if each video is too long. Additionally, as a dialogue act is attributed for each phrase unit, utterances of 10–20 s usually contain more than 10 phrases; i.e., some context information is still present in the dialogue passage.

For each conversation passage, head rotation angles were computed by the head motion generation model described in the previous subsection "**NOD&TILT**".

For comparison, we prepared two types of motion. One is a reproduction of the head rotations extracted from the original motion capture data "**ORIGINAL**", and the other is the nod-only motion proposed in [4] "**NOD ONLY**".

The motions were generated in two robots: One is a female android robot (Geminoid F) and the other is a humanoid robot (Robovie R2), as shown in Fig. 7.6.

Robovie R2 has three degrees of freedom for its head, enabling the rotation angles to be directly mapped to the actuator commands by a linear mapping. Geminoid F also has three actuators for the head (as shown in Fig. 7.6), but their correspondence to the three rotation angles is not as straightforward as for Robovie

**Fig. 7.6** External appearance and actuators of Geminoid F and Robovie R2

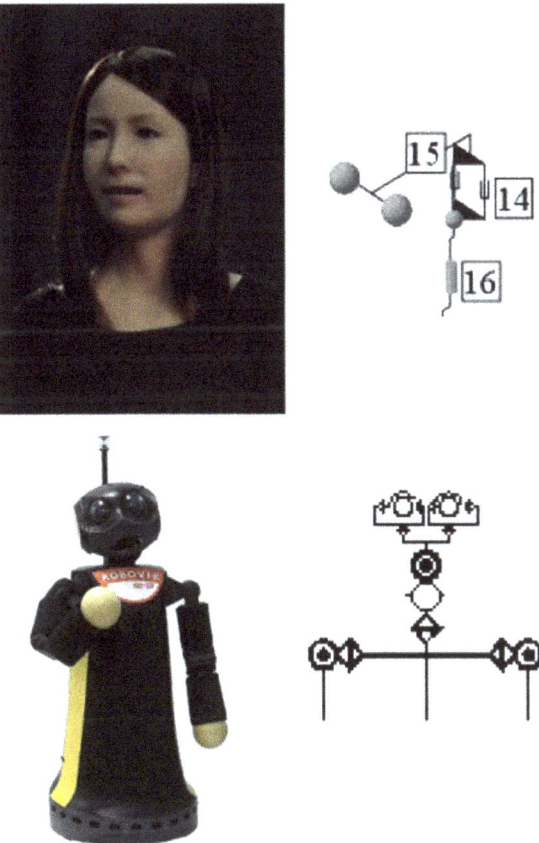

R2. Therefore, the mapping function between the rotation angles and android actuator commands proposed in our previous work was applied [16].

Commands were sent every 20 ms for Geminoid F and every 100 ms for Robovie R2 according to the constraints in each robot's hardware, but these rates are thought to be sufficient for head motion control purposes. Furthermore, for the android, the lip motion (jaw lowering motion) was reproduced from the distances between the nose and chin markers in the original motions.

Video clips were recorded for each motion type, resulting in 33 videos (11 conversation passages and three motion types) for each robot type. In each trial, subjects were shown a sequence of two videos with different head motion control types for the same conversation passage, allowing them to compare the effects of changing the head motion control.

Pairs of videos were presented to subjects in the following order:

- NOD ONLY versus NOD&TILT
- NOD&TILT versus ORIGINAL
- NOD ONLY versus ORIGINAL

The video pairs were sorted in random order.

Subjects were asked to rate the naturalness of the motion for each video and preference scores for each pair of video according to the following questionnaire. Only one response could be selected for each question.

- Is the motion of the robot natural?

Clearly unnatural (1) | unnatural (2) | slightly unnatural (3) | difficult to decide (4) | slightly natural (5) | natural (6) | clearly natural (7)

Thirty-eight paid subjects (18 male and 20 female, aged 18–60 with a mean of 34 and s.d. of 13) participated in the experiment. None of the subjects was involved in robotics research.

## 7.4.3  Experimental Results

Figure 7.7 shows the average subjective naturalness scores for individual motion types in each robot. Subjective naturalness was quantified using a scale of 1–7.

To understand the significance of these scores, a repeated measures analysis of variance (ANOVA) was conducted. A significant main effect was found ($F(2, 36) = 26.152$, $p < 0.0005$ for Geminoid F and $F(2, 23) = 19.109$, $p < 0.0005$ for Robovie R2).

Regarding the naturalness of individual motion, "NOD&TILT" was judged to be the most natural, with "NOD ONLY" and "ORIGINAL" obtaining lower scores in both robot types. Possible reasons for these results are discussed in Sect. 7.4.4.

Comparing the two graphs in Fig. 7.7, the scores for Geminoid F are higher than those for Robovie R2 for all three motion types. As the motion reproduced in the

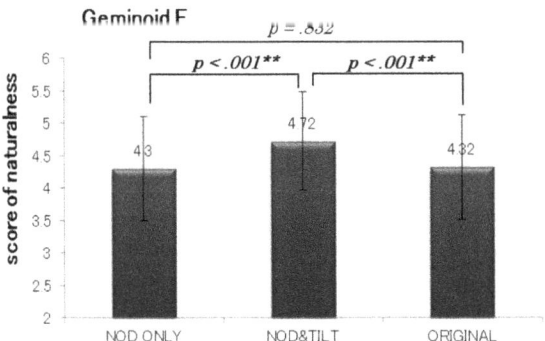

**Fig. 7.7** Subjective naturalness for each motion type (standardized value) for Geminoid F (top) and Robovie R2 (bottom). Error bars indicate standard deviation

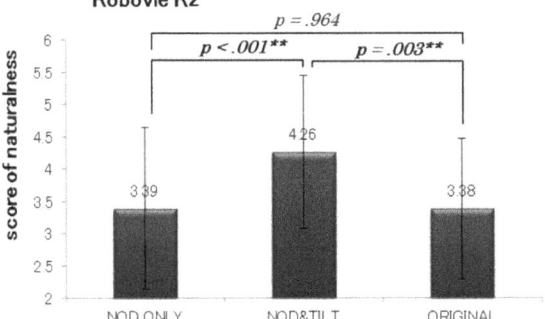

two types of robots was quite similar, we believe this difference is caused by the external appearance of the robots. This is discussed in more detail in Sect. 7.4.4.

For Geminoid F, a comparison between "NOD ONLY" and "NOD&TILT" shows that "NOD&TILT" (single nods plus head tilts with adequate timing) is significantly more natural than "NOD ONLY" (p < 0.0005), indicating the effectiveness of the head tilt motion generated in the specific weak phrase boundaries. A comparison between "NOD&TILT" and "ORIGINAL" shows that the "NOD&TILT" model is more natural than "ORIGINAL" (p < 0.0005). It was expected that "ORIGINAL" would have higher preference scores than "NOD ONLY." However, the results do not show a significant difference between these two cases. This means that simply trying to reproduce the original motion does not necessarily imply that natural motion will be generated.

For Robovie R2, we can observe similar overall tendencies as in the results for Geminoid F. "NOD&TILT" scores significantly higher than "NOD ONLY" and "ORIGINAL" ("NOD&TILT" vs. "NOD ONLY": p < 0.0005; vs. "ORIGINAL": p = 0.003).

We think that better control of the other rotation angles and the intensity of nodding may improve naturalness. However, the present results show that a "slightly natural" motion can be achieved using a very simple timing control model.

### 7.4.4  Discussions

Regarding the differences in the subjective naturalness scores between the two robot types, the results in Sect. 7.4.3 indicate that, for the same motion in both robots, the Robovie R2 scores lower than Geminoid F. One possible reason is that Robovie R2 does not have movable lips. For the android, lip motion was reproduced from the distances between the nose and chin markers, so that the subjects recognize the intervals of the speech utterance through both visual information and auditory information. However, for Robovie R2, as head motion is not strongly related to phonetic features, it is difficult to believe that the speech is being uttered by the robot from watching the robot's head motion alone. We think that this lack of visual information causes the lower subjective naturalness scores for Robovie R2, as it does not have movable lips.

## 7.5  Evaluation of Face-up Motion for Utterance Signature

To address the problem discussed in Sect. 7.4.4 of increasing perceived naturalness for robots without movable lips (such as Robovie R2), we evaluated the effects of including a slight "face-up" motion (3 degrees) as an indication of utterance intervals, similarly to the experiments in Sect. 7.3. This upward motion is often observed in natural speech and should improve motion naturalness in a robot that does not have movable lips, such as Robovie R2.

### 7.5.1  Experimental Setup

The "NOD ONLY" and "NOD&TILT" head motion generation models described in Sect. 7.4.1 were combined with a face-up motion during speech utterance intervals. We call these new-generation models "NOD ONLY+" and "NOD&TILT+".

Eight conversation passages were randomly selected from the previous experiment. For each conversation passage, six types of motion were generated: "NOD ONLY+" and "NOD&TILT+" for Robovie R2, as well as "NOD ONLY" and "NOD&TILT" for both Robovie R2 and Geminoid F (the latter were used as baseline conditions for comparison).

Video clips were recorded for each speech segment accompanied by motion, resulting in 48 videos for Robovie R2 and Geminoid F. As in the previous experiment (Sect. 7.4), subjects were shown a sequence of two videos with different head motion control types for the same conversation passage and were asked to rate the naturalness of the motion for each segment.

Ten Japanese subjects (five male and five female) participated in the experiment. All subjects were in their 20 s (college students) and were not involved with robotics research.

## 7.5.2  Experimental Results

Figure 7.8 shows the subjective naturalness scores for each motion type. As in the previous experiment, subjective naturalness was quantified using a scale of 1–7, where "1" denoted "clearly unnatural" and "7" denoted "completely natural."

We compared the scores for motion types with upward motion (Robovie R2) against those without upward motion in both robots (the two baseline conditions). The middle bar in Fig. 7.8 shows the subjective naturalness scores for the proposed "NOD&TILT+" model (with face-up motion) for Robovie R2, whereas the left and right bars show the naturalness scores for "NOD&TILT" (without the face-up motion) model for Robovie R2 and Geminoid F, respectively.

In place of lip movements synchronized to speech utterances, the face-up motion during the speech is expected to have a similar effect.

Analysis reveals a significant difference between "NOD ONLY+" and "NOD ONLY" for Robovie R2. However, no significant difference was found between "NOD&TILT+" and "NOD&TILT" for the same robot. The reason for this is that the head tilt during the speech utterance serves much the same purpose as the face-up motion by attracting the listener's attention before the final nod. In future, it could be interesting to further clarify the cause of these results and investigate

**Fig. 7.8** Subjective naturalness for "NOD&TILT+" compared with "NOD ONLY" and "NOD&TILT". Error bars indicate standard deviation

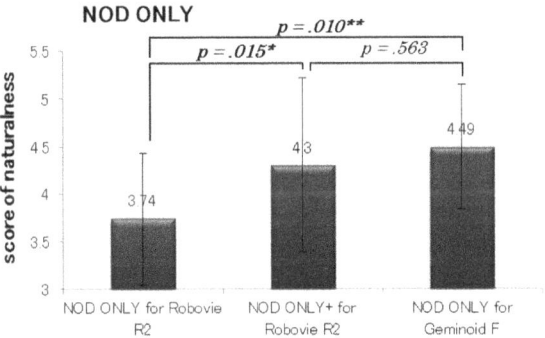

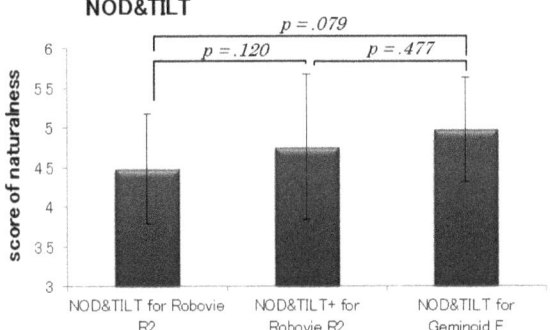

whether there are other motions that can be used in this way to improve perceived naturalness.

However, the face-up motion was still not sufficient to make the subjective naturalness scores for Robovie R2 as high as those for Geminoid F. As Geminoid F is a very humanlike anthropomorphic robot, which is at first glance indistinguishable from real humans, it may be the external appearance that enhances the subjective naturalness. We plan to evaluate the effect of this face-up motion in other types of robots in the future.

## 7.6 Evaluation of Head Motion and Eye Gazing During Human–Robot Dialogue Interaction

The results from the experiment in the previous sections left some unanswered questions. First, how do subjects evaluate the models when they are interacting with a real robot and not merely watching video footage? Second, why was the "ORIGINAL" motion, which we thought would be perceived as the most natural, judged to be as unnatural as the "NOD ONLY" model and less natural than the proposed "NOD&TILT" model? For the latter question, the literature strongly points to the importance of gaze movements for a wide variety of interaction schemes, both in human science [19, 20] and in human–robot interaction [21, 22]. However, the "ORIGINAL" motion only uses head movements, with no gaze movement reproduced.

Therefore, we conducted an experiment with two purposes: to verify the effectiveness of our models when actual robots are used and determine whether the addition of gaze movements would give the expected results for the original motion.

### 7.6.1 Analysis of Eye Gazing

We first analyzed eye gazing in human–human dialogue conversations. Thirteen free conversations from our multimodal dialogue speech database were analyzed. The following tag set was used to annotate eye gazing based on the displayed video information from the speaker's viewpoint.

- **u**: upwards gaze;
- **d**: downwards gaze;
- **l**: to the left (from speaker's viewpoint);
- **r**: to the right;
- **ul**: up and to the left;
- **ur**: up and to the right;
- **dl**: down and to the left;

- **dr**: down and to the right;
- **no**: no gaze movement (looking at conversation partner).

Figure 7.9 shows the distribution of eye gaze tags for each dialogue act tag (described in Sect. 7.2), i.e., where conversation partners looked during the dialogue when speaking. The intensity of each square indicates how often the speakers looked in that direction. The darker the background of the square, the higher the frequency of that eye gaze act.

In Fig. 7.9, it can be observed that the person speaking mostly looks directly toward the listener (especially for "bc," "g," and "k"). However, the speaker looks away in about 50% of the cases when lengthening their previous utterance and thinking about what to say next ("f" and "k3"). In such cases, the speaker has a

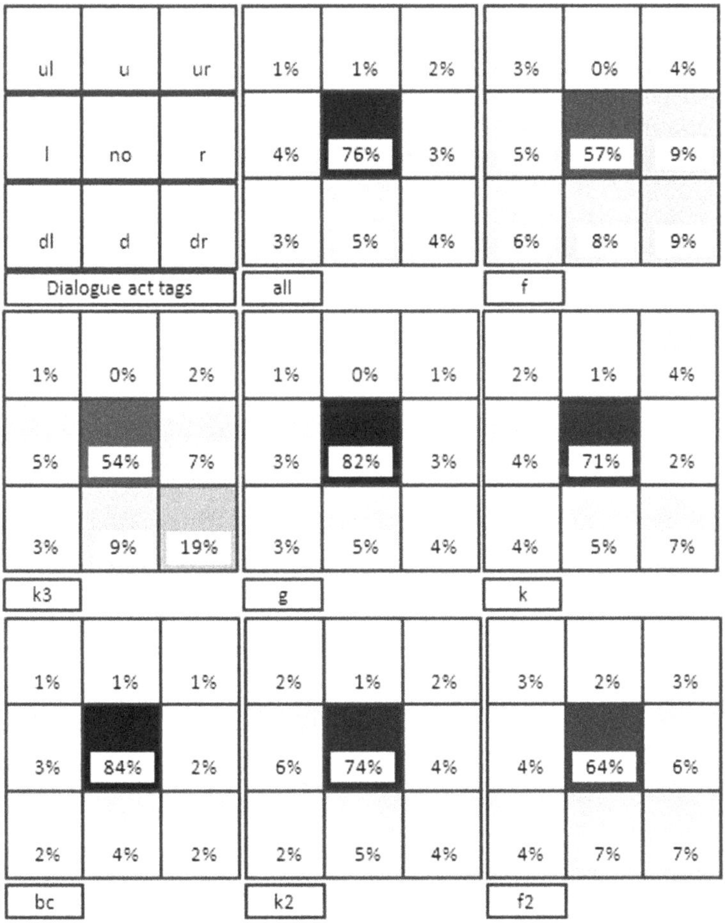

| ul | u | ur | 1% | 1% | 2% | 3% | 0% | 4% |
|---|---|---|---|---|---|---|---|---|
| l | no | r | 4% | 76% | 3% | 5% | 57% | 9% |
| dl | d | dr | 3% | 5% | 4% | 6% | 8% | 9% |
| Dialogue act tags | | | all | | | f | | |
| 1% | 0% | 2% | 1% | 0% | 1% | 2% | 1% | 4% |
| 5% | 54% | 7% | 3% | 82% | 3% | 4% | 71% | 2% |
| 3% | 9% | 19% | 3% | 5% | 4% | 4% | 5% | 7% |
| k3 | | | g | | | k | | |
| 1% | 1% | 1% | 2% | 1% | 2% | 3% | 2% | 3% |
| 3% | 84% | 2% | 6% | 74% | 4% | 4% | 64% | 6% |
| 2% | 4% | 2% | 2% | 5% | 4% | 4% | 7% | 7% |
| bc | | | k2 | | | f2 | | |

**Fig. 7.9** Distribution of eye gaze tags for different dialogue acts: Intensity for each square indicates how often speakers looked in that direction

**Fig. 7.10** Gaze direction
during speaker head tilting:
"Same direction" ("opposite
direction") means gaze
direction coincides (does not
coincide) with the direction of
the head tilt

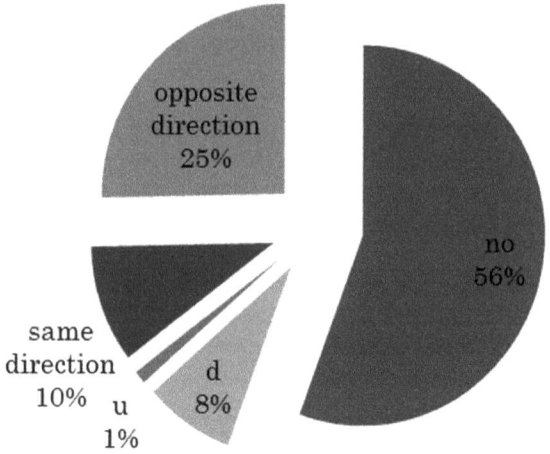

tendency to look downward rather than upward when looking away. For "k3," the
speakers also frequently look down and to the right (19% of the time).

The results in Fig. 7.9 indicate that eye gazing has similarities with head tilting.
In both cases, behavior reflects weak phrase boundaries when the speaker lengthens
the end of the phrase or pauses in the middle of a sentence because of disfluencies
(f, k3). Figure 7.10 shows the physical relationship (interdependency) between
head tilting and eye gazing.

During head tilting, the speaker most often looks directly at the listener (56%),
but also often looks in the opposite direction to that in which they tilt their head
(25%) (e.g., if the speaker tilts their head left, they often look to the right).
Sometimes, the speaker looks in the same direction as the head tilt (10%) or
downward (8%). The speaker rarely looks upward (1%).

The above results indicate that by altering its gaze direction during a head tilt by
looking in the opposite direction, a robot could be perceived as more natural. Based
on this, we evaluated the control of gaze in addition to the head motion in the
human–robot interaction experiment of the present section.

## 7.6.2 Experimental Setup

To allow for a comparison of different motion generation strategies in face-to-face
human–robot interaction, we created a scenario in which subjects act as visitors to
an information desk and robots play the role of receptionists. In this scenario, we
can ask subjects to pose the same questions to the robots, enabling the effects of
different motion generation strategies to be fairly compared using the same utterances with the same lip motion and changing only the head motion and gazing
strategies.

We first simulated a human human interaction for this scenario and recorded audio and motion capture data for a female speaker playing the role of the receptionist (collaborator). During the interaction, another speaker playing the role of the visitor asked several questions to induce the receptionist to naturally produce thinking behavior. The following items appeared in the interaction:

1. directions to IRC lab;
2. what kind of research is done at IRC lab;
3. how many staff work at IRC lab;
4. directions to the bathroom;
5. places to eat;
6. Japanese restaurants;
7. timetable for the bus;
8. directions to Kobe;
9. sightseeing spots in the vicinity;
10. sightseeing spots in Nara.

Two robots were used for the experiment: the Geminoid F android and a child-sized minimally designed robot Telenoid R2 (see Fig. 7.11). Like Robovie R2, Telenoid R2 has a head that can rotate about three independent axes and is less humanlike in appearance than Geminoid F, but it also has movable lips, which allow for a clearer comparison of motion types.

At the start of the experiment, subjects were given instructions and told they would be playing the role of a visitor to an information desk where the two robots assumed the role of receptionists. The subjects then entered a room in which the two robots were positioned around a desk, and met and conversed with each robot one at a time. Specifically, the subjects asked each robot the questions in the above list in a natural manner. After each question, the robot responded according to a remote control by the experimenters. This procedure was repeated five times, once for each motion type. After each session, subjects were asked to fill out a brief questionnaire. The answers for all questions were prepared ahead of time by asking

**Fig. 7.11** External appearance of Telenoid R2

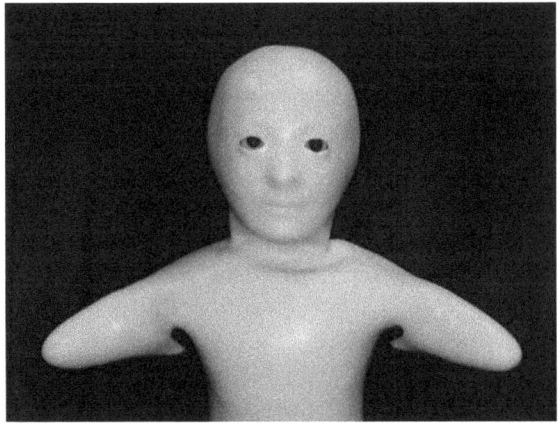

a collaborator to act out the scenario; audio, video, and motion data of our collaborator were recorded and used as the original data to create the robot's answers.

Five motion control types were reproduced on the two robots:

- NOD ONLY
- NOD&TILT
- NOD&TILT with GAZING
- ORIGINAL
- ORIGINAL with GAZING

The motion control types "NOD ONLY," "NOD&TILT," and "ORIGINAL" are the same as described in Sect. 7.4 (the latter is generated by reproducing the original motion recorded from the motion capture data).

During the analysis of eye gaze movements in Sect. 7.6.1, we found that when people tilt their heads, they often turn their gaze in the opposite direction. We applied this rule to generate eye gaze movements for the "NOD&TILT" model, thereby creating a new condition, "**NOD&TILT with GAZING**."

Eye gaze control for the "**ORIGINAL with GAZING**" model was based on annotated eye gazing tags corresponding to each point in the dialogue.

For all five motion control types, lip motion was generated using the method proposed in [23], which is based on a rotation of the vowel space given by the first and second formants around the center vowel and a mapping to the degree of lip opening.

Motion types were presented in a nearly random fashion, as we ensured that all subjects compared the three types directly: "NOD&TILT," "NOD&TILT with GAZING," and "ORIGINAL with GAZING." Therefore, these three motion types were always presented together as a group, either in this order or reversed. The order of this group and the other two motion types was then determined randomly.

Subjects were asked to fill out a questionnaire for each motion control type to evaluate the naturalness of the robot's motion. The questionnaire had the following items:

- **Is the motion of the robot natural?**
- **Compared with the previous one, which motion is more natural?**

The options for the answers to these questions were the same as those described in Sect. 7.4.2.

A total of 22 paid Japanese speakers (11 male and 11 female, aged 18–50 with a mean of 29 and s.d. of 11) participated in the experiment.

## 7.6.3 Experimental Results

Figure 7.12 shows the subjective naturalness scores for individual motion types in each robot. These results were quantified on a 1–7 scale, where "1" denoted "clearly unnatural" and "7" denoted "completely natural."

Fig. 7.12 Subjective
naturalness for each motion
type and robot: Geminoid F
(top) and Telenoid R2
(bottom). Error bars indicate
standard deviation

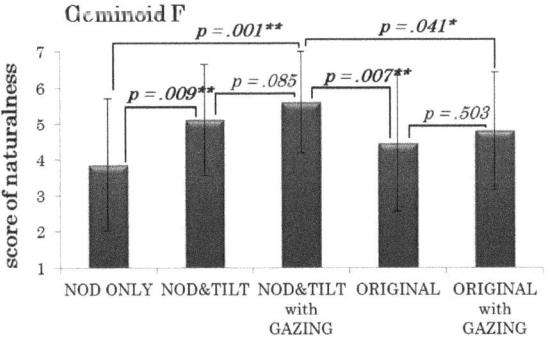

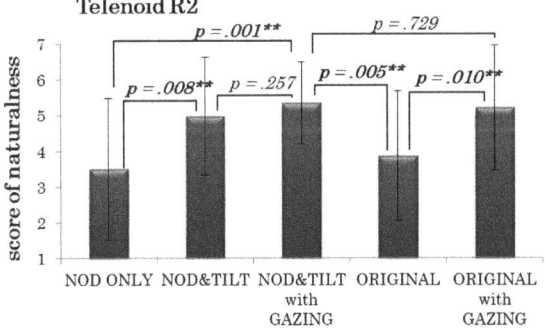

Comparing the performance of the motion control types using repeated measures ANOVA gave $F(4, 18) = 4.451$, $p < 0.011$ for Geminoid F and $F(4, 15) = 4.558$, $p < 0.013$ for Telenoid R2.

We can see that "NOD ONLY," "NOD&TILT," and "ORIGINAL" obtained similar results as in the video-based evaluation in Sect. 7.4.

Both Geminoid F and Telenoid R2 achieved significantly higher subjective scores for "NOD&TILT with GAZING" than for "NOD ONLY" and "ORIGINAL" ("NOD&TILT with GAZING" vs. "NOD ONLY": $p = 0.001$ for Geminoid F, $p = 0.001$ for Telenoid R2; "NOD&TILT with GAZING" vs. "ORIGINAL": $p = 0.007$ for Geminoid F, $p = 0.005$ for Telenoid R2), but a comparison with "NOD&TILT" only shows non-significant or almost-significant differences ($p = 0.085$ for Geminoid F, $p = 0.257$ for Telenoid R2). This result is discussed in the next section.

For the Telenoid, "ORIGINAL with GAZING" scored as highly as "NOD&-TILT with GAZING" and significantly higher than "ORIGINAL" ($p = 0.010$), as expected. However, for the android, this was not the case: "NOD&TILT with GAZING" performed best. A comparison between "ORIGINAL with GAZING" and "NOD&TILT with GAZING" shows a significant difference ($p = 0.041$). We suspected that imperfect reproduction of the original motions was again the culprit. Specifically, Geminoid F cannot tilt its head about the roll axis without also rotating in the pitch axis because of the position of its head actuators (see Fig. 7.6), causing the chin to move forward. To confirm this, we recorded the robot's reproduced motion data using the motion capture system and compared it with the original motion data captured from a human speaker. We found that pitch and yaw motions were reproduced well, but that roll motions were not accurately reproduced, causing a potential source for perceived unnaturalness (see Fig. 7.13).

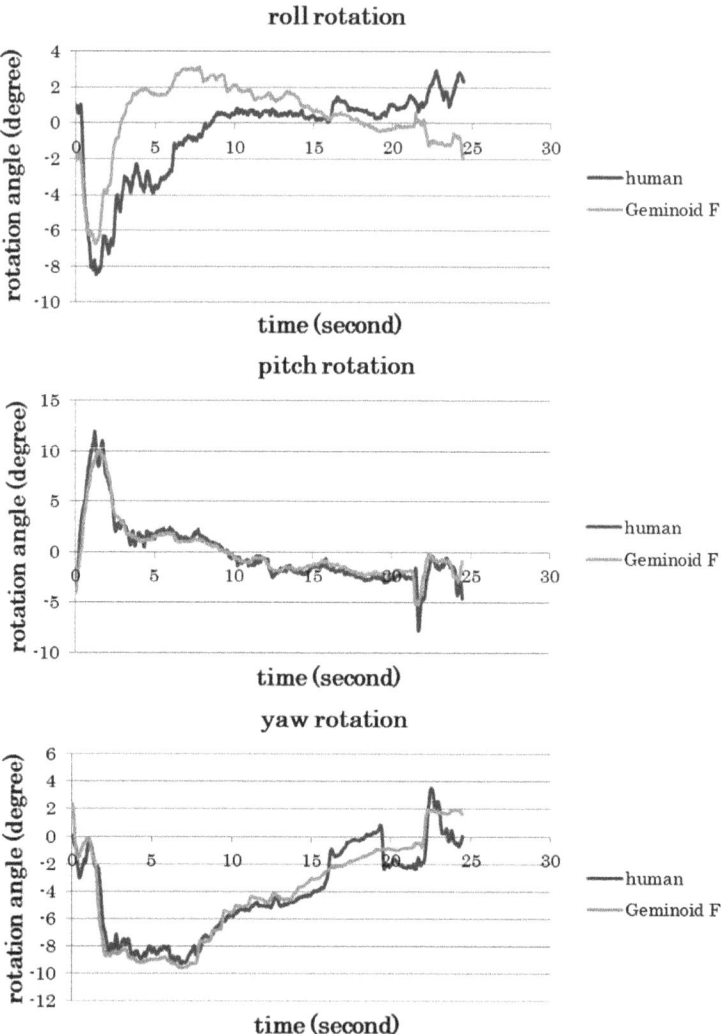

**Fig. 7.13** Motion data from human and Geminoid F

### 7.6.4 Discussions

The effect of adding gaze control to the developed generation models was not as great as we had anticipated. We believe there may be two causes for this. First, people may communicate what they are thinking in different ways: by looking to one side as if recalling something, tilting their head to one side, or both. We think that doing both is not necessarily perceived as more natural, because it may convey only the same information (i.e., the person is thinking). Additionally, robots tend to be less complicated in terms of their communicative capabilities than people; because of this perception, it may be acceptable for some information (such as eye movements) that is normally present with humans to be abstracted or missing in robots. This highlights another interesting possibility: By removing extraneous, misleading, or undesired communicative cues that can be observed in humans (such as involuntary twitching), robots could one day become capable of conveying information more clearly than actual humans. Thereby, robots would be able to contribute to service tasks in which such natural communication is an asset, e.g., teaching or theater performances.

This study has focused on exploring how generated nodding, head tilting, and gazing may contribute to the perceived naturalness of a communicating robot. Evaluations were performed both with video footage and face to face with three different humanoid robots.

However, the generality of the current work is necessarily limited by the target group, choice of robot, cues for generating head motions, and the investigated motions. In this study, Japanese people of various ages evaluated our robots for naturalness, but it remains to be shown whether the results will be exactly the same in different countries or for different cultures. Likewise, it would be interesting to investigate mechanisms for increasing the perception of naturalness in non-humanoid communicative robots. Finally, we imagine that other motions during conversation, such as posture shifts or hand movements, could also have an interesting effect on naturalness.

## 7.7  Conclusions

A rule-based nodding and head tilting motion generation model was proposed based on dialogue act function tags. Nodding generation was evaluated in two humanoid robots for utterances with confident behaviors. Subjective evaluation showed that the proposed model can generate head motion that appears more natural than reproducing the original head motion.

The model including head tilting was then evaluated for utterances including thinking behaviors. Subjective scores showed that the proposed model including head tilting and nodding can generate head motion with increased naturalness compared to nodding only or directly mapping people's original motions.

A "face-up" motion was proposed for robots without movable lips, as an alternative to explicitly signal utterance motions. It was shown that the inclusion of such a motion improves the perceived naturalness of robots that do not have a mouth, such as the humanoid Robovie R2.

An experiment in which participants conversed with the robots face to face confirmed the validity of the previous video-based evaluation. The inclusion of eye gazing control was also shown to be effective for increasing naturalness when head tilting motion is accompanied.

Remaining topics for future work include the development of methods for extracting linguistic and paralinguistic information from speech (such as phrase boundaries, disfluencies, and dialogue acts) in order to automate the generation of head motion commands.

**Acknowledgements** This work was supported by JST CREST.

# References

1. Ishi, C.T., C. Liu, H. Ishiguro, and N. Hagita. 2010. Head motion during dialogue speech and nod timing control in humanoid robots. In *Proceedings of the 5th ACM/IEEE international conference on human-robot interaction (HRI 2010)*, 293–300.
2. Liu, C., C. T. Ishi, H. Ishiguro, and N. Hagita. 2012. Generation of nodding, head tilting and eye gazing for human-robot dialogue interaction. In *Proceedings of the 7th ACM/IEEE international conference on human-robot interaction (HRI 2012)*, 285–292.
3. Ishi, C.T., H. Ishiguro, and N. Hagita. 2013. Analysis of relationship between head motion events and speech in dialogue conversations. *Speech Communication* 57 (2014): 233–243.
4. Ishi, C.T., C. Liu, H. Ishiguro, and N. Hagita. 2010. Head motion during dialogue speech and nod timing control in humanoid robots. In *Proceedings of IEEE/RSJ human robot interaction (HRI 2010)*, 293–300.
5. Liu, C., C. Ishi, H. Ishiguro, and N. Hagita. 2013. Generation of nodding, head tilting and gazing for human-robot speech interaction. *International Journal of Humanoid Robotics (IJHR)* 10(1).
6. Sidner, C., C. Lee, L.-P. Morency, and C. Forlines. 2006. The effect of head-nod recognition in human-robot conversation. In *Proceedings of IEEE/RSJ human robot interaction (HRI 2006)*, 290–296.
7. Morency, L.-P., C. Sidner, C. Lee, and T. Darrell. 2007. Head gestures for perceptual interfaces: The role of context in improving recognition. *Artificial Intelligence* 171(8–9): 568–585.
8. Yehia, H.C., T. Kuratate, and E. Vatikiotis-Bateson. 2002. Linking facial animation, head motion and speech acoustics. *Journal of Phonetics* 30: 555–568.
9. Sargin, M.E., O. Aran, A. Karpov, F. Ofli, Y. Yasinnik, S. Wilson, E. Erzin, Y. Yemez, and A.M. Tekalp. 2006. Combined gesture-speech analysis and speech driven gesture synthesis. In *Proceedings of IEEE international conference on multimedia*.
10. Munhall, K.G., J.A. Jones, D.E. Callan, T. Kuratate, and E. Vatikiotis-Bateson. 2004. Visual prosody and speech intelligibility—Head movement improves auditory speech perception. *Psychological Science* 15(2): 133–137.
11. Graf, H.P., E. Cosatto, V. Strom, and F.J. Huang. 2002. Visual prosody: Facial movements accompanying speech. In *Proceedings of the IEEE international conference on automatic face and gesture recognition (FGR'02)*.

12. Beskow, J., B. Granström, D. House. 2006. Visual correlates to prominence in several expressive modes. In *Proceedings of interspeech 2006—ICSLP*, 1272–1275.
13. Busso, C., Z. Deng, M. Grimm, U. Neumann, and S. Narayanan. 2007. Rigid head motion in expressive speech animation: Analysis and synthesis. *IEEE Transactions on Audio, Speech and Language Processing*.
14. Iwano, Y., S. Kageyama, E. Morikawa, S. Nakazato, K. Shirai. 1996. Analysis of head movements and its role in spoken dialogue. In *Proceedings of the international conference on spoken language processing (ICSLP'96)*, 2167–2170.
15. Foster, M.E., J. Oberlander. 2007. Corpus-based generation of head and eyebrow motion for an embodied conversational agent. *Language Resources and Evaluation* 41(3–4): 305–323.
16. Ishi, C.T., J. Haas, F.P. Wilbers, H. Ishiguro, and N. Hagita. 2007. Analysis of head motions and speech, and head motion control in an android. In *Proceedings of IEEE/RSJ international conference on intelligent robots and systems (IROS 2007)*, 548–553.
17. Ishi, C.T., H. Ishiguro, and N. Hagita. 2006. Analysis of prosodic and linguistic cues of phrase finals for turn-taking and dialog acts. In *Proceedings of the interspeech'2006—ICSLP*, 2006–2009.
18. Minato, T., M. Shimada, H. Ishiguro, and S. Itakura. 2004. Development of an android robot for studying human-robot interaction. *Innovations in applied artificial intelligence*, 424–434. Springer.
19. DeBoer, M., and A.M. Boxer. 1979. Signal functions of infant facial expression and gaze direction during mother-infant face-to-face play. *Child Development* 50(4): 1215–1218.
20. Langton, S.R.H., R.J. Watt, and V. Bruce. 2000. Do the eyes have it? Cues to the direction of social attention. *Trends in Cognitive Sciences* 4(2): 50–59.
21. Kaplan, F., and V. Hafner. 2004. The challenges of joint attention. *Interaction Studies* 67–74. http://cogprints.org/4067/.
22. Nagai, Y., M. Asada, and K. Hosoda. 2006. Learning for joint attention helped by functional development. *Advanced Robotics* 20: 1165–1181(17).
23. Ishi, C.T., C. Liu, H. Ishiguro, and N. Hagita. 2011. Speech-driven lip motion generation for tele-operated humanoid robots. In *International Conference on Auditory-Visual Speech Processing*.

# Chapter 8
# Uncanny Valley of Androids and the Lateral Inhibition Hypothesis

Michihiro Shimada, Takashi Minato, Shoji Itakura and Hiroshi Ishiguro

**Abstract** From the viewpoint of designing a robot for communication, it is important to avoid the 'uncanny valley,' although this is an essential phenomenon for discovering the principles relevant to establishing and supporting social interaction between humans and robots. Studying the uncanny valley allows us to explore the boundary of humanlike-ness. We have empirically and experimentally obtained evidence for the uncanny valley effect, which has thus far only been hypothesized through the development of androids that closely resemble human beings. We have also obtained experimental evidence to suggest that the uncanny valley varies owing to the development of individuals. We refer to this variable uncanny valley as the *age-dependent uncanny valley*. We assume that the uncanny valley is induced by a lateral inhibition effect, which is the same mechanism observed in sensory cells, and is referred to herein as the *lateral inhibition hypothesis of the uncanny valley*. The present paper presents evidence concerning the uncanny valley and describes the likelihood of the present hypothesis.

**Keywords** Uncanny valley · Lateral inhibition · Development on perception of humanlike-ness

This chapter is a modified version of a previously published paper [1], edited to be comprehensive and fit with the context of this book.

M. Shimada · T. Minato (✉) · H. Ishiguro
Asada Project, ERATO, Japan Science and Technology Agency, 2-1 Yamada-oka,
565-0871 Suita, Osaka, Japan
e-mail: minato@atr.Jp

M. Shimada · H. Ishiguro
Graduate School of Engineering, Osaka University, 2-1 Yamada-oka,
Suita, Osaka 565-0871, Japan

S. Itakura
Department of Psychology, Graduate School of Letters, Kyoto University,
Yoshida Honmachi, Sakyo-ku, Kyoto 606-8501, Japan

© Springer Nature Singapore Pte Ltd. 2018
H. Ishiguro and F. Dalla Libera (eds.), *Geminoid Studies*,
https://doi.org/10.1007/978-981-10-8702-8_8

## 8.1 Introduction

It is important to avoid the uncanny valley [2, 3] from the viewpoint of designing a communication robot. However, the uncanny valley is an essential phenomenon for discovering the principles relevant to establishing and supporting social interaction between humans and robots. Examination of the uncanny valley offers a clue for exploring the boundary of humanlike-ness and investigating human cognitive activities in communication processes.

The uncanny valley was first described by Mori [2, 3], who discussed the relationship between the similarity of a robot to a human and the perception of familiarity of the subject. The familiarity of a robot increases with its similarity until a certain point, at which imperfections cause the robot to appear repulsive (Fig. 8.1). This sudden drop in familiarity is referred to as the uncanny valley.

The original uncanny valley was based on Mori's empirical intuition rather than any experimental data. In recent years, a number of researchers have investigated issues on natural human–robot communication through experiments in order to confirm the existence of the uncanny valley [4–7]. However, no very humanlike robots have yet been developed. We have developed an android robot that closely resembles a human being in order to clarify the principles relevant to establishing and maintaining social interaction between humans and robots. This android enables us to investigate the principle of the uncanny valley through psychological experiments.

Thus far, we have developed several androids and studied methods to overcome the uncanny valley, implement natural movement in the android, and evaluate human–android communication. Although the first androids were situated deep within the uncanny valley, improved technology has allowed them to climb out. Moreover, through psychological experiments, we revealed that the uncanny valley varies with the development of the individual in early childhood.

We assume that this evidence is likely to be explained by the lateral inhibition effect, which is the same mechanism observed in sensory cells. Lateral inhibition

**Fig. 8.1** The uncanny valley [2, 3]

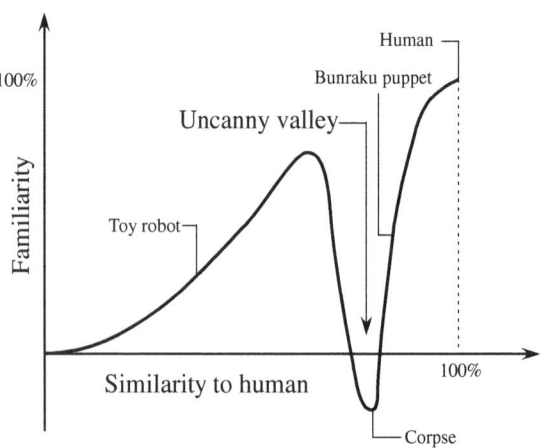

helps a sensory cell to make the origin of a stimulus more salient [8]. It is likely that lateral inhibition occurs not only in sensory cells, but also at many levels of recognition processes, such as face and body recognition. This appears to be necessary for individuals to have a high sensitivity in recognizing human beings, which is important for living in a human society. Based on the assumption that the pattern recognition system of the human brain makes use of the lateral inhibition effect, a discriminant circuit for humans is excessively inhibited when individuals see a robot that is slightly different from a human being, and the robot is categorized as an object, which is quite different from a human being. This phenomenon induces the uncanny valley. The change in the uncanny valley described above can also be explained by the formation process of the lateral inhibition effect. We call this assumption the *lateral inhibition hypothesis*. Proposing a hypothesis on human brain information processing and human cognitive activity from the viewpoint of robotics is a significant step.

The present paper reports the impression (familiarity) of three developed androids as well as techniques to improve their humanlike-ness. Psychological experiments were conducted to evaluate the humanlike-ness of the androids. Furthermore, the present paper describes experiments to clarify changes in the uncanny valley with respect to infant development. Finally, the paper proposes a lateral inhibition hypothesis that explains the uncanny valley and discusses a process that yields the uncanniness.

## 8.2  Familiarity of the Developed Androids

The appearance of an android is of considerable importance. After making the appearance of an android as humanlike as possible, it is necessary to implement a motion mechanism to realize humanlike motion. Here, the material used to construct the skin is important. It should be moldable enough to emulate the human shape, sufficiently flexible to allow joint movement, and have a soft humanlike texture. Concerning the motion mechanism, the configuration of the degrees of freedom (DoFs) is important. It is difficult to develop actuators and joints that can perfectly replicate human muscles and joints. The development of a motion mechanism is required to make the apparent motion of the android similar to that of humans.

We have developed three androids to investigate the humanlike-ness of their appearance and their motion mechanisms. In this section, the development of the androids and empirical evidence regarding their familiarity are described.

### 8.2.1  Repliee R1

To examine the humanlike-ness of the appearance, we developed the android *Repliee R1* shown in Fig. 8.2. Repliee R1 is based on an actual five-year-old girl. We used a

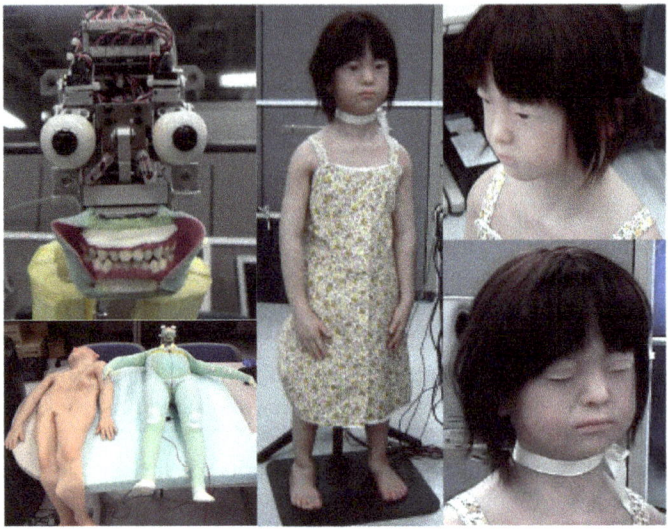

**Fig. 8.2**  Repliee R1

cast of the girl's body to mold the skin of the android, which is composed of a type of silicone that has a humanlike feel. The skin color is replicated in detail, including features such as blood vessels. Prosthetic eyeballs were used for the eye. Cameras were not embedded in the eyes to maintain a more humanlike appearance. The mechanical parts of the head are covered with fiber reinforced plastic (FRP) skull-like parts so that the bone under the skin can be felt when touching the head of the android. Repliee R1 has nine DoFs in the head (two for the eyelids, three for the eyes, one for the mouth, and three for the neck) and several free joints. All actuators (electrical motors) are embedded within the body. The main limitations of the head of this android are as follows:

- Repliee R1's range of motion is limited by the low elasticity of the silicone skin.
- Only the part of the skin that covers the eyeball moves during blinking; however, the expansion and contraction of the skin is not natural.
- Saccadic eye movements are not realized because of limitations in motor speed.
- The facial expression cannot be changed by facial elements other than the mouth and eyelids.
- The head oscillates through quick neck movements because of the low stiffness of the supporting frame.

Regarding the impression of Repliee R1, the main problems are the head oscillation, eyelid movement, and facial expression. The closed eyes appear different from those of humans. The appearance of the face closely resembles that of a human. However, the imperfect eyelids and reduced facial expression produce a negative impression. Although no statistical data were collected, several visitors to our laboratory reported a 'strange' or 'eerie' feeling with respect to Repliee R1.

## 8.2.2 *Repliee Q1*

Next, we developed *Repliee Q1*, shown in Fig. 8.3, to investigate the humanlikeness of the movement. The form of the body was not copied from a person, and the face is based on an averaged face of women from all over the world. Details such as wrinkles were reduced by the averaging process. The realism of the eyelid movement was improved by moving the area of skin below the eyebrows when blinking.

The silicone skin covers the neck, head, and forearms, with clothing covering other body parts. Unlike Repliee R1, the silicone skin does not cover the entire body, facilitating flexibility and increasing the range of motion. The android is driven by air actuators (air cylinders and air motors) that provide 31 DoFs from the waist up. The legs and feet are not powered, and the android can neither stand up nor move from a chair. The number of DoFs in the head is the same as that of Repliee R1. A high power-to-weight ratio is necessary for the air actuators in order to enable multiple actuators to be mounted in the human-sized body.

The eyelid movement and head oscillation problems of Repliee R1 have been improved, but the problems related to facial expressions remain. Repliee Q1 has been shown to the general public at several exhibitions (Fig. 8.6), and people have reported a better impression of Repliee Q1 than Repliee R1. In future studies, the hypothesis that an average face is more attractive than individual faces [9] may be examined for a 3-D face.

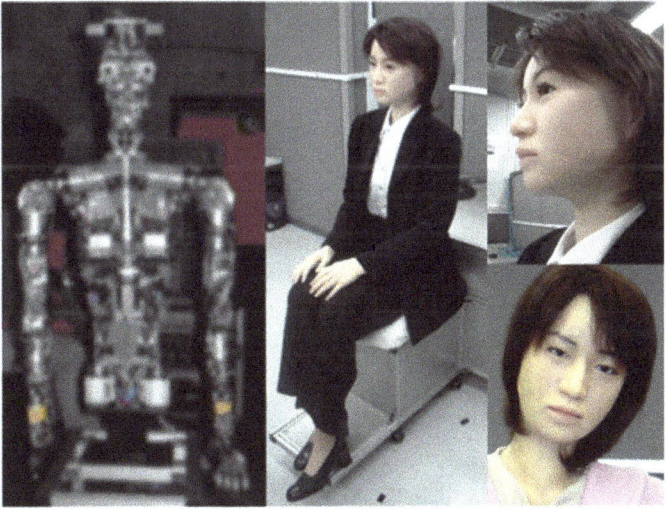

**Fig. 8.3** Repliee Q1

## 8.2.3 Repliee Q2

Repliee Q1 has been upgraded to *Repliee Q2*, shown in Fig. 8.4. The face is now modeled on a specific Japanese woman in order to realize a more humanlike appearance. Repliee Q2 has a total of 42 DoFs and can generate facial expressions and finger motions in addition to the movements of Repliee Q1. Sixteen actuators are embedded in the head (one for the eyebrows, one for the eyelids, three for the eyes, seven for the mouth, one for the cheeks, and three for the neck). Several facial expressions are shown in Fig. 8.5. As shown in the figure, only subtle changes in facial expression can be realized because of the number of actuators and limitations in the range of motion. Symbolic facial expressions are not generated, making the facial expressions natural.

The humanlike-ness of the appearance and facial expressions of Repliee Q2 represent an improvement over the previous model. We have also shown Repliee Q2 to the general public at several exhibitions (Fig. 8.6). Many people evaluated its humanlikeness very highly, with most reporting a positive impression. By improving the motion mechanism and facial expressions, it seems that the uncanniness of Repliee R1 can be overcome.

**Fig. 8.4** Repliee Q2

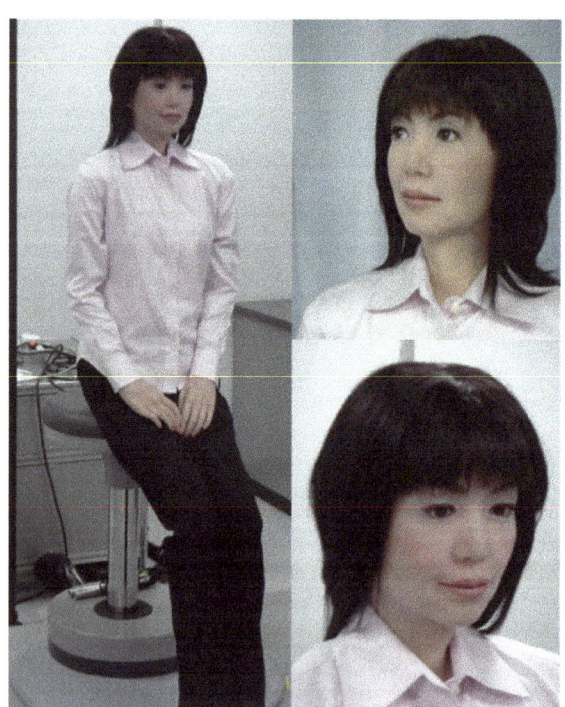

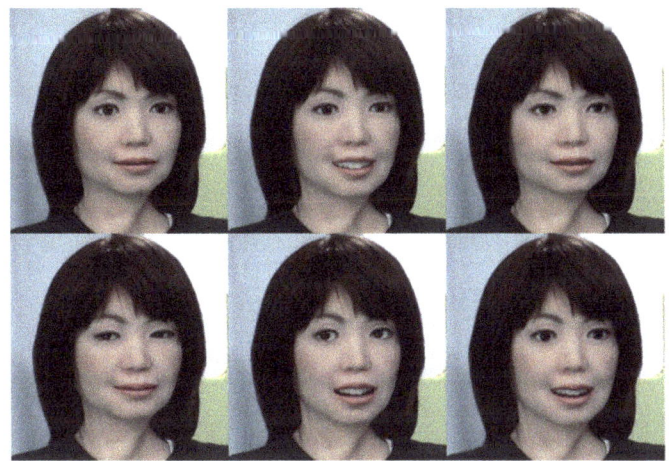

**Fig. 8.5**   Examples of facial expressions of Repliee Q2

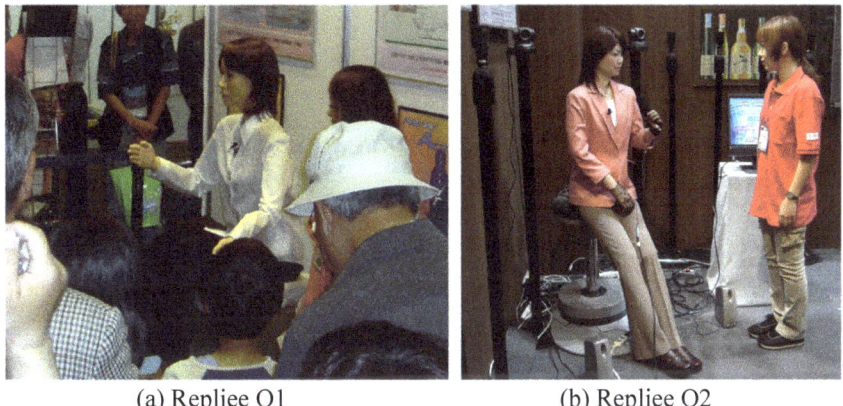

(a) Repliee Q1                           (b) Repliee Q2

**Fig. 8.6**   Androids at exhibitions

## *8.2.4   Empirically Induced Familiarity of the Androids*

Although there are no statistical data, it is likely that the order of familiarity of the
three androids is as follows:

$$\text{Repliee R1} < \text{Repliee Q1} < \text{Repliee Q2}$$

Concerning the facial appearance, a slight fault in the duplication of a person produces a marked strangeness (Repliee R1), but a created face does not produce this
feeling of strangeness (Repliee Q1). Therefore, a partial fault in the face (e.g., the
eyelids of Repliee R1) is considered to produce a stronger strangeness than a fault in

the entire face. Even a slight change in facial expression is important for improving the impression toward an android.

To realize a humanlike face, difficulties remain in imitating moving parts with large skin deformations, such as the eyelids and lips. The development of more flexible material for the skin is required. Furthermore, natural-looking deformations are also necessary.

## 8.3 Evaluation of the Humanlike-ness of the Androids

In the previous section, the familiarity and humanlike-ness of the androids were described. To evaluate these factors quantitatively and qualitatively, psychological experiments should assess the subjective impressions of various people. We have conducted a number of experiments to measure people's subconscious response toward robots with the aim of evaluating differences between robots and humans. This section describes an experiment to evaluate the humanlike-ness of the three androids [10].

### 8.3.1 Evaluation of an Android Through Subconscious Recognition

*Purpose*: The purpose of this experiment is to investigate the humanlike-ness of androids by measuring the gaze behavior that represents the subconscious mental state. Social signal theory states that gaze behavior is a type of social signal. Thus, we would expect gaze behavior to be influenced by the interlocutor in a conversation.

*Procedure*: Participants were asked to have a conversation with a questioner. The eye movements of the subjects were measured while the subjects were thinking about the answers to the questions posed by the questioner. There were two types of questions: thinking questions and knowledge questions. The former type compels the subject to derive the answer. The knowledge questions were used as the control condition.

The subjects were asked 10 knowledge questions and 10 thinking questions in random order. The faces of the subjects were videotaped, and their gaze direction was coded from the end of the question to the beginning of the answer. The eye movements of the subjects were analyzed to determine the average duration of gaze in eight directions. Four types of questioners were used: a Japanese person, Repliee R1, Repliee Q1, and Repliee Q2. The androids were made to appear as humanlike as possible. At the beginning of the experiment, the experimenter was seated beside the android and explained the experiment to the subject in order to allow the subject to become accustomed to the android.

The human questioner, Repliee Q2, Repliee Q1, and Repliee R1 asked questions to 29, 28, 8, and 17 subjects, respectively. The average ages of the subjects were 33.1 years, 30.2 years, 27.5 years, and 20.5 years, respectively. The subjects were recruited from a temporary employment agency. Most of the subjects were unfamiliar with the androids. Each subject was asked questions by only one questioner.

*Result*: The polar plot in Fig. 8.7 illustrates the average percentage of time that the subjects looked in each eye direction. To examine the effect of the questioner on the duration of breaking eye contact, a repeated-measures three-way ANOVA with one between-subject factor (questioner) and two within-subject factors (question type and eye direction) was conducted. The effects of the questioner, question type, eye direction, interaction between questioner and question type, and interaction between questioner and eye direction were significant. The most significant result is that the eye direction changes with the questioner. To investigate the interaction between questioner and eye direction, a Tukey's HSD (Honestly Significant Difference) test

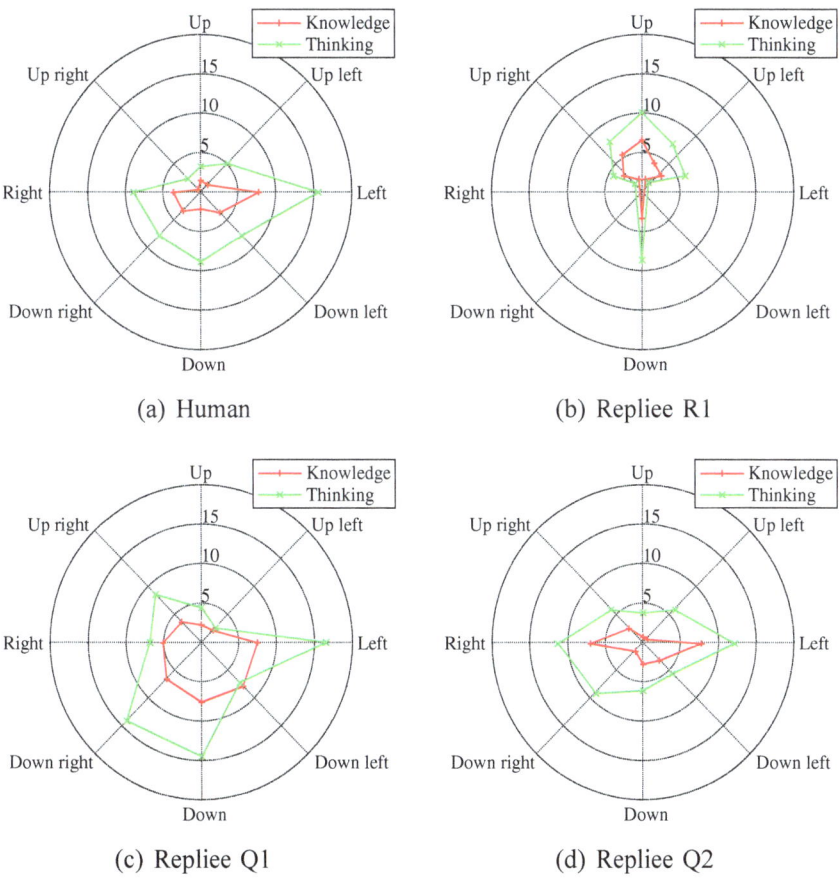

(a) Human                              (b) Repliee R1

(c) Repliee Q1                         (d) Repliee Q2

**Fig. 8.7**   Average duration of gaze in eight averted directions (%)

was conducted. As a result, significant differences were observed between looking downward to the right and downward to the left in the case of Repliee R1 and Repliee Q1, looking upward and downward to the left in the case of the human questioner and Repliee R1, and looking downward to the right in the case of Repliee R1 and Repliee Q2.

Figure 8.7 shows that, when asked thinking questions, the subjects tended to avert their eyes to the left or right (Repliee Q2 and the Japanese human questioner), to the left, downward to the right, and downward (Repliee Q1), and downward (Repliee R1). To investigate the similarity of the graph shapes, we calculated the inner product between the human and Repliee Q2, human and Repliee Q1, and human and Repliee R1. The elements of the vector were the average percentage of gaze duration in each averted direction. The results are 0.96, 0.94, and 0.75 for Repliee Q2, Repliee Q1, and Repliee R1, respectively. Therefore, we concluded that the gaze behavior toward Repliee Q2 is most similar to that toward a human. Moreover, if the gaze behavior indicates humanlike-ness, the androids are found to be humanlike in the order of Repliee Q2, Repliee Q1, and Repliee R1.

### 8.3.2  Climbing the Uncanny Valley

The results in the previous section show that Repliee Q2 was subconsciously treated as a human. In contrast, Repliee Q1 and Repliee R1 were not treated as human, because gaze behavior toward them differed from that toward the human questioner. The following describes the gaze directions. In the case of the human questioner or Repliee Q2, the subject looked left or right. In the case of Repliee Q1, the subjects looked left, downward to the right, and downward. In the case of Repliee R1, the subjects looked right and downward. Thus, the gaze direction changed from downward to sideways as the questioner became more humanlike.

The results suggest that the order of humanlike-ness of the three androids is as follows:

$$\text{Repliee R1} < \text{Repliee Q1} < \text{Repliee Q2}$$

By taking into account the relationship described in Sect. 8.2.4, the androids can be placed in the uncanny valley as shown in Fig. 8.8. In this figure, the abscissa denotes the similarity in appearance and movement. Thus, we obtained evidence for the existence of the uncanny valley that had previously been hypothesized through the development of androids.

## 8.4  Another Uncanny Valley

Does the uncanny valley appear in infant cognitive processes? Infants do not feel strangeness toward a robot that is slightly different from a human (e.g., Repliee R1).

**Fig. 8.8** Familiarity of the androids

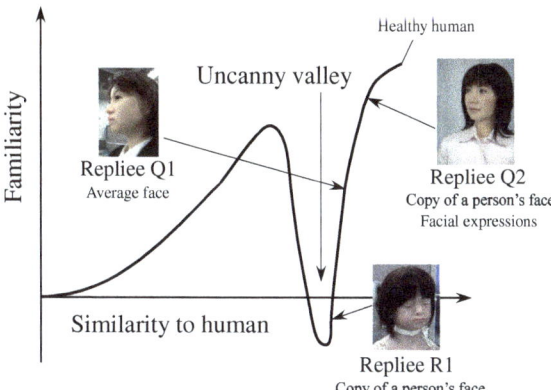

However, preschool children express strong fear [11], which suggests that the impression toward most human beings, i.e., the interpersonal cognition model, changes with infant development. This section describes an experiment to clarify this hypothesis.

### 8.4.1 Evaluation of Infant Cognitive Development

*Purpose*: We investigated how an infant recognizes the androids using the preferential looking method.

*Procedure*: Infants were seated 1 m away from a monitor while being held by their mothers from behind. After attracting the attention of the infants, a stimulus (a pair of movies) was exhibited for 10 s. Then, once the infant was looking at the monitor, the next stimulus was presented. The movies were displayed side by side, and the gaze time was measured for each trial. In the experiment, there were three movies: human, Repliee R1, and humanoid robot Robovie (mechanical-looking robot). The positions of the movies were interchanged for the purpose of counterbalancing. Thus, a total of six stimuli were displayed. Each movie showed only the upper body, and only the neck moved. The trajectory and speed of the neck movement were the same in all stimuli. The subjects were 27 infants ranging in age from 10 to 25 months.

*Results*: We classified the infants into three groups: (A) 12-month-old (6 infants), (B) 18-month-old (13), and (C) 24-month-old (7). Each gaze time is shown in Fig. 8.9.

The gaze time was found to change at 18 months of age. For example, in the case of Repliee R1 versus Robovie, the 12-month-old infants showed interest in Repliee R1. However, the 18-month-old infants showed more interest in Robovie. In addition, the 24-month-old infants showed an interest in Repliee R1. The same pattern appeared for the case of Repliee R1 versus the human questioner. Younger infants appeared to have an interest in Repliee R1, but this was not found in group B infants. However, the older group C infants exhibited a renewed interest in Repliee R1.

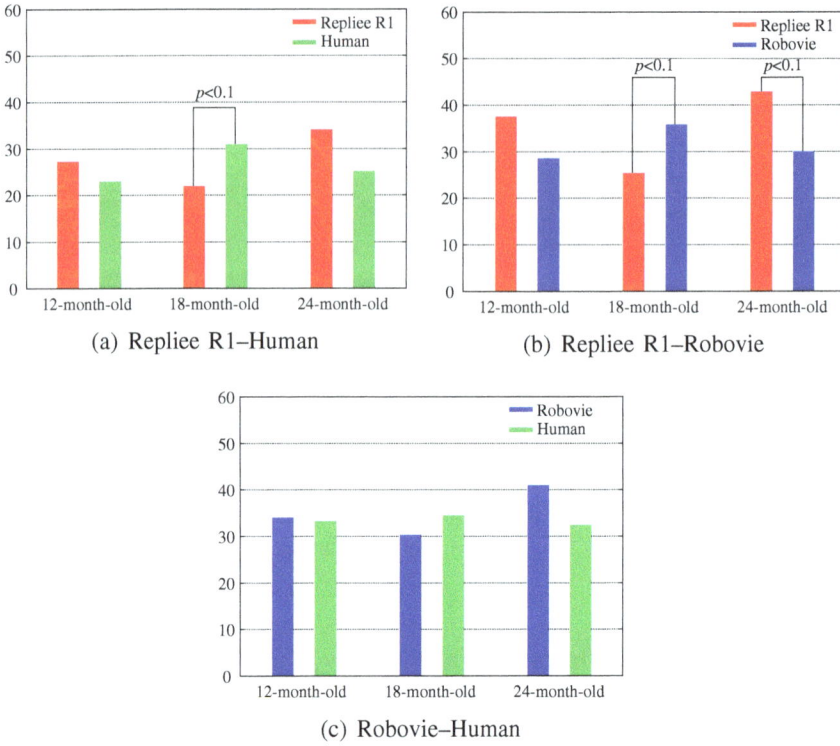

**Fig. 8.9** Average duration of gaze toward the stimuli (%)

## 8.4.2 Age-Dependent Uncanny Valley

The 12-month-old and 24-month-old infants expressed a great deal of interest in Repliee R1, and the 18-month-old infants expressed less interest. This is thought to be because the 12-month-old infants made a new model of humans when they observed the android. Therefore, they looked at the android for longer. However, the 18-month-old infants felt fear because the androids did not match the human model they had already built. The 24-month-old infants accepted various beings in their environment. Thus, they looked at the android for longer. This suggests another valley structure (U-shape structure) for the impression that indicates the development of an interpersonal cognition model.

From these results, we hypothesize that the uncanny valley varies with the development of the individual. The familiarity with a robot that is slightly different from a human being varies as shown in Fig. 8.10. Up to a certain point, the familiarity increases with the age of the individual. At this point, imperfections result in a feeling of strangeness. We assume that the change is a result of the development of the model of the human being. We refer to this relationship as the *age-dependent*

**Fig. 8.10**  Hypothesis of the age-dependent uncanny valley

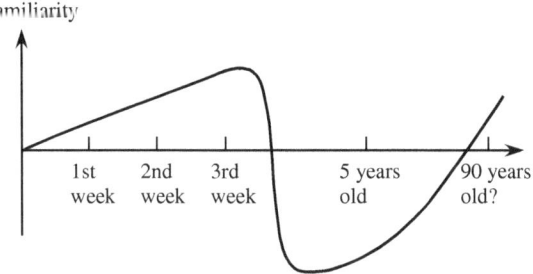

*uncanny valley.* In the next section, we propose a hypothesis to explain the uncanny valley and the age-dependent uncanny valley.


## 8.5  Lateral Inhibition Hypothesis of the Uncanny Valley

### 8.5.1  Lateral Inhibition of Sensory Cells

Lateral inhibition helps sensory cells to determine the origin of a stimulus more precisely [8]. One example is a retinal ganglion cell, which forms a receptive field that receives inputs from multiple cone cells (photoreceptors). The receptive field forms a ring shape, as shown in Fig. 8.11, in which the central disk part has excitatory connections and the surrounding ring part has inhibitory connections. Light shining on the central part excites the ganglion cell, whereas light on the surrounding part inhibits it. This competitive effect is referred to as lateral inhibition. The response of the ganglion cell varies depending on the light stimulus pattern, as shown in Fig. 8.11. Light pattern D increases the activity of the ganglion cell, whereas light pattern B decreases the activity of the ganglion cell. The lateral inhibition effect can enhance the response to the boundary of a pattern. Other ganglion cells detect color, and sensory cells such as auditory and tactile cells also exhibit the lateral inhibition effect. Furthermore, the Mexican-hat function, which is a model of lateral inhibition, is used for pattern recognition in computer information processing studies.


### 8.5.2  Lateral Inhibition Hypothesis

Inputs from sensory cells result in perception through multiple information processing in the brain. Lateral inhibition is likely to occur not only in sensory cells, but also at all levels of the recognition processes. Thus far, we have observed the following:

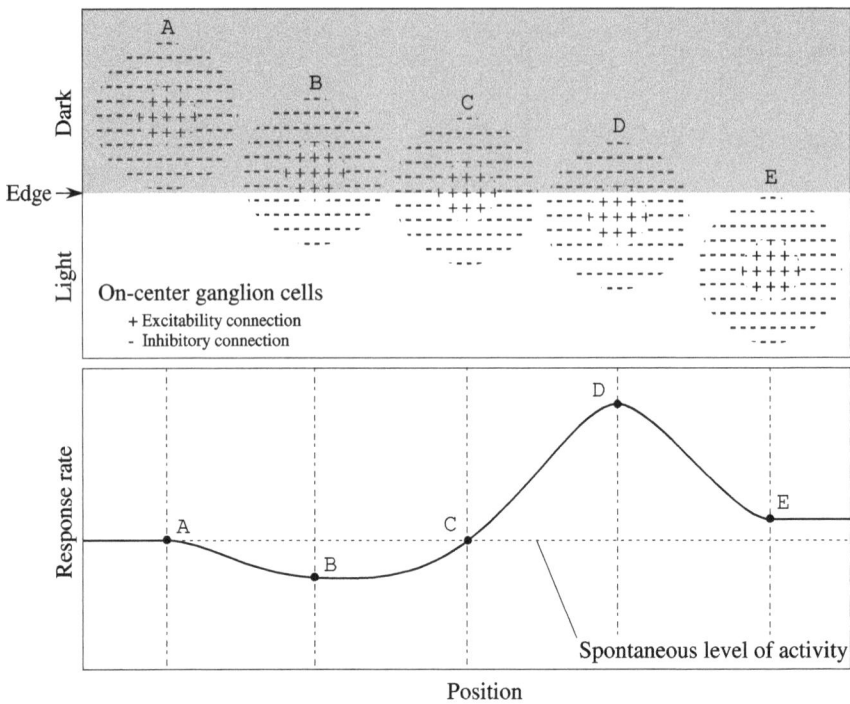

**Fig. 8.11** Detecting differences in luminance [12]

- Individuals need high-accuracy recognition of human beings in a human society.
- Lateral inhibition helps sensory cells to determine the origin of a stimulus more precisely.
- Lateral inhibition occurs at all levels of recognition processes.
- A model of lateral inhibition is used for pattern recognition in computer information processing.

This evidence indicates that it is reasonable to assume a lateral inhibition model in the recognition process of humanlike-ness. Based on the assumption that the human brain makes use of the lateral inhibition effect to recognize human beings, the humanlike-ness of robots that seem almost human is excessively inhibited. We also assume that there is another recognition model in which the humanlike-ness is proportional to the similarity to human beings. This is intuitively reasonable. By integrating this model with the lateral inhibition model, the valley structure shown in Fig. 8.12 appears. This is the valley of recognition of humanlike-ness derived from the lateral inhibition hypothesis, and we expect this graph to be a feature of the original uncanny valley hypothesis. This is the *lateral inhibition hypothesis* of the uncanny valley. The relationship between the recognized humanlike-ness and familiarity is discussed in Sect. 8.5.3.

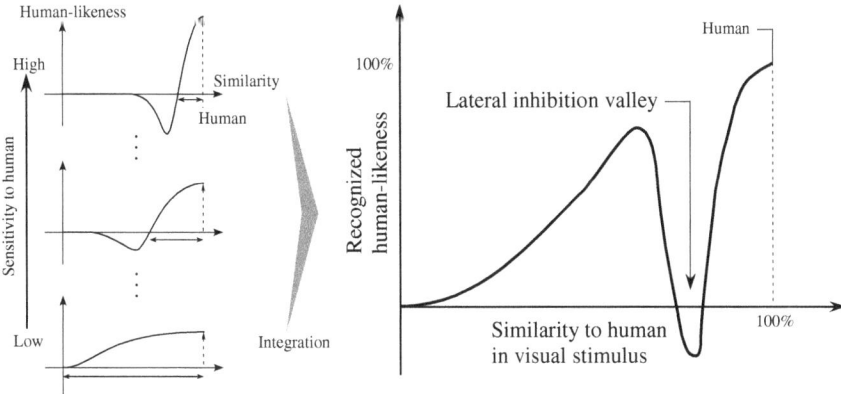

**Fig. 8.12**  Humanlike-ness based on lateral inhibition hypothesis

Based on the lateral inhibition hypothesis, we can also explain the age-dependent uncanny valley shown in Fig. 8.10. The mechanisms for the formation of the receptive field with lateral inhibition have been studied previously. The basic principle can be seen in Turing diffusion-driven instability [13]. Consider a two-component (X, Y) system in which the two diffusion rates are different (Fig. 8.13). Activator X increases X and Y, and inhibitor Y decreases X. (a) Initially, the concentration distributions of the two components are stable and relatively constant with small fluctuations. (b) Once the concentration of X is biased by the fluctuations, the concentrations of X and Y increase under activation by X. (c) If the diffusion rate of Y is greater than X, then the distribution of lateral inhibition is formed. The concentration of X after point A increases at first, before decreasing after a certain time. If the distribution of X is regarded as the activity of a receptive field, this process means that the discriminant boundary gradually becomes distinct.

We have hypothesized that the age-dependent uncanny valley (Fig. 8.10) is caused by the development of the model for recognizing human beings. Based on the lateral inhibition hypothesis, the transition of the discriminant boundary of the model can be explained by the formation process of lateral inhibition. Thus, the transition of familiarity toward Repliee R1 is shown in Fig. 8.14. An infant with immature lateral inhibition does not experience a feeling of strangeness toward Repliee R1. In contrast, preschool children with mature lateral inhibition feel a strong degree of strangeness. The discriminant performance declines with age, and the impression of strangeness disappears again in old age.

### 8.5.3  Where Does Uncanniness Come From?

In the present hypothesis, we have argued that the valley in the recognition of humanlike-ness is induced by the lateral inhibition effect. In this case, why does

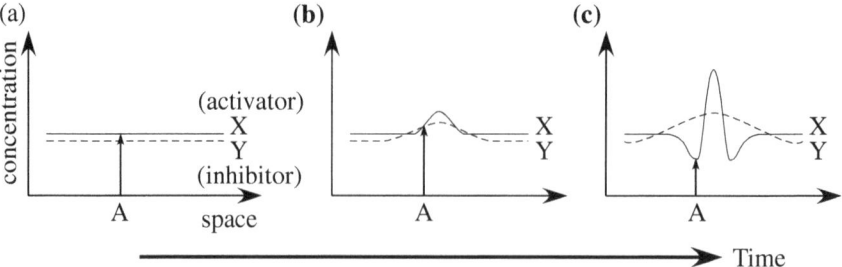

**Fig. 8.13** Pattern formation in Turing's reaction-diffusion systems

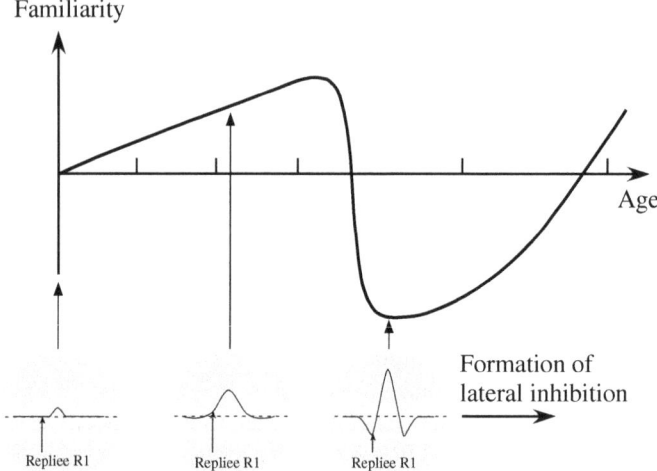

**Fig. 8.14** The age-dependent uncanny valley owing to the formation process of lateral inhibition

'uncanniness' or 'eeriness' occur? In this section, we discuss the relationship between the valleys of humanlike-ness and familiarity.

Empirically, it appears that most human beings feel eeriness in advance of their recognition. In fact, the human brain is innately programmed to generate fear in the presence of particular objects (e.g., certain animals and high places) and after observing the fear response in others [14]. However, it is not clear whether the brain is innately programmed to generate fear toward a robot that is slightly different from a human being.

In the brain, there are two (or more) pathways from sensory organs (thalamus) to the emotional system (amygdala). One is a direct pathway, and the other runs through the perception and recognition system (cortex), as shown in Fig. 8.15 [15]. The amygdala receives rough information on visual stimuli through the direct (low-level) path because of the absence of the perception process. When an individual sees an object that is similar to a human being, the amygdala receives a signal indicating that the object is human through the low-level path and a signal indicating that the

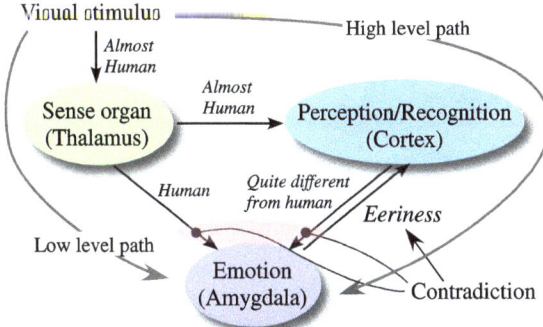

**Fig. 8.15** Uncanniness comes from a contradiction between sensory and recognized information

object is quite different from a human through the high-level path. This competitive state resulting from contradictory information may produce the feeling of eeriness. By taking account of the fact that infants have an interest in moving or complex-looking objects, it is likely that the familiarity is proportional to the humanlike-ness at the left side of the valley in Fig. 8.12. In any case, in future studies, it will be necessary to consider the model that produces this emotion of eeriness.

### 8.5.4   Extension of the Uncanny Valley

The original uncanny valley hypothesis should be reconsidered based on the lateral inhibition effect. We present the following two hypotheses on the appearance and behavior of robots:

- If the appearance of a robot is very humanlike, humans will attempt to relate to the android as if it were a human being. Therefore, subtle differences create a strangeness induced by lateral inhibition.
- Humans expect balance between appearance and behavior when they recognize an animate object.

Based on the lateral inhibition hypothesis, the uncanny valley is plotted in three dimensions in the upper-left image of Fig. 8.16. Subtle differences in any aspect increase the 'strangeness' of the android. On the contrary, the second hypothesis suggests that the familiarity increases for well-balanced appearance and behavior. We refer to this as the *synergy effect*. For example, a robot should have robot-like behaviors and a human should have humanlike behaviors [16]. This differs from the concept of the uncanny valley, because humans do not have sensitive mental models for recognizing robots and other toys. By fusing these hypotheses, we have the extended uncanny valley shown in the right-hand image of Fig. 8.16. The axes in the extended uncanny valley are not clearly defined. How do we quantify similarity and how do we evaluate human–robot interaction? A study to answer these questions will help to reveal the essence of human–robot communication.

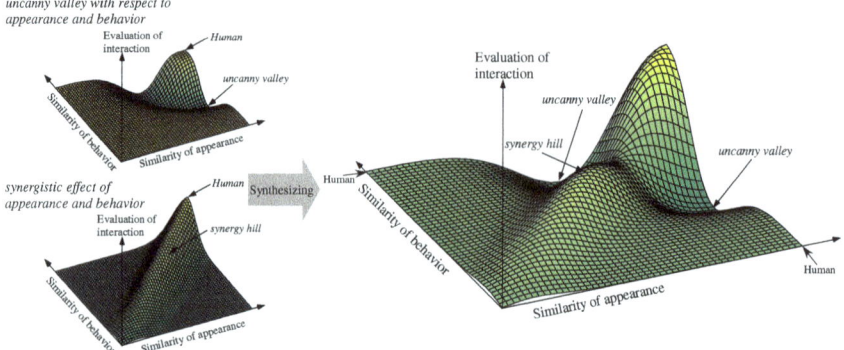

**Fig. 8.16**  Extended uncanny valley [17]

## 8.6  Conclusion

The present paper has proposed the lateral inhibition hypothesis to explain the uncanny valley phenomenon. The lateral inhibition hypothesis is induced from evidence obtained through the development of androids and psychological experiments using these androids. This hypothesis seems reasonable in light of various lateral inhibition effects in the perception and recognition processes, as well as the high discriminant performance of perception and recognition. In addition, this hypothesis can explain the age-dependent uncanny valley, whereby the uncanny valley changes with infant development. We expect that evidence supporting this hypothesis will be found through brain science and psychology studies. Such android studies help to clarify human cognitive activities and develop a design methodology for communication robots through an interdisciplinary framework of engineering and science [18]. The contributions of robotics to the elucidation of human cognitive development are not yet sufficiently advanced. Thus, our proposal of a cognitive model, i.e., the lateral inhibition hypothesis, is significant from the standpoint of robot development.

**Acknowledgements**  The android robots Repliee R1, Repliee Q1, and Repliee Q2 examined in the present study were developed in collaboration with Kokoro Company, Ltd.

## References

1. Michihiro Shimada, Takashi Minato, Shoji Itakura, and Hiroshi Ishiguro. (2007). Uncanny valley of androids and its lateral inhibition hypothesis. In *16th IEEE international symposium on robot and human interactive communication (RO-MAN 2007)*, 374–379.
2. Mori, M. 1970. Bukimi no tani [the uncanny valley] (in Japanese). *Energy* 7 (4): 33–35.
3. MacDorman, K.F. (2005). Androids as an experimental apparatus: Why is there an uncanny valley and can we exploit it? In *Proceedings of CogSci-2005 workshop: Toward social mechanisms of android science*, 106–118.

4. Oztop, E., D.W. Franklin, T. Chaminade, and G. Cheng. 2005. Human humanoid interaction: Is a humanoid robot perceived as a human? *International Journal of Humanoid Robotics* 2 (4): 537–559.
5. Ramey, C.H. (2006). An inventory of reported characteristics for home computers, robots, and human beings: Applications for android science and the uncanny valley. In *Proceedings of ICCS/CogSci-2006 long symposium: Toward social mechanisms of android science*, 21–25.
6. Hanson, D. (2006). Exploring the aesthetic range for humanoid robots. In *Proceedings of ICCS/CogSci-2006 long symposium: Toward social mechanisms of android science*, 16–20.
7. MacDorman, K.F., and H. Ishiguro. 2006. The uncanny advantage of using androids in cognitive science research. *Interaction Studies* 7 (3): 297–337.
8. Squire, L.R., F.E. Bloom, S.K. McConnell, J.L. Roberts, N.C. Spitzer, and M.J. Zigmond. 2002. *Fundamental Neuroscience*, 2nd ed. San Diego, California: Academic Press.
9. Langlois, J.H., and L.A. Roggman. 1990. Attractive faces are only average. *Psychological Science* 1: 115–121.
10. M. Shimada, T. Minato, S. Itakura, and H. Ishiguro. (2006). Evaluation of android using unconscious recognition. In *Proceedings of the IEEE-RAS international conference on humanoid robots*, 157–162.
11. Itakura, S., N. Kanaya, M. Shimada, T. Minato, and H. Ishiguro. (2004). Communicative behavior to the android robot in human infants. In *Proceedings of the international conference on development and learning*, 44.
12. Purves, D., G.J. Augustine, D. Fitzpatrick, L.C. Katz, A. LaMantia, J.O. McNamara, and S.M. Williams. 2000. *Neuroscience*, 2nd ed. Sunderland, Massachusetts: Sinauer Associates Inc.
13. Turing, A.M. 1952. The chemical basis of morphogenesis. *Philosophical Transactions of the Royal Society B* 237 (641): 37–72.
14. Carter, R. 2002. *Consciousness*. London: Weidenfeld Nicolson Illustrated.
15. LeDoux, J. 1996. *The emotional brain: The mysterious underpinnings of emotional life*. New York: Simon & Schuster.
16. Chaminade, T., and J. Decety. 2001. A common framework for perception and action: Neuroimaging evidence. *Behavioral and Brain Sciences* 24: 879–882.
17. Minato, T., M. Shimada, S. Itakura, K. Lee, and H. Ishiguro. 2006. Evaluating the human likeness of an android by comparing gaze behaviors elicited by the android and a person. *Advanced Robotics* 20 (10): 1147–1163.
18. Ishiguro, H. (2005). Android science-toward a new cross-interdisciplinary framework. In *Proceedings of the international symposium of robotics research*.

# Chapter 9
# Evaluation of Robot Appearance Using a Brain Science Technique

Goh Matsuda, Kazuo Hiraki and Hiroshi Ishiguro

**Abstract** We evaluate the humanlike-ness of humanoid robots using electroencephalography (EEG). As the activity of the human mirror-neuron system (MNS) is believed to reflect the humanlike-ness of observed agents, we compare the MNS activity of 17 participants while observing certain actions performed by a human, an extremely humanlike android, and a machine-like humanoid. We find the MNS to be significantly activated only when the participants observe actions performed by the human. Despite the participants' rating of the android appearance as more humanlike than that of the robot, the MNS activity corresponding to each of the three agents does not differ. These findings suggest that appearance does not crucially affect MNS activity, and that factors such as motion should be targeted for improving the humanlike-ness of humanoid robots.

**Keywords** Humanoid robot · Android · EEG · Mirror neuron
Mu suppression

This chapter is a modified version of a previously published paper [1], edited to be comprehensive and fit with the context of this book.

K. Hiraki (✉)
The University of Tokyo, 3-8-1 Komaba, Meguro, Tokyo 153-8902, Japan
e-mail: khiraki@idea.c.u-tokyo.ac.jp

G. Matsuda · K. Hiraki
Graduate School of Arts and Sciences, University of Tokyo, Tokyo, Japan
e-mail: matsuda@ardbeg.c.u-tokyo.ac.jp

H. Ishiguro
Graduate School of Engineering Science, Osaka University, Osaka, Japan

© Springer Nature Singapore Pte Ltd. 2018
H. Ishiguro and F. Dalla Libera (eds.), *Geminoid Studies*,
https://doi.org/10.1007/978-981-10-8702-8_9

## 9.1   Introduction

To develop effective, communicative humanoid robots, it is important to know what aspects of humanoids make them more humanlike. This is because the very nature of human communication involves human communicative partners. In the present study, we examine the effect of appearance on the perceived humanlike-ness of robots using electroencephalography (EEG). The measurement of brain activity is known to reveal cognitive processes and mechanisms beyond the scope of traditional psychological methods.

One possible neurological index of humanlike-ness is the activity of the human mirror-neuron system (MNS). The MNS consists of brain regions located in the premotor cortex and the inferior parietal lobule, which are activated by self-actions as well as the observation of similar actions performed by others [2]. The MNS is thought to play an important role in understanding the actions and intentions of others, which is essential for social interactions. Therefore, it is likely that MNS activity reflects the humanlike-ness of agents serving as social partners. In the present study, we compared MNS activity during the observation of actions performed by three different agents, namely a human, an android (a humanoid robot with significant human resemblance), and a machine-like humanoid. If a humanlike appearance improves the social capability of a humanoid, MNS activation should be higher when an observer views actions performed by an android as compared with actions performed by a machine-like humanoid. As the suppression of the mu-rhythm (8–13 Hz) power of EEG in the sensorimotor area is considered to reflect the activation of the premotor cortex [3], we focused on changes in mu power as a measure of MNS activity.

## 9.2   Materials and Method

### 9.2.1   Participants

The participants comprised eight males and nine females (mean age 21.0 years; range 18–26); all were right handed and had normal or corrected-to-normal vision. Informed consent was obtained prior to their participation in the study, which was approved by the ethics committee of the University of Tokyo.

### 9.2.2   Stimuli

The stimuli were seven different black-and-white video clips (see Fig. 9.1) presented on a 17-in CRT monitor (9.7 cm × 9.4 cm). There were three experimental conditions. In the first condition (human), a woman performed two actions, namely

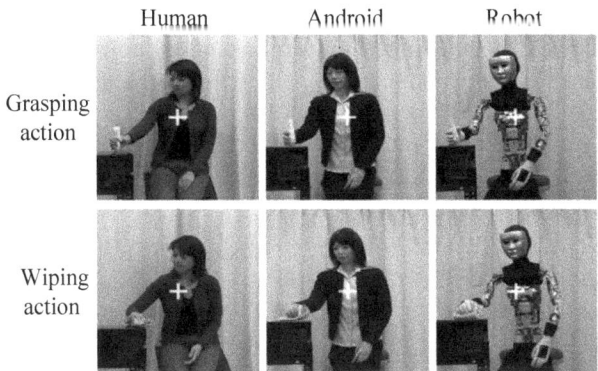

**Fig. 9.1** Example frames of the experimental video clips

grasping a tube or wiping a table with a cloth. The duration of each action was approximately 3.5 s. In the second (android) and third (robot) conditions, a female android (Repliee Q2 [4]) and a mechanical humanoid robot, respectively, performed the same actions. Because the android and robot have an identical inner architecture, the physical size and motion of both were almost the same. Visual white noise was presented as a baseline condition. A white cross was displayed at the center of each video clip as a fixation point.

### 9.2.3  Procedure

EEG measurements were made across 14 blocks of stimuli. In each block, participants observed one of the seven video clips 24 times with no inter-stimulus interval. The same video clip was presented in two blocks. A total of 174 s of EEG data were recorded per video clip. The order of presentation was randomized across participants.

After EEG recordings had been made, participants evaluated the humanlike-ness of the appearance and motion of each agent. A 7-point rating scale was used, with ratings of 1 and 7 representing "not humanlike at all" and "extremely humanlike," respectively.

### 9.2.4  EEG Recordings and Data Analysis

EEGs were acquired with a 64-channel high-density sensor net system (Net Station, EGI Inc.). The sampling rate was 500 Hz. Data from three electrode sites representing the sensorimotor area (C3, Cz, and C4) were selected for analysis. The mu-rhythm power at those electrodes was calculated using a fast Fourier transform and then averaged across the electrodes. Mu suppression for each experimental condition was defined as the log-transformed ratio of mu power relative to the mu power during the baseline condition. As a log ratio below zero indicates power suppression, each log ratio of mu power was compared with zero using a $t$-test. We also performed a repeated-measures analysis of variance (ANOVA) with the agent type as the factor.

A separate repeated-measures ANOVA with the agent type as the factor was performed for the subjective appearance and motion ratings. Bonferroni corrections were applied to multiple comparisons.

## 9.3 Results

### 9.3.1 Mu Suppression

Figure 9.2 shows the mu suppression recorded during each experimental condition. Significant suppression was observed only in the human condition ($t_{16} = 2.67$, $p < 0.05$, corrected). Although the mean log ratio of mu power was also negative in the android and robot conditions, they were not significantly less than zero. ANOVA revealed no significant effect of agent type ($F_{2,32} = 0.27$, $p = 0.77$).

### 9.3.2 Subjective Rating

There were significant differences in humanlike-ness for appearance ($F_{2,32} = 197.13$, $p < 0.01$) and motion ($F_{2,32} = 58.46$, $p < 0.01$) among the agents. Multiple comparisons in both ratings revealed significant differences between all pairs ($p < 0.01$, corrected); the human ranked the highest and the android was ranked the second highest for both ratings (see Fig. 9.3).

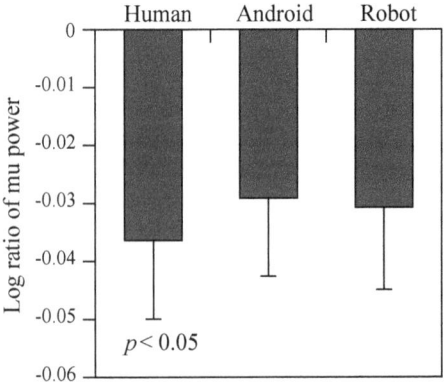

**Fig. 9.2** Log ratio of mu power during each experimental condition

**Fig. 9.3** Ratings of
human-likeness for
appearance and motion

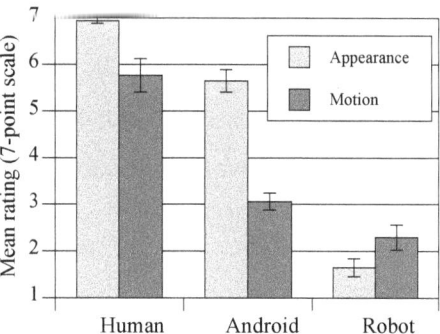

## 9.4   Discussion

Although there was no significant difference in mu power among the experimental
conditions, significant mu suppression was only observed in the human condition.
This indicates that human actions activated the MNS slightly more than the
android's and robot's actions.

Despite the android being rated as having a more humanlike appearance than the
robot, these agents produced similar MNS activity. This suggests that the extent of
humanlike-ness in appearance is not a crucial modulator of MNS activity. The
motion of the android was also rated as being more humanlike than that of the
robot. However, the mean motion ratings for both agents were less than 4, which
implies that the participants did not consider the motion of the android or the robot
to be humanlike.

From our results, it appears that MNS response may be influenced by
humanlike-ness of motion rather than appearance. A previous study reported that
point-light biological motion, which has an extremely simple appearance, activates
the MNS [5], highlighting the importance of motion for MNS activity. We hope to
investigate these phenomena in the near future using the methodology applied in
the present study.

**Acknowledgements**  This work was supported by KAKENHI (20220002 and 18200018).

## References

1. Matsuda, G., K. Hiraki, and H. Ishiguro. 2010. Evaluation of robot appearance by using a brain
   science technique. In *Proceedings of 2010 IEEE/RSJ international conference on intelligent
   robots and systems (IROS 2010) workshop: Human-robot symbiosis: Synergistic creation of
   human-robot relationships*.
2. Rizzolatti, G., L. Fogassi, and V. Gallese. 2001. Neurophysiological mechanisms underlying
   the understanding and imitation of action. *Nature Reviews Neuroscience* 2: 661–670.

3. Pineda, J.A. 2005. The functional significance of mu rhythms: translating "seeing" and "hearing" into "doing". *Brain Research Reviews* 50: 57–68.
4. Chikaraishi, T., T. Minato, and H. Ishiguro. 2008. Development of an android system integrated with sensor networks. In *Proceedings of 2008 IEEE international conference intelligent robots and systems (IROS 2008)*, Nice, Sep. 2008, 326–333.
5. Ulloa, E.R., and J.A. Pineda. 2007. Recognition of point-light biological motion: mu rhythms and mirror neuron activity. *Behavioural Brain Research* 183: 188–194.

# Chapter 10
# Persistence of the Uncanny Valley

**Jakub A. Złotowski, Hidenobu Sumioka, Shuichi Nishio, Dylan F. Glas, Christoph Bartneck and Hiroshi Ishiguro**

**Abstract** In recent years, the uncanny valley theory has been heavily investigated by researchers from various fields. However, the videos and images used in these studies do not permit any human interaction with the uncanny objects. Therefore, in the field of human–robot interaction, it is still unclear what impact, if any, an uncanny-looking robot will have in the context of an interaction. In this paper, we describe an exploratory empirical study using a live interaction paradigm that involves repeated interactions with robots that differ in embodiment and their attitude toward humans. We find that both components of uncanniness investigated here (likeability and eeriness) can be affected by an interaction with a robot. The likeability of a robot is mainly affected by its attitude, and this effect is especially prominent for a machine-like robot. Merely repeating interactions is sufficient to reduce the degree of eeriness, irrespective of a robot's embodiment. As a result, we urge other researchers to investigate the uncanny valley theory in studies that involve actual human–robot interactions in order to fully understand the changing nature of this phenomenon.

**Keywords** Uncanny valley · Anthropomorphism · Human–robot interaction
Multiple interactions · Eeriness · Likeability · Dehumanization

---

This chapter is a modified version of a previously published paper [1], edited to be comprehensive and fit with the context of this book.

---

J. A. Złotowski
CITEC Center of Excellence Cognitive Interaction Technology, Bielefeld University, Bielefeld, Germany
e-mail: jakub.zlotowski@pg.canterbury.ac.nz

H. Sumioka · S. Nishio (✉) · D. F. Glas · H. Ishiguro
Advanced Telecommunications Research Institute International, Keihanna Science City, Kyoto, Japan
e-mail: nishio@botransfer.org

C. Bartneck
Human Interface Technology Laboratory New Zealand, University of Canterbury, Christchurch, New Zealand

H. Ishiguro
Department of Systems Innovation, Graduate School of Engineering Science, Osaka University, Osaka, Japan

© Springer Nature Singapore Pte Ltd. 2018
H. Ishiguro and F. Dalla Libera (eds.), *Geminoid Studies*,
https://doi.org/10.1007/978-981-10-8702-8_10

## 10.1 Introduction

The uncanny valley theory was originally presented by Mori [2] in relation to a prosthetic arm. In recent years, it has gathered considerable attention in the fields of robotics, virtual agents, and cognitive sciences, as well as in mass media. The uncanny valley hypothesis suggests a nonlinear relationship between a robot's anthropomorphism and affinity. It proposes that, by increasing the humanlike appearance of a robot, we can increase our affinity for it. However, when a robot's appearance becomes sufficiently humanlike, but still distinguishable, people's emotional reaction becomes strongly negative. Once the appearance of a robot becomes indistinguishable from a real human, the affinity reaches its optimum at the same level as for human beings. Furthermore, Mori suggested that the movement of a prosthetic arm compared with a static arm will amplify the emotional response.

The uncanny valley is often used to explain people's rejection of anthropomorphic robots and virtual agents both in science and popular media, where it was given as a reason for the failure of the computer-animated movie The Polar Express. However, despite its wide adoption, there is relatively little empirical proof supporting the existence of the uncanny valley [3], e.g., the initial empirical work by [4] and [5] indicated that humanlike-ness might not be the only factor influencing our perception of an object as eerie. Rendering style could be related to the uncanny valley for virtual agents [6]. Moreover, it might be necessary to consider the effects of not only realism, but also the abnormality of artificial human appearance in investigating the uncanny valley phenomenon [7, 8]. It has been found that a mismatch between appearance and voice can result in the uncanny valley effect [9]. Furthermore, a mismatch between the appearance and movement of an android leads to stronger brain activation in the anterior portion of the intraparietal sulcus [10], which could provide a neurological explanation for the uncanny valley. In contrast, [11] reported that realistic motion can improve acceptability, especially of characters classified in the deepest point of the valley. This goes against the original theory of [2], who suggested that motion will increase the uncanny effect. The uncanny valley has been reported for other primates, with monkeys looking at real faces and unrealistic synthetic faces longer than at realistic synthetic monkey faces [12].

### 10.1.1 Related Work

Several potential explanations for the uncanny valley have been proposed. Apart from the neurological explanation [10], other factors include empathy [13], perception of experience [14], threat avoidance [2], and terror management [15]. A mathematical model using a Bayesian representation of categorical perception has been developed to explain how stimuli containing conflicting cues can give rise to a perceptual tension at category boundaries that leads to the uncanny feeling [16]. However, empirical investigations of these categorical boundaries suggest that ambiguous morphs close to the human endpoint induce a positive affect rather than the negative reaction suggested by the uncanny valley hypothesis [17, 18]. Furthermore, [19] found that images of prosthetic hands with intermediate humanlike-ness produced the strongest feelings of eeriness, whereas within different categories of images, increased humanlike-ness was related with the lowest degree of eeriness.

Vast research efforts have been dedicated to studying the dimensions of the uncanny valley. In particular, the original Japanese term used by [2]—*Shinwakan*— is difficult to translate to English. Various studies have used different translations, such as familiarity [5], likeability [20], affinity [21], eeriness [22], and empathy [23]. This variation in terms might affect the comparability of the results. Moreover, the humanlike-ness axis of Mori's graph has been the subject of empirical investigation [24].

The shape of the graph representing the uncanny valley is disputed. In one study, toy robots and humanoids were preferred over humans [25]. The authors proposed that the relationship between humanlike-ness and likeability resembles a cliff rather than a valley, where even perfectly realistic anthropomorphic robots are liked less than toy robots or mechanoids. These results imply that building highly humanlike androids might be futile, as their chances of acceptance are worse than for machine-like robots. Another study [20] reported that a highly realistic robot (android) was liked as much as a human. Furthermore, they reported that an android's realistic motion did not decrease its likeability and questioned the existence of the uncanny valley. This result is in line with a study using virtual agents [11]. However, [22] pointed out that the scales used by Bartneck and colleagues were correlated with warmth and, as a result, with each other, which might have affected the results. Overall, there is a lack of agreement between different studies regarding the dimensions and shape of the uncanny valley, with indications that Mori's theory could be too simplistic to accurately depict the relationship between humanlike-ness and perception of a robot or virtual agent. Moreover, it is not clear whether this theory has any actual consequences for interaction.

### 10.1.2   Does the Uncanny Valley Affect Human–Robot Interaction?

Despite being a common research theme, the effect of the uncanny valley hypothesis on human–robot interaction (HRI) is unknown. Previous studies on the uncanny valley used either images or videos of different targets that were supposed to induce the uncomfortable, eerie feeling (the exception is the work of [20], which involved short-term HRI). However, these studies did not permit any interaction between participants and robots or virtual agents. To understand how the uncanny valley affects HRI, it is necessary to involve physically collocated robots, as their physical presence could be an important mediating factor [26]. Previous work suggests that people's attitudes toward robots change during interaction [27], but it has never been empirically shown that the uncanny feeling will persist.

Little is known about the lasting effect of the uncanny valley. It is implicitly assumed that this negative emotional response toward anthropomorphic technology will have enduring consequences and lead people to reject androids that are distinguishable from humans. As this assumption has never been verified, it is important to consider an alternative hypothesis in which the uncanny valley only leads to a negative emotional response when the target is novel, and where the feeling of eeriness disappears during the course of HRI. It is possible that the affective habituation caused by repeated interactions will allow people to become accustomed to a machine that looks almost human, but is still not a perfect copy. Furthermore, the uncanny valley effect might decrease when an android interacts with a human in a friendly way. If this is the case, the effects of the uncanny valley on HRI might be limited to the pre-interaction phase.

### 10.1.3   Research Questions

There is some empirical evidence suggesting only a short-term effect of the uncanny valley. In a study conducted during the ARS Electronica festival, visitors who had interacted with an android were interviewed afterward. The majority did not report an uncanny feeling [28, 29]. As this study had the form of an open interview that allowed people to talk freely about their experience, only a qualitative analysis was possible. Therefore, it is important to quantitatively show whether the uncanny feeling is experienced less during and after interaction with an android. Secondly, the analysis of the uncanny valley phenomenon with virtual agents indicates that there could be a relation between knowing an agent (previous exposure) and the uncanny discomfort experienced by people exposed to it [30]. Lower levels of previous exposure to an agent were related with higher discomfort.

Moreover, there are psychological theories that can suggest a relation between repeated exposure to a stimuli and the uncanny valley hypothesis: mere exposure effect and affective habituation. It has been shown that mere exposure to a neutral stimulus leads to an increased positive affect toward it [31]. On the other hand, for strongly positive or negative stimuli, the intensity of the reaction decreases after multiple exposures. This process is called affective habituation [32].

The relationship between attraction and familiarity in interpersonal relations is well documented. Positive relationships are a result of frequent face-to-face contact [33]. However, if the person was initially disliked, greater familiarity would lead to greater dislike of that person [33]. This finding is consistent with other studies [34] that found repeated exposure to an unpleasant stimulus does not increase its likeability. Moreover, people are rated more positively by those whom they see more frequently [35] and express a stronger liking toward those whose ideas they have been exposed to for longer [36].

Four explanations have been proposed for the familiarity principle of attraction. Firstly, repeated exposure leads to increased processing fluency [37], which on its own is affectively positive [38]. Secondly, novel stimuli can produce uncertainty and negative reactions that diminish after a stimulus is found to be harmless [39]. Thirdly, as a result of classical conditioning, because most interactions are not aversive and mildly positive, others with whom people interact more often become paired with a positive affect [40, 41]. Fourthly, building on the previous explanation, repeated exposure creates an opportunity for interaction, and these interactions are more likely to lead to rewarding social experiences [41, 42].

The mere exposure effect does not require interaction, but exposure is sufficient for it to occur and it has been reported for various types of stimuli [43]. Although [44] proposed that familiarity leads to dislike in real interpersonal relations, because additional information about others makes them less similar to oneself, a live interaction paradigm showed that two previously unacquainted people exhibit a positive affect with increased familiarity [42].

In relation with the uncanny valley, it is possible that extreme stimuli weaken the affective reaction as people become more familiar because of affective habituation. However, for stimuli that are initially neutral, increased exposure could produce a more positive affect as a result of the mere exposure effect.

This study is the first exploratory work to investigate the effect of a robot's attitude and multiple interactions on the uncanny valley phenomenon by applying a live interaction paradigm in which actual HRI occurs. In particular, we focus on two aspects of interaction that could affect the uncanniness of a robot: (i) the number of interactions and (ii) the robot's attitude toward humans. Moreover, we have chosen two of the most common components representing the y-axis of the uncanny valley graph, *likeability* and *eeriness*, as they could be differently influenced by different aspects of HRI.

Likeability is an important factor affecting human–human relationships. For long-term HRI, it is expected to play an equally important role. Multiple factors affect human–human liking. One of the most important factors is a history of interaction with a specific person. In particular, we tend to like those with whom we have positive rather than negative interactions [45]. Moreover, the perception of a robot can be affected by its behavior [46]. Both positive and negative behavior have been anthropomorphized in robots, but people had more mechanistic conceptions for an impolite robot than for a positively behaving robot [27]. A robot that has a positive attitude toward humans could increase its likeability, as the classical conditioning explanation of the mere exposure effect would suggest. Similarly, an unfriendly robot could be liked less than it was before an interaction began. However, it is possible that an embodiment of a robot will play a role in affecting how strong an effect its behavior will have on its likeability. Thus, we hypothesize that:

$H_{1a}$: *A friendly robot's likeability will increase with repeated interactions.*

$H_{1b}$: *An unfriendly robot's likeability will decrease with repeated interactions.*

We also believe that previous exposure to a robot, irrespective of its behavior, is more important to its perceived eeriness. Eerie robots could produce affective habituation, and the initial strong negative emotional response will weaken with increased exposure. Similarly, for a robot that was initially perceived as neutral, repeated interactions may positively increase the affective perception according to the mere exposure effect.

In addition to looking at explicit measures such as self-reports, we investigate implicit attitudes toward humanlike robots. Implicit measures assess automatic reactions that are not consciously controllable [47] and are incrementally valid [48]. In addition, implicit measures complement rather than replace explicit measures, as they measure different aspects of the investigated attitude [49, 50]. Therefore, we also measured the perceived eeriness of the robots implicitly. Thus, our next hypotheses are:

$H_{2a}$: *Repeated interactions with a robot will reduce its explicit perceived eeriness.*
$H_{2b}$: *Repeated interactions with a robot will reduce its implicit perceived eeriness.*

Recent work in HRI indicates that it might be necessary to consider anthropomorphism as a multidimensional rather than unidimensional phenomenon [51]. These dimensions come from work on dehumanization—the process of depriving others of human qualities. It has been proposed that there are two distinct senses of humanness [52]: Human Uniqueness (HU) and Human Nature (HN). HU characteristics reflect socialization and distinguish humans from animals, e.g., intelligence, intentionality, or secondary emotions. HN covers inborn biological dispositions that distinguish humans from automata, e.g., warmth, sociability, or primary emotions. Anthropomorphism of a robot is not fixed and changes during an interaction [27]. To date, it has not been determined whether HU and HN dimensions of humanness attributed to a robot are also affected by the number of interactions or remain constant. In addition, previous work has indicated that the dimensions of mind attribution might be responsible for the uncanny valley phenomenon [14]. In particular, machines that are perceived as capable of experience but not agency are more uncanny. The dimensions of mind attribution and humanness are closely related [53]: agency reflects HU and experience reflects HN. Thus, our final hypothesis is:

$H_3$: *HN traits, but not HU traits, are related to a robot's perceived eeriness and likeability.*

## 10.2   Materials and Methods

Our study was conducted using a $2 \times 2 \times 3$ mixed experimental design in which a robot's embodiment (humanlike vs. machinelike) and attitude (positive vs. negative) were between-subjects factors and the number of interactions (Interaction I vs. Interaction II vs. Interaction III) was a within-subjects factor. We have explicitly measured a robot's perceived eeriness, anthropomorphism, likeability, and HN and HU dimensions of humanness. Furthermore, we used the Brief Implicit Association Test (BIAT) [54] as an implicit measurement tool of eeriness. This is a computer-based program that requires participants to classify a series of words into specified categories and measures the strength of the association between these concepts and attributes using participants reaction times.

## *10.2.1   Participants*

Sixty native Japanese speakers were recruited by a recruitment agency for the study. The recruitment agency for part- and full-time student jobs posted a message on its website about the possibility of participating in a study involving a robot. Participants were paid ¥2000 for their time. All participants were undergraduate students of various universities and departments located in Kansai. Participants who had previously participated in a study involving one of the robots were excluded from selection. A software failure meant that the data from two participants were corrupted. Therefore, we had to exclude these data from the analysis. Of the remaining 58 participants, 26 are female and 32 are male. Their ages ranged from 18–36 years with a mean of 21.47. The study took place at the Advanced Telecommunications Research (ATR) Institute International. Adequate ethical approval was obtained from the ATR Ethics Committee and informed consent forms were signed by the participants.

## *10.2.2   Materials and Apparatus*

All the implicit and explicit measurements were conducted using PsychoPy v1.78 running on a laptop. Participants interacted either with Geminoid HI-2 or Robovie R2. Geminoid HI-2 is the second generation of androids built as a copy of a real human (see Fig. 10.1). Geminoid is indistinguishable from a human being for several seconds. Once people realize its slight imperfections, they have a negative feeling [55, 56]. Robovie R2 is a machinelike robot that has some human features, such as a head and hands. Therefore, Geminoid HI-2 represents a robot that is near the deepest point of the uncanny valley, whereas the humanlike features of Robovie R2 should make it highly likeable [56]. Furthermore, as the uncanny valley can also be caused by a mismatch between appearance and voice or movement (e.g., [9, 10]), we ensured that the Geminoid HI-2 fell into the valley by using a synthetic childlike voice and machinelike jerky movement that does not fit the appearance of a male adult. The same movements and voice were used for Robovie R2, in which no mismatch occurs. During HRI, both robots expressed idle motions that were added to increase their degree of animation. Geminoid HI-2 exhibited movement resembling blinking and breathing, as well as idle movements of its hands and synchronization of its lips to its speech. As Robovie R2 does not have a mouth, identical idle behavior was not possible. Therefore, we implemented slight head and hand motions during speech.

   The experiment took place in a room that had been divided into two parts separated by a folding screen to block the view (see Fig. 10.2). A robot was placed in the experimental space, and all HRIs occurred there. In the measurement space, participants watched an introduction video that explained the order of the experiment and filled out all the questionnaires on a laptop. This ensured that participants did not need to judge the robot in its presence, which could have affected the results.

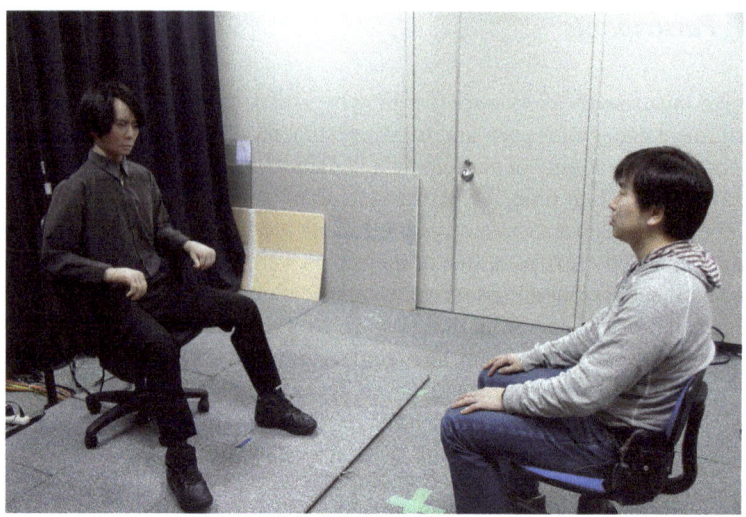

**Fig. 10.1** Geminoid HI-2 and a participant

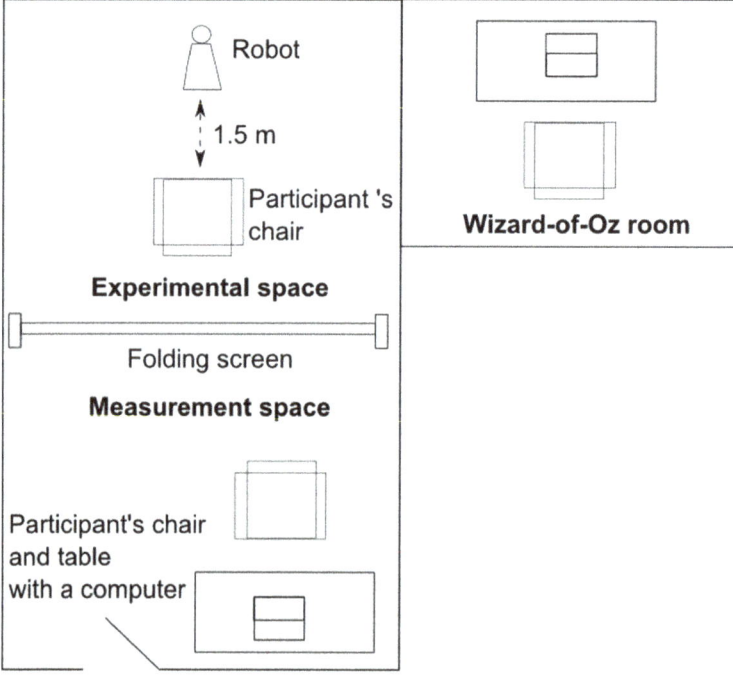

**Fig. 10.2** Diagram of the experimental and measurement spaces and Wizard-of-Oz room

The experimental space was equipped with cameras and the robot's behaviors were controlled by a Wizard-of-Oz sitting in another room.

## 10.2.3 Procedure

We used a live interaction paradigm. Participants were first shown an introduction video that explained the experimental procedure. They were told that the study involved creative and persuasive talking and that they would need to convince a robot to give them a job based on a CV (an identical CV was given to all participants). The experimenter ensured that participants understood the instructions and sat them at a computer. During all HRIs and questionnaires, the experimenter left the participant alone in the room. The experiment was divided into four phases: pre-interaction video, Interaction I, Interaction II, and Interaction III.

Although we ensured that none of the participants had previously interacted or participated in an experiment with the specific robot to which they were assigned, it was still possible that they had seen the robot elsewhere. In particular, in Japan, it is common to see robots used in experiments on various TV programs. Therefore, to minimize the differences in potential prior exposure, participants were asked to watch a short video (~15 s) in which a robot (either Robovie R2 or Geminoid HI-2) introduced itself and its capabilities. The dialogue was identical for both robots. After the video, participants performed the BIAT and filled out all a questionnaire.

During Interaction I, participants were taken to the experimental room and sat 1.5 m in front of a robot (see Fig. 10.3). They were told to have a small conversation with the robot to become familiar before the actual job interview began. The robot was introduced as *Robo*. During this conversation, the robot asked participants three neutral questions (e.g., "Is it cold today?" or "Where do you come from?"). After this short conversation, participants asked to fill out the same questionnaire as before.

In Interaction II, the experimenter provided a short job description for which the participant was instructed to apply. Participants were asked to apply for positions as an engineer and bank manager. The order of interviews was counterbalanced between Interaction II and III. Furthermore, a participant received a CV of a person whom she was supposed to imitate during the interview. The CVs were identical for all participants, but the gender of the applicant was always the same as the real gender of the participant. Participants were asked to use this CV as the basis for their responses, but they could invent the information required to answer the questions. To motivate participants to perform the task as well as possible, they were informed by the experimenter that they would be paid an additional sum if they secured the job. They were given 5 min to prepare for the interview, and then the experimenter collected the CVs and job description sheets and brought the participant to the robot.

The interview began with the robot briefly describing the company and job position for which the participant was applying. After the introduction, the participant was asked three job interview questions. The questions were generic and common

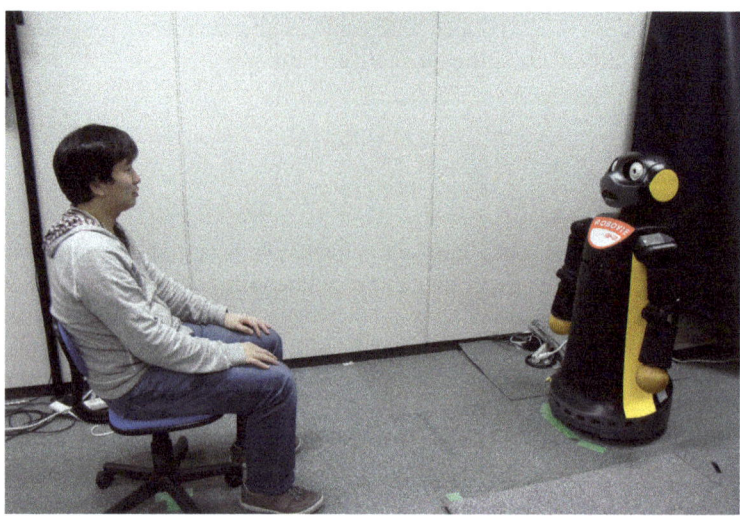

**Fig. 10.3** Experimental setup. Participant sitting in front of Robovie R2 during interaction

for job interviews, e.g., "Please tell me about yourself" or "What is your biggest weakness?" While the participant was responding, the robot provided feedback using nonlexical conversation sounds and nonverbal communication. In the positive condition, it either nodded or nodded and uttered "Un" (a Japanese expression signaling agreement with the speaker). In the negative condition, it either shook its head or nodded its head and uttered "Asso" (a Japanese expression indicating lack of interest in what the speaker says that is rather rude). This feedback was initiated by the Wizard as appropriate in the natural flow of the conversation, e.g., when a participant paused to think about her response.

After each question, the robot thanked the participant and asked the next question. After the third question, the robot informed the participant that it would announce its decision about whether to give the job to the participant later (in fact, the decision was never announced). Although the outcome was not announced directly to the participants, some hints were provided in each condition. In the positive condition, the robot hinted approval of what the participant said during the interview. In the negative condition, it was not particularly pleased with a participant's responses, suggesting they consider applying elsewhere. At that point, participants were asked to fill in a questionnaire for the third time. This time, multiple dummy questions regarding the interview were included. Interaction III was identical to Interaction II, but the CVs, job positions, and questions asked by the robot were different. participants were permitted to answer each of the questions freely and we did not measure the duration of the interactions. The whole procedure took approximately 1 h.

## *10.2.4  Measurements*

In the experiment, we used several questionnaires and BIAT [54] as dependent measures. We explicitly measured the robots' perceived eeriness and anthropomorphism on 5-point Likert scales derived from [22]. Moreover, their likeability was measured using the corresponding Godspeed scale [57] (range 1–5). To establish the relationship between multidimensional anthropomorphism and the uncanny valley, we measured two dimensions of anthropomorphism, HN and HU, on scales developed by [58]. Both dimensions had 10 items and were measured on a scale from 1 (not at all) to 7 (very much) (e.g., "The *Robo* is... shallow"). This experiment was part of a bigger study that involved additional self-report scales that were collected at the same time and are not reported here. We used a validated version of the likeability scale in Japanese. Perceived eeriness, anthropomorphism, HN, and HU were only available in English. Therefore, we conducted a back-translation process to obtain their Japanese versions. We calculated the reliability of each scale separately for each interaction round using Cronbach's $\alpha$. According to [59], Cronbach's $\alpha > 0.6$ is acceptable for newly developed scales for research purposes. Based on this threshold, all the scales (apart from HU) were adequately reliable. The lowest Cronbach's $\alpha$ values during any of the three measurements were as follows: likeability $\alpha = 0.83$, perceived eeriness $\alpha = 0.62$, anthropomorphism $\alpha = 0.88$, HN $\alpha = 0.65$, and HU $\alpha = 0.54$. The low reliability of the HU scale indicates that the HU results should be interpreted with great caution.

Furthermore, we used BIAT [54] as a computer-based implicit measurement tool of eeriness. BIATs involve participants classifying a series of words into superordinate categories. The task involved combining a concept classification (*"Robo"* vs. "Human") with an attribute classification ("Eeriness" vs. "Non-eeriness"). We were interested in measuring the strength of association between *"Robo"* and "Eeriness."

In the BIAT, only two categories at a time were displayed on the screen and a total of three categories were evaluated ("Interview Robot *Robo*," "Human," and "Eeriness"). The fourth category ("Non-eeriness") is called non-focal and was used as a distractor (attribute word that does not belong to the categories being evaluated in a specific block) for "Eeriness." The other two categories ("Interview Robot *Robo*" and "Human") were used as distractors for each other. There were two blocks of 16 trials each that were repeated four times. The following stimuli were used: "Interview Robot *Robo*" (Automaton, Machine, Robot, Artificial), "Human" (Person, Natural, Mankind, Real), "Eeriness" (Eerie, Freaky, Spine-tingling, Shocking), and "Non-eeriness" (Reassuring, Numbing, Uninspiring, Boring).

At the beginning of BIAT, participants were presented with two of the categories being evaluated (e.g., "Interview Robot *Robo*" and "Eeriness") and the words that belong to each of these categories. During the actual classification task, these categories were displayed in the top part of the screen. At the center of the screen, a series of words were shown that either belong to these categories or do not (see Fig. 10.4). Participants were asked to respond as quickly as possible by pressing "K" if the word belongs to either of the categories or "D" if not. As an example, if the

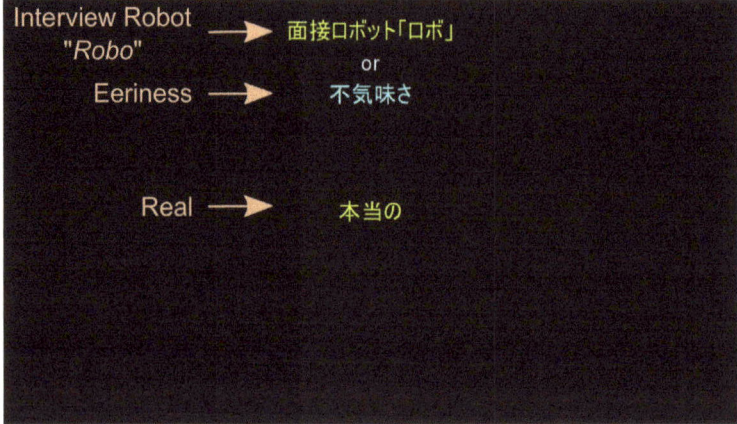

**Fig. 10.4** Screenshot from the BIAT with English annotations. Two classification concepts ("Interview Robot *Robo*" and "Eeriness") and an attribute word ("Real") are being classified by a participant

categories were "Human" and "Eeriness," a participant should press "K" if the target word is "Mankind" or "Freaky," but should press "D" if the word is "Artificial" or "Reassuring." If a participant misclassified a word, a red cross appeared on the screen until the correct key was pressed.

The total time from the word appearing until the correct key was pressed was calculated with millisecond precision. These times were used to establish the strength of association between the categories. When an association between two categories is stronger, participants should be able to make their choices faster than for a pair of categories that are implicitly not associated with each other. The order of the BIATs was randomized and the order of the blocks was counterbalanced.

## 10.3 Results

In the first step of the analyses, we looked at the explicit and implicit measures. We then looked at the relationship between these different dependent measures. To analyze the data, we conducted a series of three-way ANOVAs with embodiment and attitude as between-subjects factors and number of interactions as a within-subjects factor. The assumptions of all statistical tests were met unless otherwise specified.

## 10.3.1  Likeability

First, we looked at the likeability, and in particular how a robot's attitude can affect it during HRI (Fig. 10.5). Because the assumption of a normal distribution was violated for parametric testing for anthropomorphism, we employed a permutation test with three factors using the aovp function with 1000 iterations from the lmPerm package [60] in R [61]. Likeability was significantly affected by the robots' attitude, $p = 0.001$. Positively behaving robots (M = 3.82, $\sigma = 0.67$) were liked more than negatively behaving robots (M = 3.24, $\sigma = 0.9$). Moreover, we found a statistically significant effect of embodiment with probability $p = 0.01$. Robovie R2 (M = 3.7, $\sigma = 0.88$) was liked more than Geminoid HI-2 (M = 3.37, $\sigma = 0.78$). In addition, we found a marginally significant interaction effect between embodiment and attitude, $p = 0.07$. Robovie R2 was more liked when it behaved positively (M = 4.15, $\sigma = 0.54$) than negatively (M = 3.26, $\sigma = 0.94$), $p < 0.001$. However, the attitude of Geminoid HI-2 did not significantly affect its perceived likeability.

Furthermore, we found a statistically significant interaction effect between the robots' attitude and number of interactions, $p < 0.001$. During Interaction I, a robot's attitude did not affect its likeability. However, during Interaction II, a robot's positive attitude (M = 3.86, $\sigma = 0.66$) increased its likeability compared with the negative attitude (M = 2.93, $\sigma = 0.98$), $p < 0.001$. Similarly, during Interaction III, a robot's positive attitude (M = 3.97, $\sigma = 0.69$) resulted in higher likeability than a negatively behaving robot (M = 3.2, $\sigma = 0.94$), $p < 0.001$. The interaction effect between embodiment and measurement was significant with $p < 0.001$. The difference was only observed during Interaction I, when Robovie R2 (M = 3.9, $\sigma = 0.56$) was liked more than Geminoid HI-2 (M = 3.34, $\sigma = 0.61$).

**Fig. 10.5** Effect of three factors on likeability. The rating of likeability based on attitude and interaction round, grouped by robot type

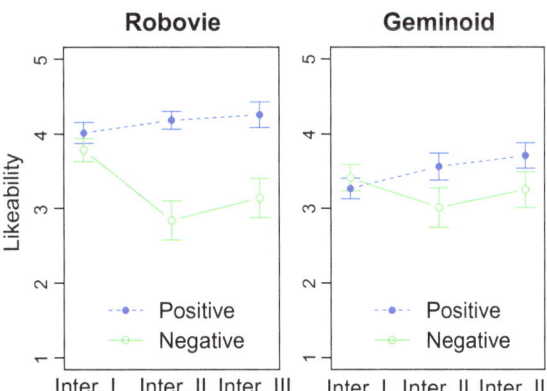

**Fig. 10.6** Effect of three
factors on explicit eeriness.
The rating of explicit
eeriness based on attitude
and interaction round,
grouped by robot type

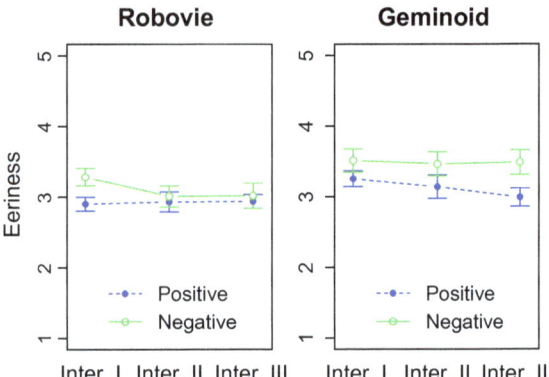

## 10.3.2 Eeriness

The second component of the uncanny valley—eeriness—was measured explicitly
and implicitly (Fig. 10.6). We are interested in establishing the effect of repeated
interactions on a robot's perceived eeriness. The explicit measure of eeriness showed
the main effect of embodiment to be statistically significant, $F(1,54) = 5.14, p = 0.03$,
$\eta_G^2 = 0.07$. Geminoid HI-2 ($M = 3.31, \sigma = 0.62$) was perceived as significantly more
eerie than Robovie R2 ($M = 3.01, \sigma = 0.51$). Moreover, there was a significant main
effect relating to attitude, $F(1,54) = 4.27, p = 0.04, \eta_G^2 = 0.06$. A robot behaving negatively ($M = 3.3, \sigma = 0.64$) was perceived as more eerie than one behaving positively
($M = 3.03, \sigma = 0.49$). In addition, there was a main effect relating to the number of
interactions, $F(2,108) = 3.1, p = 0.05, \eta_G^2 = 0.01$. Post-hoc tests using the Bonferroni
correction revealed that, with marginal significance, participants rated the robots as
more eerie after Interaction I ($M = 3.25, \sigma = 0.52$) than after Interaction III ($M =
3.11, \sigma = 0.6$), $p = 0.08$.

Apart from the explicit eeriness, we also measured implicit eeriness. In the BIAT,
shorter response times indicate a stronger association between categories. Thus, an
increased time would indicate a weaker association between a robot and eeriness.
However, the reduced response time with the increased number of interactions could
also be due to participants improving at the task itself. Therefore, we have transformed the reaction times to z-scores within each interaction round, enabling the
comparison of results between interactions. A three-way ANOVA with embodiment
and attitude as between-subjects factors and the number of interactions as a within-
subjects factor did not indicate any statistically significant main or interaction effects.

### 10.3.3   Anthropomorphism

We then looked at one- and two-dimensional measures of anthropomorphism. It was expected that there would be a main effect relating to a robot's embodiment, and in particular that Geminoid HI-2 would be perceived as more humanlike than Robovie R2. As the assumption of a normal distribution was violated for parametric testing for anthropomorphism, we employed a permutation test with three factors using the aovp function with 1000 iterations from the lmPerm package [60] in R [61]. We found a marginally statistically significant main effect of embodiment with probability $p = 0.08$ (see Fig. 10.7). Geminoid HI-2 (M = 2.47, $\sigma = 1.1$) was more anthropomorphic than Robovie R2 (M = 2.17, $\sigma = 0.92$). Moreover, we found a significant interaction effect between the robots' attitude and number of interactions with probability $p < 0.001$. Only during Interaction III did a robot's positive attitude (M = 2.63, $\sigma = 1.07$) result in higher likeability compared with a negatively behaving robot (M = 2.11, $\sigma = 1.02$), $p = 0.05$.

We then proceeded to the two-dimensional measurement of anthropomorphism to investigate its relation with the uncanny valley. The results related to the model of anthropomorphism proposed by [51] will be discussed in another paper. In line with previous research, we did not find statistically significant main or interaction effects for the HU dimension (see Fig. 10.8).

However, we found a main effect relating to embodiment, with F(1,54) = 5.13, $p = 0.03$, $\eta_G^2 = 0.07$ on the HN dimension (see Fig. 10.9). Robovie R2 (M = 3.16, $\sigma = 0.77$) was attributed with more HN traits than Geminoid HI-2 (M = 2.74, $\sigma = 0.85$). In addition, there was a significant main effect of attitude, F(1,54) = 8.46, $p = 0.005$, $\eta_G^2 = 0.12$. Robots with positive attitude (M = 3.21, $\sigma = 0.74$) were attributed as having more HN than those with a negative attitude (M = 2.67, $\sigma = 0.85$). There was also a significant main effect relating to the number of interactions, F(2,108) = 7.39, $p = 0.001$, $\eta_G^2 = 0.02$. Post-hoc tests using the Bonferroni correction for the family-wise error revealed that the robots were attributed more HN traits after Interaction I (M = 3.4, $\sigma = 0.77$) than after Interactions II (M = 2.88, $\sigma = 0.87$), $p = 0.02$, or

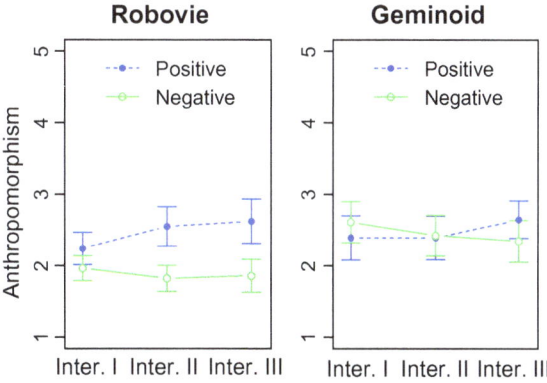

**Fig. 10.7** Effect of three factors on anthropomorphism. The rating of anthropomorphism based on attitude and interaction round, grouped by robot type

**Fig. 10.8** Effect of three
factors on Human
Uniqueness. The rating of
Human Uniqueness based on
attitude and interaction
round, grouped by robot type

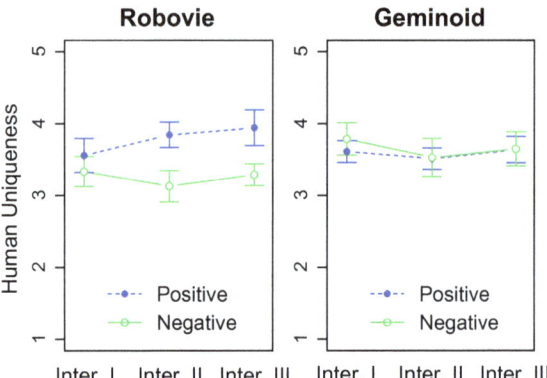

**Fig. 10.9** Effect of three
factors on Human Nature.
The rating of Human Nature
based on attitude and
interaction round, grouped
by robot type

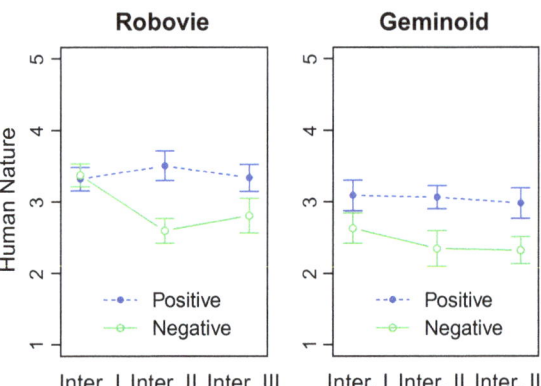

III (M = 2.86, $\sigma$ = 0.86), $p$ = 0.02. Furthermore, there was a significant interaction effect between attitude and number of interactions, $F(2,108) = 9.8$, $p < 0.001$, $\eta_G^2 = 0.03$. Attitude only produced a significant effect in Interactions II [$F(1,56) = 15.82$, $p < 0.001$, $\eta_G^2 = 0.22$] and III [$F(1,56) = 7.75$, $p = 0.007$, $\eta_G^2 = 0.12$].

### 10.3.4 Relationship Between the Uncanny Valley and HRI Factors

Next, we looked at the relationship between the different dependent variables used in this study to establish how the uncanny valley is related to factors that are important for HRI. The correlations between likeability, eeriness, one- and two-dimensional anthropomorphism are presented in Table 10.1.

The following convention was used to determine the effect size of Pearson's r coefficient: small ($0.1 \leq |r| < 0.3$), medium ($0.3 \leq |r| < 0.5$), large ($0.5 \leq |r|$).

**Table 10.1** Correlation between dependent measures using Pearson's r coefficient

|  | Likeability | Eeriness | Anthropomorphism | HU | HN |
|---|---|---|---|---|---|
| Likeability |  | −0.11 | 0.4*** | 0.31*** | 0.51*** |
| Eeriness | −0.11 |  | 0.05 | 0.15* | 0.13 |
| Anthropomorphism | 0.4*** | 0.05 |  | 0.15* | 0.15* |
| HU | 0.31*** | 0.15* | 0.15* |  | 0.35*** |
| HN | 0.51*** | 0.13 | 0.15* | 0.35*** | • |

$^*p < 0.05$, $^{***}p < 0.001$

A correlation with a large effect size was observed between likeability and HN. Furthermore, likeability had a medium effect size correlation with anthropomorphism and HU. Eeriness and likeability were not correlated. Eeriness was significantly correlated with HU with a small effect size.

## 10.4 Discussion

In this study, we investigated the effect of repeated interactions and a robot's attitude on the uncanny valley phenomenon using a live interaction paradigm. In particular, we investigated the impact of these factors on a robot's likeability, as well as explicit and implicit measures of perceived eeriness. Explicit eeriness and likeability were not significantly correlated, which indicates that they measure different aspects of the uncanny valley. Although that might initially seem like an unexpected and counterintuitive finding, there are examples showing that a negative correlation between eeriness and likeability is not necessary. People can dislike other people, but at the same time do not perceive them as eerie. However, there are also cases when eeriness is desirable, e.g., people who like to watch horror movies that might involve eerie creatures. Therefore, measuring both of aspects offers a richer picture than considering only one.

The analysis of likeability showed that the more machinelike robot (Robovie R2) was liked more than the highly humanlike Geminoid HI-2. Moreover, a robot's attitude toward a human interaction partner could be used to affect its likeability, with friendly robots being liked more than unfriendly robots. However, the effect of a robot's attitude is not independent of its embodiment. The interaction effect between embodiment and attitude is especially profound in the case of a more machinelike robot. Although Robovie R2's positive behavior resulted in a small increase in likeability, its negative attitude resulted in a drop in likeability to a level similar to that observed for the negatively behaving Geminoid HI-2. In case of the latter robot, its attitude did not significantly affect its likeability. Thus, $H_{1a}$ and $H_{1b}$ are not supported.

These results seem to indicate that a robot that is perceived as uncanny is not able to affect its likeability through positive or negative interactions. In that sense, its

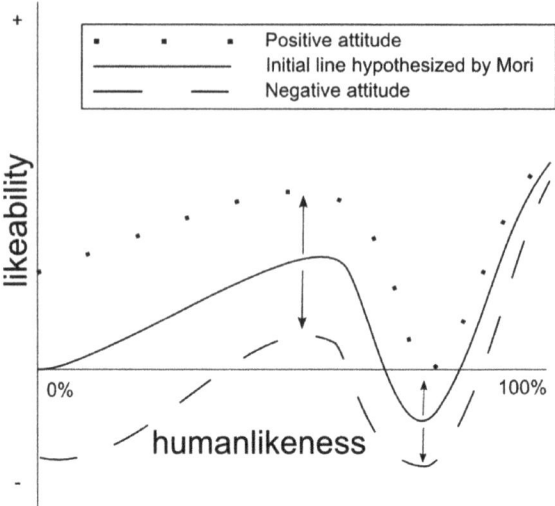

**Fig. 10.10** Hypothesized effect of robots attitude on the uncanny valley. Likeability of a robot will increase with its positive attitude toward a human interaction partner or decrease with its negative attitude. The less humanlike the robot, the stronger the effect

lower likeability is persistent. On the other hand, the impact of a machinelike robot's attitude is much greater, especially when it behaves negatively. The less humanlike a robot is, the stronger that effect could be. A hypothetical relationship between humanlike-ness and the effect of a robot's attitude on its likeability is presented in Fig. 10.10.

These findings on likeability provide a new perspective on the psychological theories related with the effect of familiarity. In particular, the results are more consistent with the mere exposure effect than affective habituation. As suggested by the work of [33, 34], greater familiarity with an unpleasant stimulus did not enhance the likeability of Geminoid HI-2, which contradicts the affective habituation theory. However, in the case of the more neutral stimulus (Robovie R2), its behavior during interactions affected its likeability. This supports the explanation of the familiarity effect proposed by [41, 42], in which repeated exposure creates opportunities for interaction and those interactions that are positive will lead to a favorable impression of a person, or in this case a robot. Therefore, in live HRI, mere exposure to a robot is insufficient to induce a positive affect toward it, and a positively toned interaction is required. However, in the case of a strongly unpleasant robot, even positive behavior can be insufficient to enhance its likeability.

Looking at the eeriness aspect of the uncanny valley, we found that Geminoid HI-2 was rated as more eerie than Robovie R2. However, more interestingly, we observed that after the last interaction, both robots were perceived as being less eerie than after the first interaction. This indicates that perceived eeriness decreases with exposure to a robot. Moreover, this reduction is the same between robots that initially had different levels of eeriness, thus supporting $H_{2a}$. Therefore, although perceived eeriness of a highly anthropomorphic robot can be decreased by merely increasing

**Fig. 10.11** Hypothesized effect of repeated HRIs on the uncanny valley. The reduction of a robot's eeriness is relatively constant, regardless of the level of its humanlike-ness

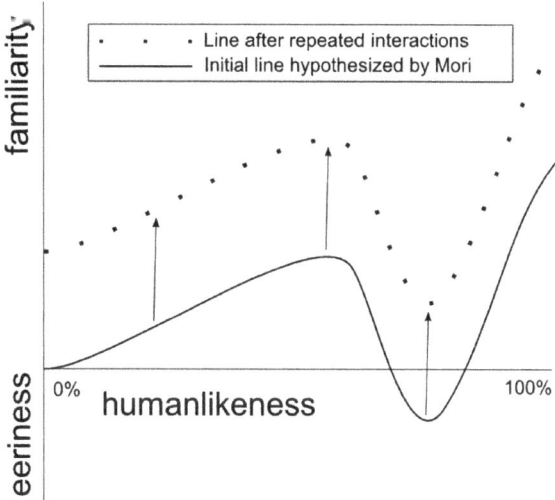

the number of HRIs, the gap between machinelike and humanlike robots remains relatively constant (see Fig. 10.11).

As both robots were perceived as less eerie after multiple interactions, it is possible that the mere exposure effect [31] and affective habituation [32] were both involved. Geminoid HI-2 was initially perceived as an extremely eerie robot. In this case, it is possible that affective habituation occurred and the affective reaction became weaker with increased exposure. On the other hand, for an initially neutral-looking robot (Robovie R2), additional exposure could not decrease its eeriness, irrespective of behavior. Therefore, the effect of familiarity on the perceived eeriness worked differently than for likeability, where a robot's positive behavior led to a favorable impression. If the familiarity effect of attraction also affects perceived eeriness, the explanation that a positive interaction is required is not supported. The more probable explanations for the results obtained for Robovie R2 are that a novel stimulus that initially fosters wary reactions is found to be benign after repeated interactions [39] or that additional exposure increases a robot's processing fluency [37] as its appearance becomes more familiar. As increased processing fluency has a positive effect, it is possible that this processing affect is then transferred to the robot, leading to a decrease in perceived eeriness. However, as the goal of this study was not to investigate which processes are responsible for the familiarity impact on attraction, future studies should verify whether any of these explanations are correct.

Our findings for both likeability and perceived eeriness are relevant for HRI designers. The likeability of robot is affected by its behavior. However, the effect is much stronger in the case of a more machinelike robot. In particular, a machinelike robot can swiftly stop being liked, despite its appearance, by employing negative behavior. It is much harder to increase the likeability of a robot that initially falls into the uncanny valley, as a friendly attitude is not sufficient make it likeable.

However, people quickly become used to the unfamiliar appearance of a robot. In our study, three short interactions were sufficient to reduce its perceived eeriness. However, this reduction was not found to be stronger for the more anthropomorphic robot. Therefore, the relative difference in perceived eeriness between the robots remained at the same level. Nevertheless, in this study, we enhanced the eeriness of Geminoid HI-2 by creating a mismatch between its appearance, speech, and movement. If the only source of eeriness was its embodiment, it is possible that the effect of multiple interactions with Geminoid HI-2 would be more profound. It is also noteworthy that the perceived eeriness of Geminoid HI-2 after Interaction III reached the level of Robovie R2 after Interaction I. Therefore, Geminoid HI-2 only remained more eerie because the perceived eeriness of Robovie R2 had decreased. It is possible that a higher number of interactions, after the machinelike robot had reached its maximum familiarity, may allow a highly humanlike robot to reach the same level.

We also found that a negatively behaving robot was rated as more eerie than a positively behaving robot. However, this finding could be explained as a result of the HRI context used in this experiment. In Japanese culture, it is not typical for an interviewer to express lack of interest during a job interview in such an explicit and rude way as the robot did in this experiment. Such an attitude could have led the robot to be perceived as more eerie than if it had behaved in a way that is common during human–human job interviews.

The analysis of implicit eeriness using BIAT did not show any significant differences. Thus, $H_{2b}$ is not supported. Therefore, in the current form, BIAT might not be optimally suited to measuring eeriness. We speculate that this result could be due to the weak association between a robot's category ("Interview Robot *Robo*") that was displayed on a screen and the specific robot with which the participants interacted. As implicit attitudes tend to change more slowly than explicit attitudes, it is possible that our manipulation was too weak to modify attitudes toward a specific robot. As a result, participants might have responded to the robot's category as being merely a representation of robots in general rather than their specific robotic interaction partner. In future studies, it might be beneficial to use a picture of a robot instead of a name as a representation of its category.

In line with previous research, the HU dimension of anthropomorphism was not significantly affected by the embodiment of a robot. Furthermore, the attribution of HN traits was affected by the embodiment and is therefore more relevant to the uncanny valley. Thus, $H_3$ is supported. However, in contrast with previous work [14], it was the less uncanny robot (Robovie R2) that was attributed with more HN. Despite this dimension having more impact on the uncanny valley, the relationship looks to be more complex than initially proposed. The biggest difference between the work of [14] and ours is the robots used in the experiments. In [14], a single robot that either had the back of its head visible or a humanlike face cover was used. The HN dimension is closely related with emotions, and a robot that has no face is not capable of expressing emotions with facial expressions. Therefore, it will be attributed with less capability to experience (HN). In our experiment, the default and fixed appearance of Robovie R2's face could be perceived as a smile. However,

Geminoid HI-2 has a highly humanlike face that suggests that it can exhibit facial expressions. As a result, participants might have had higher expectations, but the robot's facial expression remained the same (and was rather stern) during the interactions. This could have been perceived as emotional coldness in the robot, leading participants to attribute less HN to Geminoid HI-2. Nevertheless, more research is needed to establish the relationship between HN and the uncanny valley. Furthermore, considering the inadequate reliability of the HU dimension, it is necessary to interpret these results with special care. It is possible that HU is a different construct in Japan than in Western cultures.

## 10.4.1 Limitations and Future Work

In our experiment, we used only two robots that differed in their level of anthropomorphism. An alternative explanation for our results could be that it is a robot's friendliness in appearance that is more important than humanlike-ness in fostering likeability. We cannot exclude the possibility that there are differences along some other dimensions reflected by appearance. It is possible that the interaction between embodiment and attitude would be reversed if we used a different pair of robots. In particular, Geminoid HI-2 has a stern-looking facial expression, whereas the design of Robovie R2 could be perceived as cute and friendly with its big, childlike head. The appearance of Robovie R2 could invoke expectations for it to behave positively, and the mismatch between these expectations and the actual behavior of the robot could result in a strong decrease in likeability. If a friendlier-looking android, e.g., Geminoid F, was used in the experiment instead of Geminoid HI-2, it is possible that we would have observed a similar pattern of reactions to its unfriendly behavior as for Robovie R2. However, the question remains as to why the opposite trend was not observed in the case of Geminoid HI-2's mismatched positive attitude. Therefore, future studies should also include qualitative data that could help to understand why people perceive robots as eerie or likeable. Moreover, there could be demographic factors such as age, gender, or educational background that work as moderators. The role of these factors on the uncanny valley has not yet been explored in sufficient detail.

The scale used for measuring anthropomorphism [22] in the uncanny valley experiments was developed in a study that involved only static images of robots. However, contrary to expectations, Robovie R2 and Geminoid HI-2 differ only marginally in terms of perceived humanlike-ness. Because previous work indicates that androids are perceived as more humanlike than machinelike robots (e.g., [22]), the small difference between the two robots in this study must be due to other factors than merely embodiment. To increase the uncanniness of Geminoid HI-2, we used a voice and movements that did not match its embodiment. However, the humanlike-ness scale may have been affected by this manipulation, as certain items do not only apply to the embodiment, e.g., items rated by the participants include "Artificial"-"Lifelike" or "Fake"-"Natural." As a result, our manipulation not only made Geminoid HI-2 more eerie, but also less humanlike than if only its embodiment was evaluated.

This finding also indicates that a robot's behavior can be a more important factor in anthropomorphism than its embodiment. A potential solution involves the development of a new scale of anthropomorphism that is not affected by the potential mismatch between a robot's embodiment and speech or movement. Alternatively, before investigating the uncanny valley in interactions, it would be possible to first rate a robot's humanlike-ness by presenting a static robot with no HRI.

Another limitation of this study is that participants were allowed to freely interact with the robot for as long as they wanted. Therefore, we did not consider the interaction duration in this study, but only the number of interactions. It is possible that participants who interacted with a positively behaving robot were encouraged by its positive feedback to provide more detailed answers for their questions, and interacted for longer as a result. This extended interaction could have increased the familiarity of the robot and reduced its eeriness. It is also possible that the duration of interactions was insufficient to lead to the affective habituation effect of an uncanny robot. The perceived eeriness of both robots was reduced as a result of repeated interactions. However, it is still possible that, after a higher number of interactions, the affective habituation effect would become stronger for the more eerie robot. A long-term study with highly anthropomorphic robots could answer this question. In particular, future experiments should involve longer interactions with a robot, with sessions spread over multiple days.

Future work should also consider the dynamic nature of anthropomorphism. The complexity and multifaceted nature of anthropomorphism highlights a potential challenge for investigating the uncanny valley in actual, long-term HRI, rather than using images or videos that focus only on a robot's embodiment. Previous work on the uncanny valley treated it as a static feature of a robot or virtual agent. However, [27] showed that a robot's anthropomorphism changes during HRI. The results of this study also illustrate that, at least in the case of Robovie R2, attitude affects perceived humanlike-ness. Mori's hypothesis does not accommodate for such a finding. Studies of the uncanny valley should recognize that both anthropomorphism and uncanniness can vary during HRI, and they might consider whether the uncanny valley should be investigated using the pre-interaction level of anthropomorphism based only on a robot's appearance or the level of anthropomorphism measured in HRI at the same point of time as measures of uncanniness.

This study was an exploratory work that, for the first time, investigated the uncanny valley in repeated HRIs. Our results show the potential benefits of researching the complexity of this phenomenon in studies that involve human interaction with a collocated robot. Nevertheless, at the same time, our results indicate that, to understand the impact of the uncanny valley on HRI, future research must go beyond picture- and video-based studies and enable people to interact with robots. The great majority of studies have tried to find the origin of this phenomenon. This is a worthy goal. However, until we can show that Mori's theory has any significant (long-term) impact on HRI, we risk spending resources on research into an artificial problem. In the end, it matters very little whether a picture of a robot is perceived as eerie or disliked if, during an actual interaction with the robot, this effect vanishes as a result of behavior or interaction-context factors being more prominent.

**Acknowledgements** This work was partially supported by Grant-in Aid for Scientific Research (S), KAKENHI (25220004) and JST CREST (Core Research of Evolutional Science and Technology) research promotion program "Creation of Human-Harmonized Information Technology for Convivial Society" Research Area. The authors would like to thank Kaiko Kuwamura, Daisuke Nakamichi, Junya Nakanishi, and Kurima Sakai for their help with data collection.

# References

1. Złotowski, J.A., H. Sumioka, S. Nishio, D.F. Glas, C. Bartneck, and H. Ishiguro. 2015. Persistence of the uncanny valley: The influence of repeated interactions and a robot's attitude on its perception. *Frontiers in Psychology* 6: 883.
2. Mori, M. 1970. The uncanny valley. *Energy* 7 (4): 33–35.
3. Blow, M., K. Dautenhahn, A. Appleby, C. Nehaniv, and D. Lee. 2006. Perception of robot smiles and dimensions for human-robot interaction design. In *The 15th IEEE international symposium on robot and human interactive communication, 2006. ROMAN 2006*, 469–474.
4. Hanson, D. 2006. Exploring the aesthetic range for humanoid robots. In *Proceedings of the ICCS/CogSci-2006 long symposium: Toward social mechanisms of android science*, 39–42.
5. MacDorman, K.F. 2006. Subjective ratings of robot video clips for human likeness, familiarity, and eeriness: An exploration of the uncanny valley. In *ICCS/CogSci-2006 long symposium: Toward social mechanisms of android science*, 26–29.
6. McDonnell, R., M. Breidt, and H.H. Bälthoff. 2012. Render me real?: Investigating the effect of render style on the perception of animated virtual humans. *ACM Transactions on Graphics* 31 (4): 91:1–91:11.
7. MacDorman, K.F., R.D. Green, C.-C. Ho, and C.T. Koch. 2009. Too real for comfort? uncanny responses to computer generated faces. *Computers in Human Behavior* 25 (3): 695–710.
8. Seyama, J., and R.S. Nagayama. 2007. The uncanny valley: Effect of realism on the impression of artificial human faces. *Presence: Teleoperators and Virtual Environments* 16 (4): 337–351.
9. Mitchell, W.J., K.A. Szerszen, A.S. Lu, P.W. Schermerhorn, M. Scheutz, and K.F. MacDorman. 2011. A mismatch in the human realism of face and voice produces an uncanny valley. *i-Perception* 2 (1): 10–12.
10. Saygin, A.P., T. Chaminade, H. Ishiguro, J. Driver, and C. Frith. 2012. The thing that should not be: Predictive coding and the uncanny valley in perceiving human and humanoid robot actions. *Social Cognitive and Affective Neuroscience* 7 (4): 413–422.
11. Piwek, L., L.S. McKay, and F.E. Pollick. 2014. Empirical evaluation of the uncanny valley hypothesis fails to confirm the predicted effect of motion. *Cognition* 130 (3): 271–277.
12. Steckenfinger, S.A., and A.A. Ghazanfar. 2009. Monkey visual behavior falls into the uncanny valley. *Proceedings of the National Academy of Sciences* 106 (43): 18362–18366.
13. MacDorman, K.F., P. Srinivas, and H. Patel. 2013. The uncanny valley does not interfere with level 1 visual perspective taking. *Computers in Human Behavior* 29 (4): 16711685.
14. Gray, K., and D. Wegner. 2012. Feeling robots and human zombies: Mind perception and the uncanny valley. *Cognition* 125 (1): 125–130.
15. MacDorman, K.F., and H. Ishiguro. 2006. The uncanny advantage of using androids in cognitive and social science research. *Interaction Studies* 7 (3): 297–337.
16. Moore, R.K. 2012. A bayesian explanation of the 'Uncanny valley' effect and related psychological phenomena. *Scientific Reports* 2.
17. Cheetham, M., P. Suter, and L. Jancke. 2014. Perceptual discrimination difficulty and familiarity in the uncanny valley: More like a happy valley. *Frontiers in Psychology* 5.

18. Looser, C.E., and T. Wheatley. 2010. The tipping point of animacy how, when, and where we perceive life in a face. *Psychological Science* 21 (12): 1854–1862.
19. Poliakoff, E., N. Beach, R. Best, T. Howard, and E. Gowen. 2013. Can looking at a hand make your skin crawl? peering into the uncanny valley for hands. *Perception* 42 (9): 998–1000.
20. Bartneck, C., T. Kanda, H. Ishiguro, and N. Hagita. 2009. My robotic doppelganger—A critical look at the uncanny valley theory. In *18th IEEE international symposium on robot and human interactive communication, RO-MAN2009*, 269–276. IEEE.
21. Mori, M., K.F. MacDorman, and N. Kageki. 2012. The uncanny valley. *IEEE Robotics and Automation Magazine* 19 (2): 98–100.
22. Ho, C., and K. MacDorman. 2010. Revisiting the uncanny valley theory: Developing and validating an alternative to the godspeed indices. *Computers in Human Behavior* 26 (6): 1508–1518.
23. Misselhorn, C. 2009. Empathy with inanimate objects and the uncanny valley. *Minds and Machines* 19 (3): 345–359.
24. Cheetham, M., P. Suter, and L. Jancke. 2011. The human likeness dimension of the "Uncanny valley hypothesis": Behavioral and functional MRI findings. *Frontiers in Human Neuroscience* 5.
25. Bartneck, C., T. Kanda, H. Ishiguro, and N. Hagita. 2007. Is the uncanny valley an uncanny cliff? In *Proceedings—IEEE international workshop on robot and human interactive communication*, 368–373, Jeju, Republic of Korea.
26. Kiesler, S., A. Powers, S.R. Fussell, and C. Torrey. 2008. Anthropomorphic interactions with a robot and robot-like agent. *Social Cognition* 26 (2): 169–181.
27. Fussell, S.R., S. Kiesler, L.D. Setlock, and V. Yew. 2008. How people anthropomorphize robots. In *HRI 2008—Proceedings of the 3rd ACM/IEEE international conference on human-robot interaction: Living with robots*, 145–152, Amsterdam, Netherlands.
28. Becker-Asano, C., K. Ogawa, S. Nishio, and H. Ishiguro. 2010. Exploring the uncanny valley with geminoid HI-1 in a real-world application. In *Proceedings of the IADIS international conference on interfaces and human computer interaction 2010, IHCI, proceedings of the IADIS international conference on game and entertainment technologies 2010, part of the MCCSIS 2010*, 121–128, Freiburg, Germany.
29. von der Pütten, A.M., N.C. Krämer, C. Becker-Asano, and H. Ishiguro. 2011. An android in the field. In *Proceedings of the 6th international conference on Human-robot interaction, HRI '11*, 283–284, New York, NY, USA. ACM.
30. Dill, V., L.M. Flach, R. Hocevar, C. Lykawka, S.R. Musse, and M.S. Pinho. 2012. Evaluation of the uncanny valley in CG characters. In *12th international conference on intelligent virtual agents, IVA 2012, September 12, 2012–September 14, 2012*, vol. 7502 LNAI, 511–513. Springer.
31. Zajonc, R.B. 1968. Attitudinal effects of mere exposure. *Journal of Personality and Social Psychology* 9 (2p2): 1.
32. Dijksterhuis, A., and P.K. Smith. 2002. Affective habituation: Subliminal exposure to extreme stimuli decreases their extremity. *Emotion* 2 (3): 203.
33. Ebbesen, E.B., G.L. Kjos, and V.J. Konečni. 1976. Spatial ecology: Its effects on the choice of friends and enemies. *Journal of Experimental Social Psychology* 12 (6): 505–518.
34. Perlman, D., and S. Oskamp. 1971. The effects of picture content and exposure frequency on evaluations of negroes and whites. *Journal of Experimental Social Psychology* 7 (5): 503–514.
35. Brockner, J., and W.C. Swap. 1976. Effects of repeated exposure and attitudinal similarity on self-disclosure and interpersonal attraction. *Journal of Personality and Social Psychology* 33 (5): 531–540.
36. Brickman, P., P. Meyer, and S. Fredd. 1975. Effects of varying exposure to another person with familiar or unfamiliar thought processes. *Journal of Experimental Social Psychology* 11 (3): 261–270.
37. Bornstein, R.F., and P.R. D'Agostino. 1994. The attribution and discounting of perceptual fluency: Preliminary tests of a perceptual fluency/attributional model of the mere exposure effect. *Social Cognition* 12 (2): 103–128.

38. Reber, R., P. Winkielman, and N. Schwarz. 1998. Effects of perceptual fluency on affective judgments. *Psychological Science* 9 (1): 45–48.
39. Lee, A.Y. 2001. The mere exposure effect: An uncertainty reduction explanation revisited. *Personality and Social Psychology Bulletin* 27 (10): 1255–1266.
40. Clark, L.A., and D. Watson. 1988. Mood and the mundane: Relations between daily life events and self-reported mood. *Journal of Personality and Social Psychology* 54 (2): 296–308.
41. Denrell, J. 2005. Why most people disapprove of me: Experience sampling in impression formation. *Psychological Review* 112 (4): 951–978.
42. Reis, H.T., M.R. Maniaci, P.A. Caprariello, P.W. Eastwick, and E.J. Finkel. 2011. Familiarity does indeed promote attraction in live interaction. *Journal of Personality and Social Psychology* 101 (3): 557–570.
43. Bornstein, R.F. 1989. Exposure and affect: Overview and meta-analysis of research, 1968–1987. *Psychological Bulletin* 106 (2): 265–289.
44. Norton, M.I., J.H. Frost, and D. Ariely. 2007. Less is more: The lure of ambiguity, or why familiarity breeds contempt. *Journal of Personality and Social Psychology* 92 (1): 97–105.
45. Smith, E.R., and D.M. Mackie. 2007. *Social psychology*, 3rd ed, Jan. New York: Psychology Press.
46. Goetz, J., S. Kiesler, and A. Powers. 2003. Matching robot appearance and behavior to tasks to improve human-robot cooperation. In *ROMAN 2003. The 12th IEEE international workshop on robot and human interactive communication*, 55–60.
47. De Houwer, J., S. Teige-Mocigemba, A. Spruyt, and A. Moors. 2009. Implicit measures: A normative analysis and review. *Psychological Bulletin* 135 (3): 347–368.
48. Steffens, M.C., and S. Schulze. 2006. König. Predicting spontaneous big five behavior with implicit association tests. *European Journal of Psychological Assessment* 22 (1): 13–20.
49. Admoni, H., and B. Scassellati. 2012. A multi-category theory of intention. In *Proceedings of COGSCI 2012*, 1266–1271, Sapporo, Japan.
50. Gawronski, B. 2002. What does the implicit association test measure? a test of the convergent and discriminant validity of prejudice-related IATs. *Experimental Psychology (formerly Zeitschrift für Experimentelle Psychologie)* 49 (3): 171–180.
51. Złotowski, J., E. Strasser, and C. Bartneck. 2014. Dimensions of anthropomorphism: From humanness to humanlikeness. In *Proceedings of the 2014 ACM/IEEE international conference on human-robot interaction, HRI '14*, 66–73, New York, NY, USA. ACM.
52. Haslam, N. 2006. Dehumanization: An integrative review. *Personality and Social Psychology Review* 10 (3): 252–264.
53. Haslam, N., B. Bastian, S. Laham, and S. Loughnan. 2012. Humanness, dehumanization, and moral psychology.
54. Sriram, N., and A.G. Greenwald. 2009. The brief implicit association test. *Experimental Psychology (formerly Zeitschrift für Experimentelle Psychologie)* 56 (4): 283–294.
55. Ishiguro, H. 2006. Android science: Conscious and subconscious recognition. *Connection Science* 18 (4): 319–332.
56. Rosenthal-von der Pütten, A.M., and N.C. Krämer. 2014. How design characteristics of robots determine evaluation and uncanny valley related responses. *Computers in Human Behavior* 36: 422–439.
57. Bartneck, C., D. Kulic, E. Croft, and S. Zoghbi. 2009. Measurement instruments for the anthropomorphism, animacy, likeability, perceived intelligence, and perceived safety of robots. *International Journal of Social Robotics* 1 (1): 71–81.
58. Haslam, N., S. Loughnan, Y. Kashima, and P. Bain. 2009. Attributing and denying humanness to others. *European Review of Social Psychology* 19 (1): 55–85.
59. Nunnally, J. 1978. *Psychometric methods*. New York: McGraw.
60. Wheeler, B. 2010. *lmPerm: Permutation tests for linear models*. R package version 1.1-2.
61. R Core Team. 2014. *R: A language and environment for statistical computing*. R Foundation for Statistical Computing, Vienna, Austria.

# Chapter 11
# Can a Teleoperated Android Represent Personal Presence?—A Case Study with Children

**Shuichi Nishio, Hiroshi Ishiguro and Norihiro Hagita**

**Abstract** Our purpose is to investigate the key elements for representing *personal presence*, which is the sense of being with a certain individual. A case study is reported in which children performed daily conversational tasks with a *geminoid*, a teleoperated android robot that resembles a living individual. Different responses to the geminoid and the original person are examined, with a special focus on the case where the target child was the daughter of the geminoid source. Our results show that children gradually adapt to conversation with the geminoid, but the operator's personal presence is not represented completely. Further research topics on the adaptation process to androids and the key elements of personal presence are discussed.

**Keywords** Personal presence · Android robot · Teleoperation

## 11.1 Introduction

What creates a person's sense of presence? When we have a conversation or watch a movie with somebody we know, we feel that particular person, rather than an anonymous individual, beside us. Fluctuating moods or emotions are factors, and personality traits are other consistent indicators of presence. These factors and how they appear have been extensively studied, mainly in the field of psychology,

---

This chapter is a modified version of previously published papers [1–3], edited to be comprehensive and fit with the context of this book.

---

S. Nishio (✉) · H. Ishiguro · N. Hagita
Advanced Telecommunications Research Institute International,
Keihanna Science City, Kyoto, Japan
e-mail: nishio@botransfer.org

H. Ishiguro
Department of Systems Innovation, Graduate School of Engineering Science,
Osaka University, Osaka, Japan

© Springer Nature Singapore Pte Ltd. 2018
H. Ishiguro and F. Dalla Libera (eds.), *Geminoid Studies*,
https://doi.org/10.1007/978-981-10-8702-8_11

through the analysis of various human behaviors. Is the combination of these factors powerful enough to describe and capture the individual differences in each person? Can current technology represent, record, and play back this individual sense of presence? Many studies have grappled with this question, including how well current transmission technologies such as telephones, TV conferencing systems, or newly developed computer-mediated communication (CMC) systems can approximate face-to-face communication [4]. For the sense of presence, much of the recent work has been based on classical studies [5]. *Co-presence,* the sense of being with somebody else in a remote environment, has been extensively examined in the field of virtual reality [6, 7]. These studies, however, have mainly focused on the typical, anonymous nature of presence rather than the details specific to each individual.

Another scheme, the *constructive approach,* builds a hypothesis and examines the issue through implementation [8]. In the field of robotics, interest continues to grow in the social aspects of human–robot interaction. One field of interest is the use of robots as a communication interface device whose main purpose is not for industry or as a carrier. These studies focus on robots as an informational interface with a physical presence in the real world and attempt to enrich their humanlike functionality, such as making gestures, eye contact, or even expressing personality and emotion. Studies have shown the importance of humanlike nonverbal channels and the superiority of the physical presence of robots to software agents or computer terminals in everyday conversation [9, 10]. Appearance remains the one difference between human beings. Recent manufacturing advances have produced android robots whose appearance is quite similar to humans, and several studies on what constitutes the remaining differences have begun [11].

Based on these studies, we seek to clarify the key elements required to represent and perceive the sense of presence that each individual holds: *personal presence.* We believe that such findings will lead to a deeper understanding of human nature and, at the same time, provide a means for building a robot that can communicate more effectively with human beings.

However, one serious issue exists when using traditional robots as a means to study human nature: intelligence. Although actively studied, it is currently impossible to build a robot that behaves and talks like a human being. This issue prevents researchers from conducting effective examinations on the characteristics of human nature that can only be seen through intelligent conversation. To overcome this 'intelligence' issue, we have developed a new android system called a *geminoid,* which is a teleoperated robotic system with an android robot that looks and behaves similarly to a person.

As a first step toward inspecting the nature of personal presence, we conducted a case study using the class of participants most sensitive to personal presence: members of one's family. As children, rather than adults, produce more direct responses, we conducted this case study with two participants who possess a special relation to the geminoid: the daughter of the human used to model the geminoid and a 4-year-old boy who did not know the model. Previous studies with androids have mainly focused on the very first impressions they make [12, 13]. In this study, we focus on two issues: (1) How the participants adapt to the geminoid, or how their attitudes change through daily conversation experiences, and (2) how well one's personal presence can be represented through the geminoid system. Our research includes an investigation of the elements that can effectively measure how well personal presence is represented.

## 11.2 The Geminoid System

Here, we briefly describe an overview of the geminoid system [14]. A geminoid is a robot that functions as a duplicate of a living person. It appears and behaves like that person and is connected to the person by a computer network. Geminoids extend the applicable field of android science. Androids are designed for studying human nature in general. With geminoids, we can study such personal aspects as presence or personality traits, tracing their origins and implementation into robots. Figure 11.1 shows the robotic part of HI-1, the first geminoid prototype. The geminoid's appearance is based on a living person and does not depend on the imagination of designers. Its movements can be constructed and evaluated by simply referring to the original person. The existence of a real person analogous to the robot simplifies comparison studies.

The robotic element has essentially an identical structure to previous androids [11]. However, efforts concentrated on making a robot that appears to be a copy of the original person, rather than simply resembling them. Silicone skin was molded by a cast taken from the original person; shape adjustments and skin textures were painted manually based on MRI scans and photographs. Fifty pneumatic actuators drive the robot to generate smooth and quiet movements, which are important attributes when interacting with humans. The allocation of actuators was determined so that the resulting robot can effectively show the necessary movements for

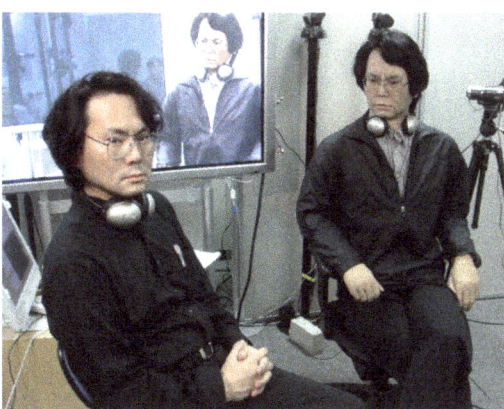

**Fig. 11.1** Geminoid HI-1 and its original person

human interaction and simultaneously express the original person's personality traits. Among the 50 actuators, 13 are embedded in the face, 15 in the torso, and the remaining 22 move the arms and legs.

As geminoids are equipped with teleoperation functionality, they are driven by more than just an autonomous program. By introducing manual control, the limitations of current AI technologies can be avoided, enabling long-term, conversational human–robot interaction experiments. Figure 11.2 shows the teleoperation interface. Two monitors show the controlled robot and its surroundings, and microphones and a headphone are used to capture and transmit utterances. The captured sounds are encoded and transmitted to the geminoid server by IP links from the interface to the robot, and vice versa. The operator's lip corner positions are measured by an infrared motion capture system in real time, converted to motion commands, and sent to the geminoid server by the network. This enables the operator to implicitly generate suitable lip movements in the robot while speaking.

The geminoid server receives robot control commands and sound data from the teleoperation interface, adjusts and merges the inputs, and sends and receives primitive control commands between the robot hardware. Figure 11.3 shows the major data flows in the geminoid system. As the robot's features become more humanlike, its behavior should also become suitably sophisticated to retain a *natural* look [13]. One thing that can be seen in every human being, and that most robots lack, is the slight body movements caused by its autonomous system, such as breathing or blinking. To increase the android's naturalness, the geminoid server emulates the human autonomous system and automatically generates these micro-movements. Such automatic robot motions are merged with explicit operation commands sent from the remote console.

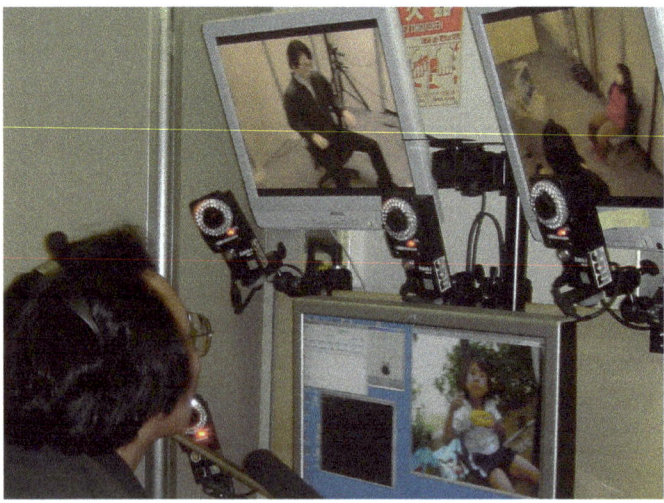

**Fig. 11.2** Geminoid teleoperation console

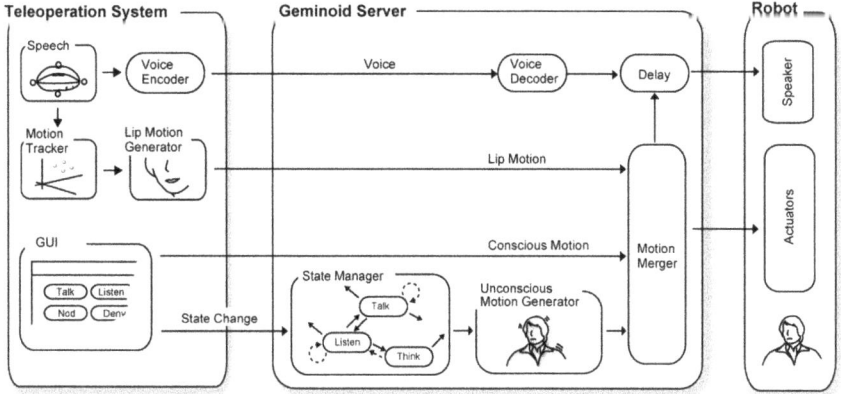

**Fig. 11.3** Block diagram of the geminoid system

## 11.3 Methods

### 11.3.1 Participants

Two children joined the experiment, a girl (R) and a boy (K). Neither child had seen or heard about the geminoid before the experiment, and they were not told that the features of geminoid HI-1 were based on Dr. Ishiguro or that he was teleoperating it.

R is a 10-year-old elementary school student. She is also the daughter of the geminoid model, Dr. Ishiguro. In the past, she modeled for a child android, *Repliee R1* [13], and had been involved in some humanoid robot experiments. Her parents describe her as shy. K is a 4-year-old boy and the son of one of the authors. He is outgoing and rarely becomes anxious even when meeting somebody for the first time. On several occasions, K has seen and played with humanoid robots at exhibitions, but had never seen androids of any kind before this experiment.

### 11.3.2 Procedures

After being led to the experimental room by the experimenter, the participants engaged in conversational tasks with the other *entity* in the room. The participants were seated in front of a 15″ LCD screen. The distance between the participants and the entity was approximately 1.4 m. Figure 11.4 shows the seat alignment.

Two conditions were compared. In the first case, the entity was a person, the original of the geminoid HI-1, Dr. Ishiguro (*P condition*). In the latter case, the participants conversed with geminoid HI-1, which was remotely operated by the same person as in the P condition (*G condition*). In the P condition, the entity

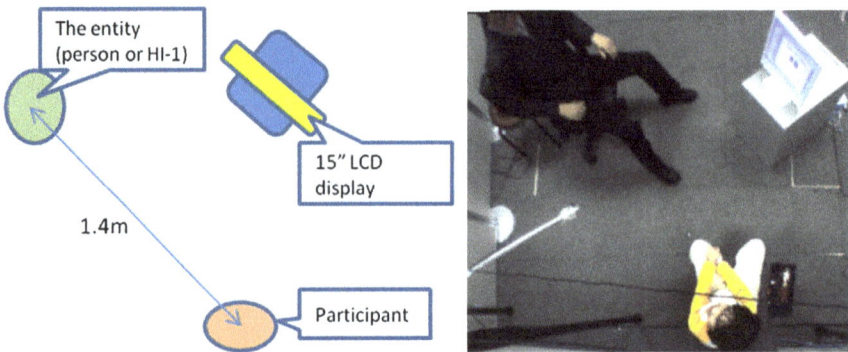

**Fig. 11.4** Experiment room setup

was told to limit his movements, so that his motion would resemble HI-1 in the G condition. Each condition was conducted twice for each participant. Thus, eight sessions were held.

Several conversational tasks were chosen, based on their parents' opinions and from pretest observations. These tasks reflected the children's interests and ages, so they could participate without becoming bored during the sessions. For R especially, some chosen tasks required conversation about family memories. Thus, a different set of tasks was chosen for each participant. R's tasks were the following:

(1) *Photo*: Viewing a series of family photographs and talking about them.
(2) *Shiritori*: Playing a Japanese word game in which players are required to say a word that begins with the last letter of the previous word.
(3) *Video*: Watching videos in which R or her father appears and talking about them. Clips from TV programs were used.
(4) *Math*: Doing simple math problems. The entity asked questions and R answered.

K's tasks were the following:

(1) *Video*: Watching family videos and talking about them.
(2) *Talk*: Talking about some recent issues related to the participant. In the experiment, several topics were chosen by K's father (one of the experimenters).
(3) *Movie*: Watching movie clips and talking about them.
(4) *English*: Counting or reciting the alphabet in English.

After these tasks, several additional tasks were performed as a trial for future experiments. As these tasks were only conducted as a trial for future experiments, they are not analyzed in this paper.

Because each task lasted until the entity decided that the participant was getting bored with the task, the duration of each task was not strictly controlled.

The average duration of each task was approximately 3 min, and the average session time, including the unanalyzed tasks, was approximately 20 min.

At the beginning of each task, the entity asked the LCD screen for the next task, and then the name of each task was shown on the display. For the '*talk*' task, a conversation topic was also displayed. During the '*photo*,' '*video*,' and '*movie*' tasks, images or movies were shown on the display. For example, in the '*photo*' task, several photographs were shown. The displayed photograph was changed by the experimenter in a separate room. In the G condition, the same contents were also shown on the teleoperation console (Fig. 11.2).

After each session, the participants were asked about their impressions of the test and the entity they were talking with. To relax the children, they were interviewed by one of their parents. At the end of each experiment day, several additional questions inquired about comparisons between the geminoid and the real person.

The main experiment was held over two days. On the first day, four sessions were conducted in the following order: R-P (participant R in the P condition), K-P, R-G, and K-G. Two weeks later, on the second day, another four sessions were conducted in the order of K-G, R-G, R-P, and K-P. One week before the first day, test sessions were conducted with identical participants, but only for the P condition, to determine the effectiveness of the tasks and help the participants become habituated to the experimental environment. To help them relax, at the beginning of the first day, the children spent some time in the experiment room with their mothers reading or playing. During this first habituation period, the android was hidden. At the beginning of the second day, the participants and their mothers spent some time talking and playing with the HI-1 (Fig. 11.5).

**Fig. 11.5** Habituation session on the second day

## 11.3.3  Measures

Although various studies have been conducted to identify suitable measures for the effect of CMC [4] or presence [15], no effective measure has been developed for personal presence. Thus, we considered some measures for capturing the participants' impression toward the entity in conversation. Each experimental session was recorded by video cameras to measure the following data:

(1)  Eloquence of conversation

The amount of conversational utterances is known to be influenced by the participant's emotional state and impression of who he/she is having a conversation with [16, 17]. The conversations in each session were transcribed from the audio recordings. As all conversations were in Japanese, we analyzed the transcripts to see how actively participants spoke in each task, which is similar to counting the words in English sentences. Here, the transcripts were morphologically analyzed and split into tokens by the *ChaSen* tokenizer [18]. The numbers of extracted tokens were counted for each participant or entity for each task. The following relative eloquence rate was derived to measure how actively each participant spoke:

$$r_{eloquence} = \frac{\text{total number of tokens in participant speech}}{\text{total number of tokens in entity speech}}$$

(2)  Gaze direction

Nonverbal behaviors, such as interpersonal distance, gestures, eye contact rate, and body movements, are also influenced by the participant's impression of who he/she is talking with [19, 20]. Thus, we selected two measures: eye contact rate and body movement. From image recordings, the gaze directions of both the entities and the participants were observed and coded into two categories: watching one another (eye contact) or not. From these data, we derived the relative eye contact rate, defined as follows:

$$r_{eye\ contact} = \frac{\text{total duration of eye contact}}{\text{total duration of the entity watching the participant}}$$

(3)  Body movement

As a simple measure to evaluate how the attitude of the participants changed, we calculated the amount of body movement of each participant from video images. The aim was to identify the broad changes in participants' nonverbal behaviors such as interpersonal distance or the number of gestures. We took the sum of the motion vector norm obtained by performing a block matching calculation between subsequent frames for the image region in which the participant appeared. The block size was set to $8 \times 8$ (pixel$^2$). Restrictions in the experimental setting meant that the camera angles differed between sessions. Additionally, the body size of the two participants differed. As the sum of the motion vector norms depends on the body area size, the obtained values were normalized by the standard area of each participant. This area was obtained from a video frame showing each participant in a neutral pose. The amount of motion was first calculated frame-by-frame (29.97 frames/s) and then totaled over 1 s intervals.

## 11.4   Results

As stated before, this experiment is a case study with only two subjects and a limited number of trials. Thus, we did not conduct any statistical analysis of the measured values. Instead, in this section, we describe the subjective tendencies observed from the results.

### *11.4.1   Eloquence of Conversation*

Figure 11.6 shows the relative eloquence measures extracted from each task. In the *'English'* task, K was asked to count or recite the alphabet in English. Thus, the amount of speech is not a meaningful value, and so the results of the *'English'* task are omitted.

For both participants, the results in the G condition seem to be lower than in the P condition. For R, the differences between each condition are rather weak. Relative eloquence in the G condition only seems to be significantly lower than in the P condition for the 'shiritori' task (R).

As for K, the overall rate is clearly lower in the G condition. In the first *'video'* task of session G-1, in particular, where the participant met the geminoid for the first time, the rate was around 1%, which indicates that the participant remained

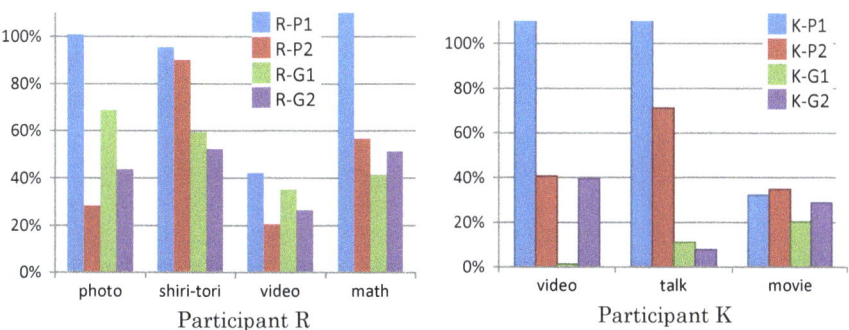

**Fig. 11.6**  Relative eloquence rate

mostly quiet throughout the task. The high value of over 100% in the same task at session P-1 shows the considerable contrast between the two conditions. In G-2, however, we can see a tendency where the value is recovering, and the differences between P-2 and G-2 have almost vanished. In the '*movie*' task for K, the differences among conditions are much smaller, perhaps reflecting the nature of the task. It seems that the participant's attention was focused on the movie, and his overall response was low, as can also be seen in other measures.

### 11.4.2   Eye Contact

The results are shown in Fig. 11.7. For R, the values of the G condition seem to be lower, except in the '*photo*' task. As for K, no clear difference between the two conditions can be found.

### 11.4.3   Body Movement

Figure 11.8 shows the average body movements. The detailed temporal changes in the first tasks for each participant are shown in Fig. 11.9. For R, the overall amount is smaller in P-1 and G-1 compared with P-2 and G-2. The values for K seem to show a clearer tendency. Obviously, the values in the G-1 session are much smaller compared with those in the other sessions, except in the '*English*' and '*movie*' tasks. This tendency matches impressions from the recorded video images. In the G-1 session, K stayed still throughout the task. In contrast, in the P condition tasks, K kept moving, spoke a lot, and showed a rich variety of facial expressions. The only exception is when he was watching movies; he concentrated on the movie and remained still in his seat.

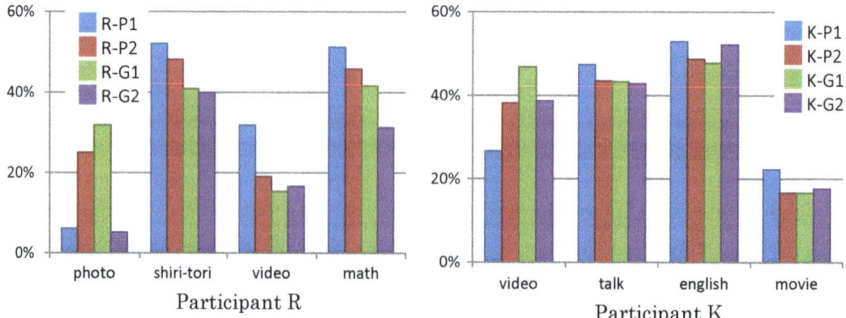

**Fig. 11.7** Relative eye contact rate

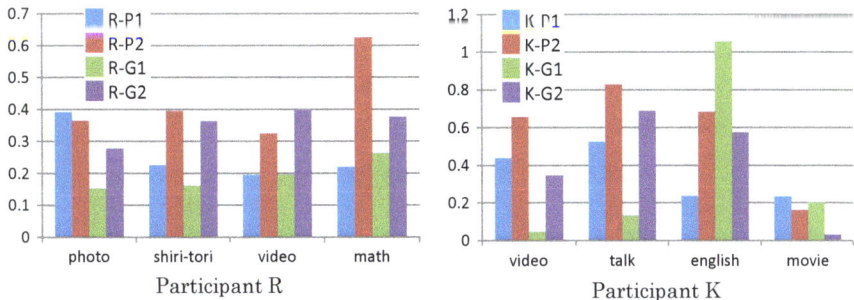

**Fig. 11.8** Average body motion

In the G-1 session, K's body motion average is largest in the *'English'* task. Watching the video, we found that K began to make large movements, frequently looking at the room exit. He might have become tired from the anxiety of the previous two tasks. In the G-2 session, the values seem to recover toward those of the P condition.

### 11.4.4  Interviews

Even though both participants were scared by the geminoid, they turned out to have quite different impressions. The answers on the first and second days were almost identical for both participants.

R seemed to notice that the geminoid was a robot controlled by her father in a different place. Most of her impressions were based on this finding. She described the geminoid as scary, mainly because its features and movements were strange. She said that the geminoid did not look like her farther, but she could not specify which part was strange or different. She preferred her real father to his geminoid because 'this robot can't play Wee (the name of a portable game player) with me, and it can't grab things.'

K had a different impression. After the first P session, K insisted that the entity (the real Dr. Ishiguro) was a robot. He thought that it was alive and breathing, and its (his) figure was normal, but he still felt that it was a robot. He described the 'robot' as very 'serious' and said that the entity listened well to his story and that he would like to play with 'it.' However, after the first G session, K said that 'I thought the first one was a robot, but that was a mistake. This one must be a robot.' K believed that the entity (HI-1) kept wearing a mask and said 'it was very scary, because it had a very thin nose.' He also mentioned that it was not breathing, and its mouth was not moving well when it was speaking.

K seemed to be rather confused by his experience. He sometimes mentioned that the 'man' (HI-1) was a robot, but later he thought it was 'a person' wearing a strange mask. 'He should take off his strange mask that he keeps wearing,' K said.

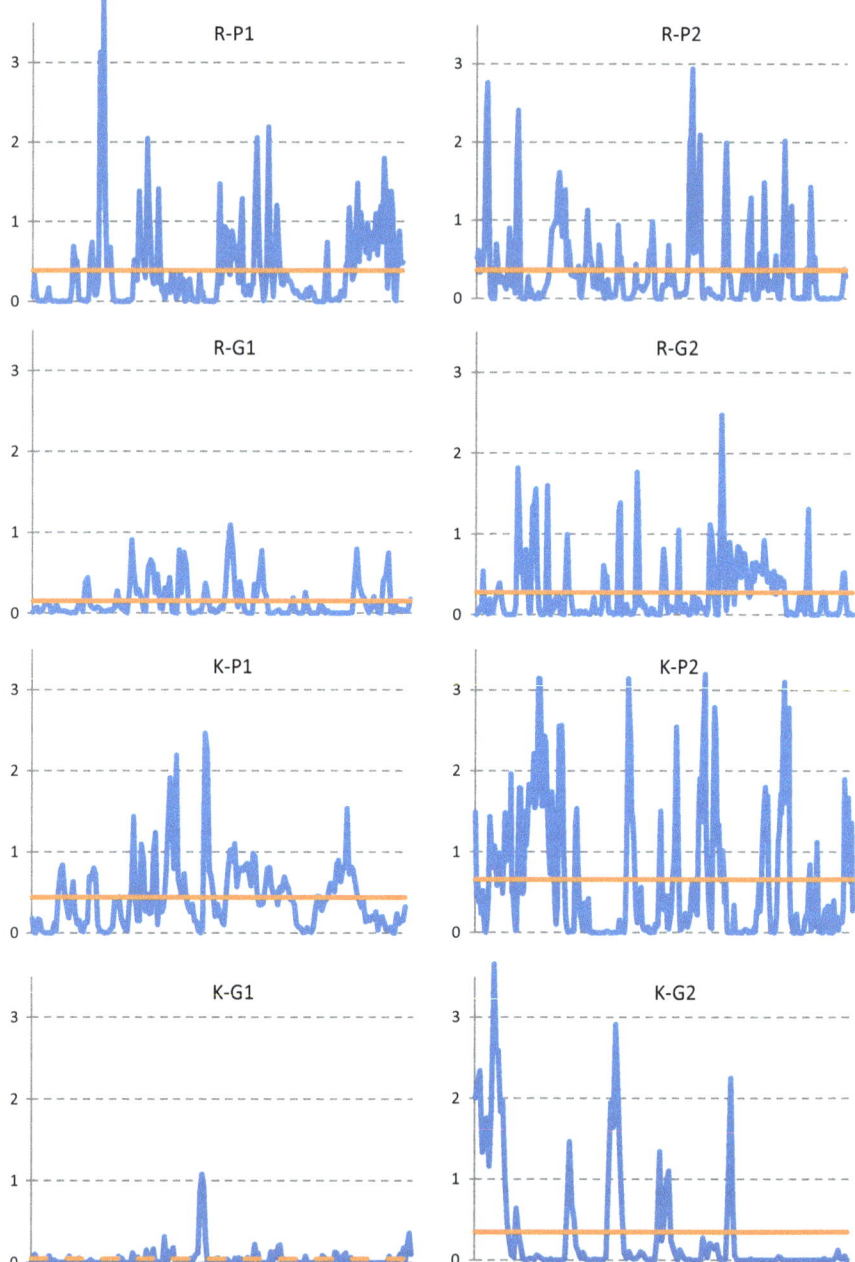

**Fig. 11.9** Body movement changes: temporal changes during the first task in each session are shown ('*photo*' task for R and '*video*' task for K). The horizontal lines indicate averages within each plot

## 11.5 Discussion

### 11.5.1 Adaptation to the Geminoid

Although both participants described the geminoid as scary in both trials, the measurement results seem to show that the participants gradually adapted to conversation through the geminoid. This was especially clear from the video recording of K, where he remained silent and still in the first half of the first G condition, and then gradually became active, as is his normal manner. The fact that the younger 4-year-old participant was scared by the geminoid seems to echo the results of a previous study [13]. The results there suggest that even a younger, more sensitive participant could adapt to the uncanny appearance of the geminoid through conversation with the entity. Although the responses of both participants toward the geminoid became similar to their response toward the real person, some differences remained, especially in the younger participant, K. One influencing factor might be the expressiveness in facial expressions. As seen in K's comment where he thought the geminoid was wearing a mask, the facial expressiveness of the android must be improved. Even if efforts are made to create a replica of an existing person, both in their appearance and behavior, many differences will still exist in the geminoid. However, even though issues remain for the geminoid before it attains the complex functionality of human beings, we can see from the measured data that both children gradually adapted to conversation with the geminoid. Indeed, even with real human beings, we sometime experience an uncanny feeling that decreases or disappears as we spend more time with that person. Similar to the developmental process in infants, where perceptual functions become optimized to frequent stimuli [21, 22], the classification function within ourselves might become 'personalized' or tuned to the behavior or expression that each specific person shows.

Does the same adaptation process work during conversations with the geminoid? Further examination of this adaptation process will elicit an understanding of the nature of human–robot interaction and may give rise to new findings in human developmental processes.

### 11.5.2 Representing Personal Presence

As for the operator's personal presence, it was not fully represented at identical levels as the real person. In the case of the daughter, the voice and content of the speech led to the quick conclusion that the geminoid was operated by her father. This is the same phenomena as seen in many adult visitors to our laboratory. When they first see the geminoid, they are surprised and experience a feeling of unease. However, after conversing for a while, they get used to the geminoid and feel it is just like talking with the real person [14]. However, even when R knew that the geminoid was operated by her father, the measurements still showed slight differences between talking with her father and with the operated geminoid.

This was much clearer in K, the participant unfamiliar with the geminoid source. Here, even stating the identification of the operator seemed to fail. In the interview after the first day, K said that the geminoid was definitely somebody that he had never met before, although he had no problem in recognizing the real entity as the same person in the two experiment days.

From K's measurement values, we believe that he also gradually adapted to the geminoid, and his attitudes became closer to those displayed toward the real person. We also think that the behavior and appearance of the geminoid displayed a different sense of personal presence to K. In the case of R, who knew the source person very well, her belief that the geminoid was operated by her father might have overcome her impression of the geminoid. As K was not familiar with the operator, what he saw for each entity might lead to a stronger impression than the content of their speech. The results with the two participants, R and K, seem to show that each focused on different aspects of the geminoid in forming their impression of its presence.

The results of this study show that the geminoid is still not perfect in representing personal presence. Although the participants gradually adapted to conversing with the geminoid, the impression they felt in the presence of the geminoid was not the same as when conversing with the original source. What are the main factors that define an individual? And what further aspects do we need to represent the presence of an individual? There are many elements that are believed to show individuality. We can often identify a person from their appearance, voice, manner of speaking, or even their gait. In the current geminoid, elements such as the voice, speech content, and memories are identical to those of the original person, who is controlling the teleoperation system. Other elements are quite close to the source, such as its appearance, and some, such as facial expression, are still not close enough, mainly because of engineering issues. Naively, it seems to be easy to express individuality when the appearance of the entity is close to the original and the speech content is exactly that of the original. However, the results of this study show that these are not sufficient. The fact that no difference was seen in the eye contact rate seems to show that the geminoid is in one aspect superior to telephone or CMC systems [4]. Overall, the current geminoid is not as good as other systems for correctly transmitting individuality. Further research is needed to identify the measurements and elements from which personal presence can be described and defined. By utilizing the function of the geminoid system, where various elements that possibly form the personal presence of an individual can be added or subtracted, further study will lead us to build robots that better represent humanlike presence, and also to clarify the key elements that make a person an individual being.

**Acknowledgements** This work was supported in part by the Ministry of Internal Affairs and Communications of Japan.

# References

1. Nishio, S., H. Ishiguro, and N. Hagita. 2007. Can a teleoperated android represent personal presence?—A case study with children. *Psychologia* 50 (4): 330–342.
2. Nishio, S., H. Ishiguro, M. Anderson, and N. Hagita. 2008. Expressing individuality through teleoperated android: A case study with children. In *HCI '08 Proceedings of the third IASTED international conference on human computer interaction*, 297–302.
3. Nishio, S., H. Ishiguro, M. Anderson, and N. Hagita. 2008. Representing personal presence with a teleoperated android: A case study with family. *AAAI Spring Symposium—Technical Report* 2008: 96–103.
4. Wainfan, L. 2005. Challenges in virtual collaboration: Videoconferencing audioconferencing and computer-mediated communications. RAND Corporation.
5. Goffman, E. 1963. *Behavior in public places*. New York: The Free Press.
6. Lombard, M. 1997. At the heart of it all: The concept of presence. *Journal of Computer-Mediated Communication* 3.
7. Zhao, S. 2003. Toward a taxonomy of copresence. *Presence: Teleoperators and Virtual Environments* 12: 445–455.
8. Ishiguro, H. 2002. Toward interactive humanoid robots: A constructive approach to developing intelligent robot. In *Proceedings of the 1st international joint conference on autonomous agents and multiagent systems*, 621–622.
9. Fong, T., I. Nourbakhsh, and K. Dautenhahn. 2003. A survey of socially interactive robots. *Robotics and Autonomous Systems* 42: 143–166.
10. Kanda, T., H. Ishiguro, M. Imai, and T. Ono. 2004. Development and evaluation of interactive humanoid robots. *Proceedings of the IEEE* 92: 1839–1850.
11. Ishiguro, H. 2005. Android science: Toward a new cross-disciplinary framework. In *Proceedings of toward social mechanisms of android science: A CogSci 2005 workshop*, 1–6.
12. MacDorman, K.F., and H. Ishiguro. 2006. The uncanny advantage of using androids in social and cognitive science research. *Interaction Studies* 7: 297–337.
13. Minato, T., M. Shimada, H. Ishiguro, and S. Itakura. 2004. Development of an android robot for studying human-robot interaction. In *Proceedings of the seventeenth international conference on industrial and engineering applications of artificial intelligence and expert systems (IEA/AIE)*, 424–434.
14. Nishio, S., H. Ishiguro, and N. Hagita. 2007. Geminoid: Teleoperated android of an existing person. In *Humanoid robotics*. Advanced Robotic Systems, ed. M. Hackel, Vienna.
15. Prothero, J, D. Parker, T. Furness, and M. Wells. 1995. Towards a robust, quantitative measure for presence. In *Proceedings of the conference on experimental analysis and measurement of situation awareness*, 359–366.
16. Leary, M.R. 1983. *Understanding social anxiety*. Beverly HiUs: Sage.
17. Nishida, K., M. Ura, T. Kuwabara, and J. Kayanno. 1988. Intermediating influence of conversation on social interaction. *Research in Social Psychology'* 3: 46–55.
18. Asahara, M., and Y. Matsumoto. 2000. Extended models and tools for high-performance part-of-speech tagger. In *Proceedings of international committee on computational linguistics (COLING)*, 21–27.
19. Feyereisen, P. 1982. Temporal distribution of co-verbal hand movements. *Ethology and Sociobiology* 3: 1–9.
20. Planalp, S. 1999. *Communicating emotion: Social, moral, and cultural processes*. New York: Cambridge University Press.
21. Kuhl, P.K., F.M. Tsao, and H.M. Liu. 2003. Foreign-language experience in infancy: Effects of short-term exposure and social interaction on phonetic learning. *Proceedings of the National Academy of Sciences* 100: 9096–9101.
22. Pascalis, O., M. Haan, and C.A. Nelson. 2002. Is face processing species-specific during the first year of life? *Science* 296: 1321–1323.

# Chapter 12
# Cues that Trigger Social Transmission of Disinhibition in Young Children

Yusuke Moriguchi, Takashi Minato, Hiroshi Ishiguro,
Ikuko Shinohara and Shoji Itakura

**Abstract** Previous studies have shown that observing the actions of a human model, but not those of a robot, can induce perseverative behaviors in young children, suggesting that children's socio-cognitive abilities may lead to perseverative errors ("social transmission of disinhibition"). This study investigates how the social transmission of disinhibition occurs. Specifically, the authors examine whether a robot with human appearance (an android) triggers perseveration in young children and compare the effects with those induced by a human model. The results reveal that the android induces the social transmission of disinhibition. Additionally, children are more likely to be affected by the human model than by the android. The results suggest that behavioral cues (biological movement) may be important for the social transmission of disinhibition.

---

This chapter is a modified version of a previously published paper [21], edited to be comprehensive and fit with the context of this book.

---

Y. Moriguchi
Joetsu University of Education, Joetsu-shi, Niigata 943-8512, Japan

T. Minato (✉)
Asada Project, ERATO, Japan Science and Technology Agency,
Suita, Osaka 565-0871, Japan
e-mail: minato@atr.Jp

H. Ishiguro
Department of Adaptive Machine Systems, Osaka University, Suita,
Osaka 565-0871, Japan

I. Shinohara
Department of Psychology, Aichi Shukutoku University, Nagakute 480-1197, Japan

S. Itakura
Department of Psychology, Graduate School of Letters, Kyoto University,
Sakyo-Ku, Kyoto 606-8501, Japan

T. Minato
Hiroshi Ishiguro Laboratories, Advanced Telecommunications Research Institute
International (ATR), 2-2-2 Hikaridai, Keihanna Science City, Kyoto 619-0288, Japan

© Springer Nature Singapore Pte Ltd. 2018         205
H. Ishiguro and F. Dalla Libera (eds.), *Geminoid Studies*,
https://doi.org/10.1007/978-981-10-8702-8_12

**Keywords** Executive function · Social cognition · Preschool children
Dimensional change card sort · Robot · Simulative process

## 12.1    Introduction

In the last decade, there has been growing interest in the development of executive
function. A search of Psyc-INFO using the key words "executive function under
12 years of age" yielded 48 records for the period 1985–1995 and 458 records for
the period 1996–2005. This expansion of the literature includes advances in brain
research (e.g., [6]), research on the developmental relationship between executive
function and other cognitive abilities (e.g., [4]), and research on developmental
disorders (e.g., [1]).

The executive function refers to the ability to plan, execute, and monitor
appropriate and relevant actions and to inhibit irrelevant and inappropriate actions
for the attainment of a specific goal. This ability develops rapidly during the preschool years, with adult-level performance being achieved during adolescence,
which is subserved by the maturation of the prefrontal cortex [5, 16, 25].

Recent studies have shown that developing cognitive control may involve more
social processes than previously considered. For example, Moriguchi et al. [18]
showed that young children's cognitive control may be affected by observing
another person's actions. They used a social modification of the dimensional
change card sort (DCCS) task, which is used to assess the executive function in
children [26]. In the standard DCCS task, children are asked to sort cards that have
two dimensions such as color and shape (e.g., yellow flowers, green houses) into
trays with target cards (e.g., a yellow house, a green flower). First, the children are
asked to sort cards according to one dimension (e.g., color) for six trials. Then, they
are asked to sort the cards according to the other dimension (e.g., shape) for six
trials. Typically, most three-year-olds fail to switch the dimension, whereas four-
and five-year-old children make the dimension switch. In the modified social DCCS
task, instead of sorting the cards by themselves, preschoolers watched an adult
model sorting the cards according to one dimension (e.g., shape), after which they
were asked to sort according to a different dimension (e.g., color). The results
confirmed that most three-year-olds fail to sort the cards according to the different
dimension and persevere with sorting according to the observed dimension, as in
the standard DCCS task (see [17]).

Interestingly, the cognitive control process in children can be affected by a
human's actions, but not by a robot's actions. Moriguchi et al. [20] showed that
children who observed a robot sorting according to one dimension had no difficulty
in sorting the cards according to a different dimension. The authors explain the
results in terms of a socio-cognitive perspective whereby children persevere with
the human model's rule because they mentally simulate the model's actions while
watching. In fact, they use the first observed rule even when asked to choose the
second rule. In contrast, the children's actions are not affected by a robot's actions

because the robot does not induce a simulative process in young children. Mor
iguchi et al. [20] concluded that children's socio-cognitive abilities can lead to
perseverative errors in the social DCCS task, and they labeled the perseverative
tendencies as the "social transmission of disinhibition."

The social understanding literature suggests that observing human actions, but
not mechanical actions, may elicit young children's and adults' imitative behaviors
[9, 11, 14]. This is consistent with Moriguchi et al. [20] explanation. However, it is
still unclear why a human's actions, but not those of a robot, may induce the social
transmission of disinhibition. The cues that trigger the social transmission of disinhibition are unknown.

Recent research regarding infants' perceptions of the goal-directed actions of
others are relevant to understanding the influence of human or robot actions: In this
field, some researchers have suggested that behavioral cues may be important for
infants' perceptions of goal-directedness, whereas others have argued that featural
cues may be relatively important [3]. The former viewpoint emphasizes that infants
are sensitive to behavioral cues, such as self-propelledness and contingent
responses [7, 10, 13, 22, 23], whereas the latter stresses that the appearance of the
agents may play a significant role in infants' perceptions of goal-directed actions
[14, 24]. Research evidence is presently inconclusive, providing some support for
each theory [3, 8, 24].

Both appearance and behavioral cues may be important for infants' social perceptions. The present study examines which cues may trigger young children's
social transmission of disinhibition. We test the hypothesis that behavioral cues
(biological movement) may affect young children's social transmission of disinhibition, and devise a new android condition for comparison with a human condition in the social DCCS task. The android has a humanlike appearance

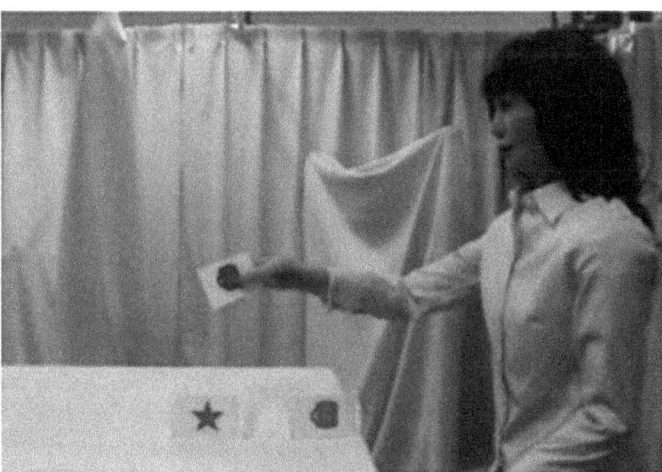

**Fig. 12.1** Android's actions in the android condition. Android Repliee Q2 was developed by
Osaka University and Kokoro Co., Ltd

(Fig. 12.1); however, its movement is mechanical, similar to the robot used in a previous study [20]. Thus, the android differs from a human in terms of its movement, but not its appearance. Using the android, we evaluate the effect of the physiological movement (as a behavioral cue) of the agents. Before the experiment, we conducted a pilot study with adult participants to verify the validity of the stimuli in the experiment.

## 12.2   Pilot Study

We conducted a pilot study to determine whether the android used in the experiment was different from a human in terms of its appearance with two groups of adult participants (N = 20 for each group, $M$ = 19.0 years of age). The first group (appearance group) judged whether the appearance of a human, an android, and a mechanical robot varied significantly. The second group (movement group) rated whether the movements of the three agents varied significantly.

To assess the differences in appearance and movement between agents, we used a scale of animacy consisting of six items evaluated on a 10-point Likert scale (Dead/Alive, Stagnant/Lively, Mechanical/Organic, Artificial/Lifelike, Inert/Interactive, and Apathetic/Responsive). This scale was developed to assess the lifelikeness of robots [2, 12]. The appearance group was asked to rate the animacy of the agents when presented with static pictures of each agent. We used pictures of the stimuli used in the present experiment (an android and a human) and a stimulus used in the previous study (a robot; [20]). The movement group was asked to rate the animacy of the agents when presented with video clips of each agent. We used video clips of card sorting from the present experiment (an android and a human) and the previous study (a robot; [20]).

The Cronbach's Alpha for the animacy scale was 0.85. We used the composite animacy scores of the six items to assess the difference in appearance and movement between agents (0–9, inanimate–animate). For the appearance group, the mean animacy scores (SD) were 4.34 (2.24) for the human, 3.97 (1.93) for the android, and 2.51 (1.57) for the robot. For the movement group, the mean scores were 6.51 (1.84) for the human, 3.89 (1.60) for the android, and 2.65 (1.46) for the robot. We conducted a group (appearance versus movement) $\times$ agent (human versus android versus robot) two-way analysis of variance (ANOVA) of the animacy scores. The results showed a significant main effect with respect to the agent, $F(2, 76) = 32.945$, $p < 0.001$, $\acute{\eta} = 0.46$, and a significant interaction between group and agent, $F(2, 76) = 6.213$, $p < 0.003$, $\acute{\eta} = 0.14$, but did not find a significant main effect with respect to the group, $F(1, 38) = 3.493$, $p > 0.06$, $\acute{\eta} = 0.08$. We conducted separate ANOVAs for each group to further examine the significant interactions between group and agent. We found a significant main effect with respect to the agent for the appearance group, $F(2, 38) = 7.262$, $p < 0.01$, $\acute{\eta} = 0.28$, and the movement group, $F(2, 38) = 33.196$, $p < 0.001$, $\acute{\eta} = 0.63$. Post hoc comparisons using the Bonferroni method revealed that the participants in the appearance group rated the human and android as more animate than the robot, whereas the participants in the movement group rated the human as more animate than the android and the robot ($p < 0.05$). The results suggest that the adult participants did not consider the android to be significantly different from the human in terms of appearance, but did consider the two to be significantly different in terms of movement.

## 12.3  Experiment

### 12.3.1  Participants

The participants were 75 three- and four-year-old children ($M = 46.0$ months, SD $= 5.2$, range $= 36$–$56$ months; 45 boys and 30 girls). They were recruited from nursery schools in Kyoto and Fukuoka, and were not reported to have developmental abnormalities. All children were from Japanese middle-class families. Informed consent was obtained from the parents of all the children prior to their involvement in the study. The children were randomly assigned to one of the following three conditions: human condition, android condition, or control condition. Mean ages (ranges) were 46.2 months (37–54 months) in the human condition, 45.9 months (39–54 months) in the android condition, and 45.8 months (36–56 months) in the baseline condition. There were no significant age differences between the conditions.

### 12.3.2  Materials

Laminated cards (9.0 × 7.5 cm) were used in the trials. Two trays (4.5 × 10.5 15 cm) were provided, one containing a target card depicting a red star and the other containing a card depicting a blue cup. There were six sorting cards, each of which depicted either a red cup or a blue star.

An android named Repliee Q2 was used as a model. Repliee Q2 was developed by Osaka University and Kokoro Co., Ltd [15]. As shown in Fig. 12.1, the face of Repliee Q2 is quite similar to that of a Japanese woman and the android wears female clothing. However, the android's movements are awkward compared to a real human; its movements are very mechanical (for details, see the stimuli at http://www.youtube.com/watch?v=JomR4cmy1-8). The android and a human female were videotaped performing the sorting task in similar ways. In the human and android conditions, the stimuli (video clips) were presented using a notebook personal computer with a 15-inch display (Dell Latitude D610).

## *12.3.3 Procedure*

There were three phases in the human and android conditions: a warm-up phase, an observation phase, and a sorting phase. The control condition included only a warm-up phase and a sorting phase. In each condition, the children were tested individually for about 5 min while seated on a chair next to an experimenter.

### 12.3.3.1 Human Condition

In the warm-up phase, the experimenter introduced the two trays with target cards and sorting cards. The child was asked to name the shape and the color on each card. Following this, the experimenter announced the general rule of the task ("There are two ways to sort the cards, color and shape. I will tell you whether you should sort the cards according to their color or shape.").

In the observation phase, the child was asked to watch a video on the computer. The video showed an adult female model and two trays and cards identical to those used with the children. The child was told that the model would sort the cards into the trays ("Now she [the model] is going to sort the cards first. Please watch carefully."). The model sorted the cards according to one dimension. Half of the children saw the model sorting the cards according to the shape dimension, and the other half saw the model sorting the cards according to the color dimension. During the observation, the children were not given any explicit rules. Instead, they were encouraged to watch the video. The model performed four trials (two blue star and two red cup cards).

In the sorting phase, the experimenter introduced the trays and sorting cards to the child again. The child was instructed to play a game ("Now, it is your turn. We are going to play a game."). In this game, the child was asked to sort the cards according to the other dimension. For example, when the model sorted the cards according to the shape dimension, the child was asked to sort the cards according to the color dimension ("Your game is a color game. In the color game, all the red ones go here and all the blue ones go there."). The child performed six trials. In each trial, the experimenter told the child the rules of the game and randomly selected a sorting card for sorting ("Where does this card go in the color game?"). The child was required to sort the cards into the two trays and was not given any feedback about the correctness of their choices.

### 12.3.3.2 Android Condition

The android condition was identical to the human condition, except that an android was shown on the video sorting the cards, rather than a human model. All other aspects of the android's actions were matched to the human model, including the speed with which the actions were performed.

In the observation phase, the instruction given in the android condition was the same as in the human condition. The child was instructed: "Now she [the android] is going to sort the cards first. Please watch carefully." The experimenter did not state that Repliee Q2 was a robot/android, and, therefore, the child did not know in advance that Repliee Q2 was a robot.

#### 12.3.3.3 Baseline Condition

The baseline condition was identical to the human condition, with the exception that there was no observation phase. After the warm-up phase, the child was instructed to sort the cards according to one dimension (e.g., "We are going to play a game. The game is a color game. In the color game, all the red ones go here and all the blue ones go there.").

### 12.3.4 Results and Discussion

Children in the human condition and the android condition watched the video clip during the observation phases and never looked away from it. This suggested that children in both conditions attended equally to the stimuli. The children's sort was scored as "correct" if they sorted a card correctly according to the dimension instructed by the experimenter in the sorting phase. As shown in previous studies (e.g., [20]), most children (63 out of 75; 84.0%) are either correct or incorrect on all six trials. Therefore, the children were classified as passing or failing the task according to whether or not they sorted at least five of the six cards correctly. Preliminary analyses using Fisher's exact tests showed no significant differences in children's performance in the shape and color games, or related to the children's sex, $p > 0.10$. Therefore, the data for these variables were combined for the subsequent analyses.

More than half of the children in the human condition failed to use the second (instructed) rule, which is consistent with previous studies (e.g., [18]). They sorted the cards according to the first rule presented by the human model. Only eight children (32%) passed the task (Fig. 12.2). However, the children in the android condition were more likely to sort the cards according to the second, instructed rule. Sixteen children (64%) in the android condition were classified as passing. In addition, as expected, almost all of the children (24 out of 25) in the baseline condition sorted the cards according to the instructed rules (Fig. 12.2).

To examine the performance differences between the human, android, and baseline conditions, we conducted chi-square tests and found a significant difference between conditions, $\chi2 (2, N = 75) = 22.222, p < 0.0001$. Post hoc analyses using Fisher's exact test (two-tailed) showed significant differences between the human and android conditions, $p < 0.05$, between the human and baseline conditions, $p < 0.0001$, and between the android and baseline conditions, $p < 0.02$. The

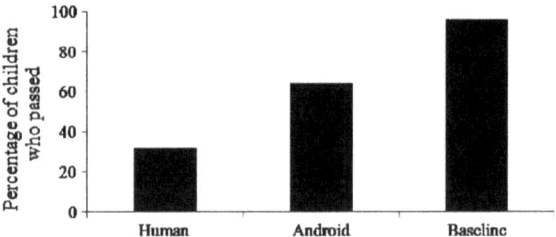

**Fig. 12.2** Percentage of children who correctly sorted cards according to the second dimension

results show that the human and android actions have a different influence on the young children's cognitive control process.

Finally, we conducted an additional analysis to compare the effect of the android to the effect of a mechanical robot in a previous study [20]. The participants in the previous robot condition (robot condition) included three-year-old children. Thus, we added six four-year-old children to the previous data (N = 26, M = 45.8 months, range = 38–56 months). In the robot condition, about 85% of the children were classified as passing. We conducted chi-square tests and found a marginally significant difference between conditions, $\chi2$ (1, N = 51) = 2.851, $p$ < 0.10.

The results are important in two aspects. First, the non-human agent with a human appearance may trigger young children's social transmission of disinhibition. Compared to the baseline condition, children in the android condition were more likely to sort the cards according to the first (modeled) dimension even though they were instructed to sort the cards according to the second dimension. These results contrast with those in the previous study using a mechanical robot [20]. In that study, the mechanical robot did not affect young children's actions; the results were as though the children had not observed any demonstrations and were not significantly different from the baseline condition. Although the differences between the android condition and the robot condition are marginally significant, the results suggest that an android triggers the social transmission of disinhibition in children. Second, the results of the present study reveal that the performance in the android condition is significantly different from that in the human condition. This is despite the fact that the participants in the android condition were told "She is going to sort the cards first." The children were not explicitly told that the android was a robot. In our pilot study, adult participants did not consider the human and android to be significantly different. The children may have detected that the android was a robot from its movements. This is interesting, because Moriguchi et al. [19] showed that a televised model triggered the social transmission of disinhibition in young children as well as a live model did. Children may identify the televised model with the live model, but discriminate the televised model from the android. The results suggest that an agent's physiological movement might play an important role in the social transmission of disinhibition.

## 12.4 Conclusion

The present study investigated whether behavioral cues may affect young children's perseverative behaviors. The results show that the android's movements did, to some extent, trigger perseverative tendencies in young children. Moreover, their performance in the android condition differed significantly from that in the human condition. The results suggest that an agent's movement (i.e., humanlike movement) might play an important role in the social transmission of disinhibition.

The present study contributes to our understanding of why young children engage in perseverative behaviors after observing another person's actions. A previous study proposed that children perseverate with a human model's actions because they mentally simulate the model's actions, and thus they execute these (mentally rehearsed) actions even when asked to choose other actions [20]. Consistent with this, we interpret the android's effect on young children's actions in terms of the simulative processes. The children expected to observe a "woman" who would demonstrate the card-sorting game, and thus initially observe the android as if she were a woman. This may trigger mental simulation. However, while watching the android's actions, they may detect that its movements are mechanical, and their simulative process may have been affected. Nevertheless, the simulative process may, to some extent, affect young children's performance in the second phase.

Our results are consistent with social understanding research. There is evidence that infants begin to detect goal-directedness in human or non-human agents during the first years of life [3]. Although the nature and emergence of infants' ability to understand others' goal-directed actions are now controversial, both appearance and behavioral cues may be critical for detecting goal-directedness [8]. In the present study, we were unable to address this controversy directly, because our research paradigm was too different from the infant goal-detection studies. Nevertheless, we believe that behavioral cues could be crucial in young children's social cognition. Further research is needed to examine whether the appearance may have a significant role in the social transmission of disinhibition and how these other cues interact with the behavioral cues observed in the present study.

## References

1. Barkley, R.A. 1997. Behavioral inhibition, sustained attention, and executive functions: Constructing a unifying theory of ADHD. *Psychological Bulletin* 121: 65–94.
2. Bartneck, C., E. Croft, and D. Kulic. 2009. Measurement instruments for the anthropomorphism, animacy, likeability, perceived intelligence, and perceived safety of robots. *International Journal of Social Robotics* 1: 71–81.
3. Biro, S., and A.M. Lesilie. 2007. Infants' perception of goal-directed actions: Development through cue-based bootstrapping. *Developmental Science* 10: 379–398.
4. Carlson, S.M., and L.J. Moses. 2001. Individual differences in inhibitory control and children's theory of mind. *Child Development* 72: 1032–1053.

5. Davidson, M.C., D. Amsoa, L.C. Anderson, and A. Diamond. 2006. Development of cognitive control and executive functions from 4 to 13 years: Evidence from manipulations of memory, inhibition, and task switching. *Neuropsychologia* 44: 2037–2078.

6. Durston, S., K.M. Thomas, Y. Yang, A.M. Ulug, R.D. Zimmerman, and B.J. Casey. 2002. A neural basis for the development of inhibitory control. *Developmental Science* 5: F9–F16.

7. Gergely, G., and G. Csibra. 2003. Teleological reasoning in infancy: The one-year-olds' naive theory of rational action. *Trends in Cognitive Sciences* 7: 287–292.

8. Gergely, G., Z. Nádasdy, G. Csibra, and S. Biro. 1995. Taking the intentional stance at 12 months of age. *Cognition* 56: 165–193.

9. Itakura, S., H. Ishida, T. Kanda, K. Lee, Y. Shimada, and H. Ishiguro. 2008. How to build an intentional android: Infants' imitation of a robot's goal-directed actions. *Infancy* 13: 519–531.

10. Johnson, S., V. Slaughter, and S. Carey. 1998. Whose gaze will infants follow? The elicitation of gaze-following in 12-month-olds. *Developmental Science* 1: 233–238.

11. Kilner, J.M., Y. Paulignan, and S.J. Blakemore. 2003. An interference effect of observed biological movement on action. *Current Biology* 13: 522–525.

12. Lee, K.M., N. Park, and H. Song. 2005. Can a robot be perceived as a developing creature? *Human Communication Research* 31: 538–563.

13. Luo, Y., and R. Baillargeon. 2005. Can a self-propelled box have a goal? Psychological reasoning in 5-month-old infants. *Psychological Science* 16: 601–608.

14. Meltzoff, A.N. 1995. Understanding of the intentions of others: Re-enactment of intended acts by 18-month-old children. *Developmental Psychology* 31: 838–850.

15. Minato, T., M. Shimada, S. Itakura, K. Lee, and H. Ishiguro. 2005. Does gaze reveal the human likeness of an android? In *Proceedings of 2005 4th IEEE international conference on development and learning,* 106–111.

16. Moriguchi,Y., and K. Hiraki. 2009. Neural origin of cognitive shifting in young children. In *Proceedings of the national academy of sciences of the United States of America,* vol. 106, 6017–6021.

17. Moriguchi, Y., and S. Itakura. 2008. Young children's difficulty with inhibitory control in a social context. *Japanese Psychological Research* 50: 87–92.

18. Moriguchi, Y., K. Lee, and S. Itakura. 2007. Social transmission of disinhibition in young children. *Developmental Science* 10: 481–491.

19. Moriguchi, Y., W. Sanefuji, and S. Itakura. 2007. Disinhibition transmits from television to young children. *Psychologia* 50: 308–318.

20. Moriguchi, Y., T. Kanda, H. Ishiguro, and S. Itakura. 2010. Children perseverate to a human's actions but not to a robot's actions. *Developmental Science* 13: 62–68.

21. Moriguchi, Y., T. Minato, H. Ishiguro, I. Shinohara, and S. Itakura. 2010. Cues that trigger social transmission of disinhibition in young children. *Journal of Experimental Child Psychology* 107 (2): 181–187.

22. Premack, D. 1990. The infant's theory of self-propelled objects. *Cognition* 36: 1–16.

23. Shimizu, Y.A., and S.C. Johnson. 2004. Infants' attribution of a goal to a morphologically unfamiliar agent. *Developmental Science* 7: 425–430.

24. Woodward, A.L. 1998. Infants selectively encode the goal object of an actor's reach. *Cognition* 69: 1–34.

25. Zelazo, P.D., and U. Müller. 2002. Executive function in typical and atypical development. In *Blackwell handbook of childhood cognitive development,* ed. U. Goswami, 445–469. Oxford: Blackwell.

26. Zelazo, P.D., D. Frye, and T. Rapus. 1996. An age-related dissociation between knowing rules and using them. *Cognitive Development* 11: 37–63.

# Chapter 13
# Effects of Observing Eye Contact Between a Robot and Another Person

Michihiro Shimada, Yuichiro Yoshikawa, Mana Asada, Naoki Saiwaki and Hiroshi Ishiguro

**Abstract** One of the common requirements for a communication robot is to be accepted by humans. Previous work has examined the effects of nonverbal factors on people's perceptions of robots for such a purpose, but always with a focus on dyadic human–robot interaction; in real human society, however, triadic interaction also plays an important role and should be considered. This paper explores a potential merit offered by the latter form of interaction; specifically, how one form of nonverbal interaction occurring between a robot and humans, eye contact, can be utilized to make the robot appear more acceptable to humans. Experiments are conducted with groups of two humans and an android. One of the humans, the "subject," is asked to communicate with a second person, the "confederate," who knows the purpose of the experiment; the confederate's role is to gaze in such a way that the subject either observes or does not observe eye contact between the confederate and the android. A post-interaction questionnaire reveals that subjects' impressions toward the robot are influenced by eye contact between the confederate and the robot. Finally, the consistency of the experimental results is discussed in terms of Heider's balance theory, and future extensions of this research are proposed.

This chapter is a modified version of a previously published paper [1], edited to be comprehensive and fit with the context of this book.

M. Shimada
Faculty of Engineering, Department of Adaptive Machine Systems,
Osaka University, 2-1 Yamada-oka, Suita, Osaka 565-0871, Japan
e-mail: shimada.michihiro@is.sys.es.osaka-u.ac.jp

Y. Yoshikawa (✉)
Asada Project, ERATO, Japan Science and Technology Agency, 2-1 Yamada-oka,
Osaka, Suita 565-0871, Japan
e-mail: yoshikawa@irl.sys.es.osaka-u.ac.jp

M. Asada · N. Saiwaki
Faculty of Human Life and Environment, Department of Clothing Environmental Science,
Nara Women's University, Kitauoyahigashi-machi, Nara 630-8506, Japan

H. Ishiguro
Faculty of Engineering Science, Department of Systems Innovation, Osaka University,
1-3 Machikaneyama, Toyonaka, Osaka 560-8531, Japan

© Springer Nature Singapore Pte Ltd. 2018        215
H. Ishiguro and F. Dalla Libera (eds.), *Geminoid Studies*,
https://doi.org/10.1007/978-981-10-8702-8_13

**Keywords**  Person's perception · Triadic interaction · Eye contact · Android
Heider's balance theory

## 13.1  Introduction

Unlike the industrial robots of the past, communicating robots are expected to communicate and work with humans in their daily life and have attracted widespread
attention because of their potential to act as an intuitive interface for humans [2],
therapeutic tools [3, 4], and mediators in communication [3]. A common requirement for such robots is to be accepted as members of society; by accept, we mean
that the robots must not be perceived negatively by the humans they are expected to
collaborate with on a daily basis. In previous work, researchers have often focused
on dyadic human–robot interaction and reported the effects of nonverbal factors on
how the robot is accepted by human users; these factors have included, for example,
appearance [5–7], movement [2], the balance between appearance and movement [8],
and the robot's responsiveness to the user [9–12]. However, communication among
more than two agents is another important consideration for communication robots,
as they may sometimes have to serve multiple users or mediate human–human communication. What kinds of factors are effective in making a robot accepted in such
triadic interactions?

Takano et al. investigated the effect of a robot's nonverbal behavior in a triadic
scenario at a real hospital [3]. They found that patients seemed more satisfied with
medical examinations when the robot appeared to smile or nod at them. However,
there have been no attempts to clarify how such nonverbal behavior could be effective
in making the robot accepted by humans in triadic interactions. Hayashi et al. [13]
found that humans receive more information from two robots that talk to each other
about a certain topic than from a single robot that speaks about the same topic. In a
triad, as implied in the previous work, how a robot is perceived by a person might
reflect that person's observations of how the robot interacts not only with the target,
but also with another agent. If we extend this idea to a scenario in which a robot tries
to join in when two humans are interacting, we might assume that the familiarity
the robot builds with one person could be shaped by letting that person observe the
interaction between the robot and the other person.

Sakamoto et al. [14] found that a person's perception in human–human–robot
communication could be influenced by whether the robot expresses positive or negative opinions of each person. Interestingly, the effects of such influence can be
predicted by Heider's balance theory [15], which explains how a person balances
their own cognitions in a triad situation, namely about the relationships between the
person and another, the person and an object, and the second person and the object.
In this theory, triadic interactions are classified as either balanced or unbalanced
depending on the perceived valence of the relationships between each pair of agents.
That is to say, given persons A, B, and C, if the number of positive relationships
is even, the interaction is seen as unbalanced, whereas if the number is odd, the

interaction is balanced. For example, if A has a positive opinion toward B and B has a negative opinion toward C, then it seems reasonable to assume, in order to balance the situation, that A will have a negative opinion toward C. In the same way, we assume that if A has a positive opinion toward C when A has a positive opinion toward B, then B is likely to have a positive opinion toward C. However, because of the limitations of language processing, we cannot assume that a robot will be able to converse with all participants. As a result, it seems difficult to utilize the effects of a participant observing verbal interaction between a robot and another participant in real situations.

In contrast, as nonverbal behavior is more easily exhibited even when another participant is talking, this would appear to be a more feasible way for a robot to give the illusion of interacting with another participant. In this paper, as one type of such nonverbal behavior, we focus on eye contact, as such synchronized behavior has been widely viewed as playing an important role in dyadic interaction [16, 17]. When a person is speaking, they avert their gaze for about 50% of the conversation [18]. Therefore, we could induce someone to feel as if the robot were establishing eye contact with another person by making it turn to the second person when that person directs their gaze toward the robot. As a result, we might expect the positive impression toward the robot to be strengthened. Once any participant's impression toward the robot has improved, the frequency with which that person looks at it, i.e., the opportunities for others to observe its nonverbal interaction, will increase. Consequently, we believe that an investigation of the effect of eye contact in triadic situations could provide an alternative, interaction-based method of preparing robots for acceptance in human society. In this paper, we construct an experiment based on a possible scenario for human–human–robot interaction to examine the effect of observing eye contact between a robot and another person in triad communication. We start from the following hypothesis:

– One's impression toward a robot can be influenced by whether another person appears to be performing nonverbal communication with it.

In the experiment, we use the scenario of a virtual job interview as an example of triadic communication, where two agents converse with each other and one other agent listens to the conversation. Human participants, a subject and a confederate, take turns playing the roles of main speaker and main listener; during this time, a robot (an android called Repliee Q2) plays the role of sub-listener. The eye contact between the confederate and the robot is controlled by operating the robot through a Wizard-of-Oz (WOZ) technique and training the confederate where to look according to two experimental conditions: the confederate establishes eye contact several times in one condition, and never does so in the other condition. We examine how the feelings of the subjects are affected by the existence of eye contact between the other agents through a post-interaction questionnaire in two different situations. The subject is biased to have a positive feeling toward the confederate in one situation (experiment 1) and a negative feeling toward him in the other situation (experiment 2). Furthermore, we examine whether Heider's theory can predict how a subject's observation of nonverbal interaction between others influences their social cognition toward them. Sakamoto et al. controlled a subject's impression of the link between others by changing the content of their verbal communication; however, we explore this issue by analyzing the results of experiments where nonverbal behavior between the interacting agents changes based on two different conditions in which the subject has a different impression of the confederate (experiments 1 and 2).

## 13.2 Experiment 1: The Effect of a Person with Positive Prepossession

In this section, we investigate how the behavior of another person (second person) affects the subject's impression of the third person (android) in the case where the second person displays communicative gaze behavior toward the subject.

### 13.2.1 Subjects

We hired thirty Japanese adults (ages: mean (M) = 21.2, standard deviation (SD) = 2.0 [y]) through a temporary employment agency. Condition 1 was performed with fifteen subjects (eight males and seven females), and condition 2 with the remaining fifteen subjects (eight males and seven females). Written informed consent was obtained from all subjects.

### 13.2.2 Apparatus

#### 13.2.2.1 Android, Repliee Q2

An android called Repliee Q2 was used in this experiment (Fig. 13.1). Repliee Q2 has a very humanlike appearance that resembles an actual Japanese woman. Features as detailed as frown lines have been duplicated, and pneumatic actuators are used to produce body movements in order to reduce motor-driving noise, which might make subjects feel uncomfortable during interactions.

Its actions were designed in advance to resemble those of a store clerk and include bowing and looking toward the person who is speaking (e.g., a customer at the store). The actions were triggered with appropriate timing by an operator who monitored the interaction from a remote room.

**Fig. 13.1** Android Repliee Q2

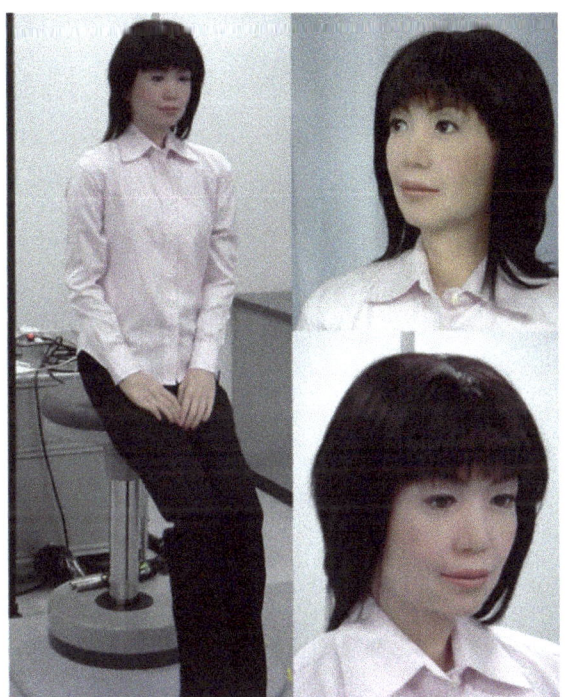

### 13.2.2.2  Experimental Environment

A sample scene from the experiment is shown in Fig. 13.2. The room used for the experiment was 3 m wide and 3.7 m long. It was surrounded by curtains and sound-proof partitions to allow subjects to concentrate on the experiment. The room contained the android (seated on a stool), a round table, and two chairs.

Two video cameras were used to record the experimental sessions. The recorded images and sounds were used not only for post-experiment analysis, but also to assist the operator, who used them during the sessions to determine when the android's actions should be executed. A microphone was placed by the confederate's chair so that the operator knew who was speaking. The distances between agents are shown in Fig. 13.2.

### 13.2.2.3  Experimental Situation

The scene of a job interview was replicated as an example of a situation involving interaction between three agents. The subject was asked to play the role of interviewer, the confederate played the role of interviewee, and the android was described to the subject as being an "Intelligent Android" (IA) whose purpose was to record the interview.

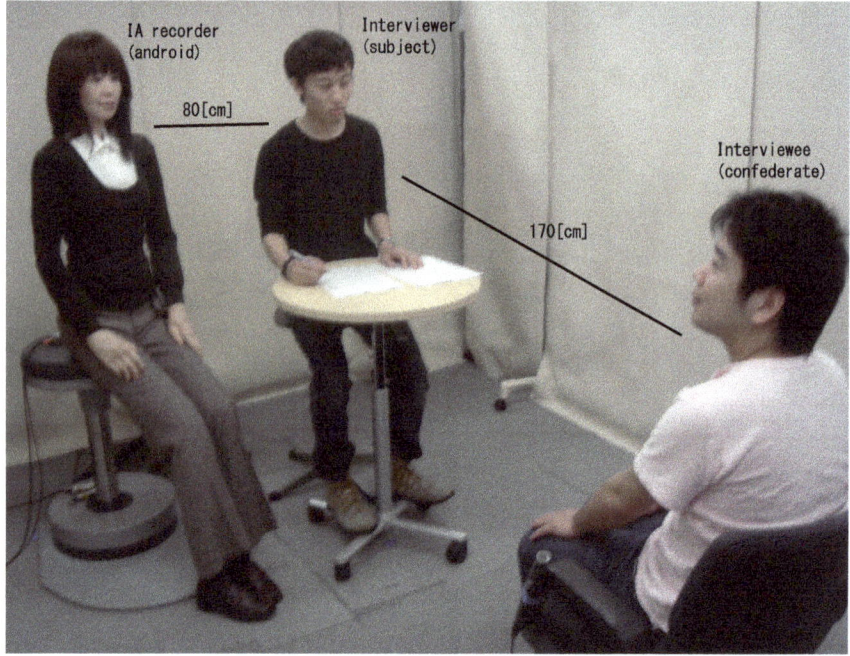

**Fig. 13.2** Experimental setup: An android plays the role of the "Intelligent Android" (IA) recorder (left), the subject plays the role of an interviewer (middle), and the confederate plays the role of an interviewee (right)

The situation of a job interview was adopted because it was considered that this would allow us to design a structured interaction where the gaze movements among subjects could be controlled to satisfy the experimental conditions. We expected the subject (i.e., the interviewer) to focus on the confederate (i.e., the interviewee), and we instructed the confederate to look not only at the interviewer, but also at the android (in condition 1) or at the wall opposite the android (in condition 2). We determined the directions in which the chairs faced and the orientation of the three agents so that the subject would not directly communicate with the android but could observe eye contact between the confederate and the android if it occurred.

## 13.2.3 Procedure

### 13.2.3.1 Task

All instructions for the experiment were given to the subject by an instructor before the subject entered the experimental room where the android was waiting. Then, the instructor asked the subject to sit on a chair next to the android. After the instructor left the experimental room, the interviewee entered the experimental room and sat down on a chair in front of the subject. After the interviewee sat in the chair, the subject started to ask questions that had been listed in a document placed on the table. The subject listened to the interviewee's answers and evaluated them on a seven-point scale. (This was a dummy task). Eight questions that could appear in a real job interview were prepared, such as "What kinds of things have you learned so far?" and "What kind of work do you want to do?" After the subject had asked all of the questions, they answered a post-interaction questionnaire designed to evaluate their feelings about the interaction, including impressions of the interviewee and the android.

#### 13.2.3.2   Instruction

Before the subject entered the experimental room, the instructor told the subject that the purpose of the experiment was to evaluate a mannequin-like device called the IA recorder. The android was described as a next-generation Integrated Chip (IC) recorder furnished with a human appearance in order to be perceived as more acceptable by humans.

The subject was told to ask the questions listed in a document on the table to another subject (the confederate) playing the role of interviewee. The subject was also told to grade the answers from the interviewee on a scale from one to seven. In this experiment, we assumed that the relationship between the subject and the interviewee was positive in both conditions. Moreover, the instructor told the subject to give the answers from the interviewee high scores (dummy task) in order to positively bias the attitude of the subject toward the interviewee. Accordingly, the subject was asked to decide whether to give five, six, or seven points for each answer.

### 13.2.4   Stimulus

Under both conditions, the android nodded to greet the subject when they entered the experimental room in order to make the subject feel as if the IA might be human. During the interview, Repliee Q2 looked at the current speaker. These motions were triggered by an operator who monitored the interview from a remote control station.

The interviewee gave the same answers to each subject. Between the two conditions, only the gaze movements of the interviewee were different. In both conditions, when the interviewee answered questions, he established eye contact with the subject. However, in condition 1, the interviewee also established eye contact with the android. The subject was assumed to feel as if a positive relationship had been established between the interviewee and the android through such behavior. The interviewee was trained to be able to establish eye contact at the same time for all subjects.

In condition 2, the interviewee looked away from the android (in the direction opposite to where the android was situated) at the same times that he had looked at the android in condition 1; this was done so that the subject would not feel a positive relationship existed between the interviewee and the android. In this way, we encouraged the subject not to conduct synchronized behavior, i.e., eye contact, with the android. Note that the interviewee was trained to answer questions and look like he was thinking in such a way that his gaze movement toward the side opposite the android would not appear strange to the subject.

## 13.2.5  Measurement

### 13.2.5.1  Questionnaire

The questionnaire consisted of questions designed to guarantee that the sessions matched the assumed experimental setting and questions used to measure the impression of the android, which was assumed to be influenced by the difference in conditions. The items that confirmed the success of the experiment are as follows: a question that asked about the relationship between the subject and the confederate, and a question that asked about the relationship between the confederate and the android based on establishing eye contact with the android.

Subjects provided a score of 1–7 for each item on the questionnaire. In the questionnaire, we used direct expressions such as "positive" and "good." Generally, direct expressions bias the results. However, we conducted a relative evaluation between conditions in this experiment. Therefore, bias is not a problem, because the same bias occurs in every condition.

### 13.2.5.2  Video Observation

We observed the behavior of the subject and the android using the recorded video. In this experiment, we sought to investigate the impact of an agent's interaction with a second agent (communication target) on the agent's impression of a third agent. Therefore, we had to exclude (as far as possible) other factors that could have influenced the formation of the impression. It is known that the amount of gaze and synchronized behavior causes an impression to change [19, 20]. Therefore, we measured the amount of time subjects spent looking at the android, the number of times subjects looked at the android, the amount of time and the number of times for which (mutual) eye contact occurred between the subject and the android, and the number of synchronized nods that took place between the subject and the android.

## 13.2.6 Results

### 13.2.6.1 Observation

An example of the experimental flow in condition 1 is shown in Fig. 13.3. The transition of gaze directions of the subject, the interviewee, and the android, as well as the transition between the interviewer and the interviewee speaking, are illustrated. The interviewer almost always looks at the interviewee, except when looking at the paper on the desk to write down the evaluation of the interviewee's answer and to check the next question. The interviewee almost always looks at the interviewer. However, the interviewee sometimes looks at the android while giving his answer. The android looks at the interviewer and the interviewee when each is speaking.

Using the recorded video (30 frames/s), we measured the time for which the subjects looked at the android, the time and the number of times that the subject and the android established eye contact, and the number of times that the subject nodded while the android was nodding. We calculated the average Cohen's kappa value from the data the author measured and the data two other observers measured in order to confirm the reliability of the author's observations. As a result, we confirmed the reliability of the measurements ($\kappa = 0.97$). Therefore, in the following test, we used the data measured by the author.

We conducted two-tailed unpaired t tests and found there was no significant difference between condition 1 and condition 2 regarding the time that the subjects were looking at the android (mean = 163.3 [frames], SD = 175.5 in condition 1, mean = 264.5[frames], SD = 322.2 in condition 2, t(26) = $-1.00$, non-significant (n.s.)), the time and the number of times that the subject and the android established eye contact (time: mean = 0.29, SD = 0.47 in condition 1, mean = 0.43, SD = 0.85 in condition 2, t(26) = $-0.64$, n.s; number of times: mean = 11.2, SD = 0.85 in condition 1, mean = 18.9, SD = 40.2 in condition 2, t(26) = $-0.54$, n.s), and the number of

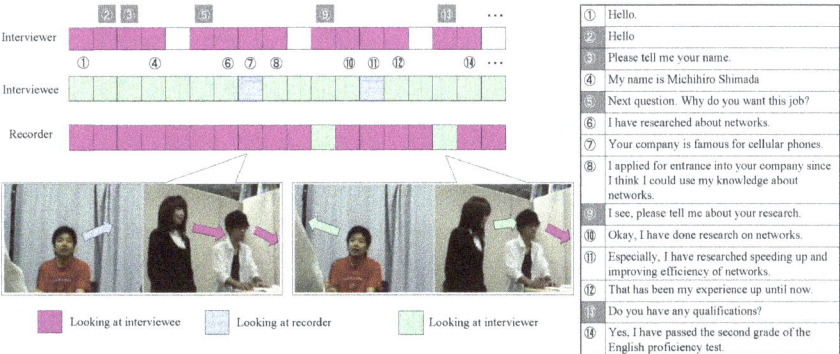

**Fig. 13.3** Sample transition of the interaction in the experiment under condition 1: what the agents were looking at is illustrated, along with who was saying what. The pictures below are the scenes viewed by the operator

times that the subjects nodded while the android was nodding (mean = 2.14, SD = 1.83 in condition 1, mean = 2.93, SD = 2.20 in condition 2, t(26) = −1.02, n.s.).

Thus, there is no significant difference between the conditions for the subject's behavior and synchronized behavior between the subjects and the android.

### 13.2.6.2 Checking Conditions for the Experiment

Before analyzing the effect of the interviewee's gaze movement on the subjects' feelings, we should confirm whether the experiments were actually administered as intended: The confederate and the android should be regarded as establishing eye contact and forming a positive relationship in condition 1, whereas they should not be regarded in that way in condition 2. Moreover, it is important to confirm that the relationship between the subject and the confederate was positive.

*Perception of eye contact.* In this experiment, we wanted to analyze the effect of observing synchronized nonverbal behavior between other agents on the observer's feelings about either agent. Therefore, we removed the data for cases where the confederate was not considered to have established eye contact with the android in condition 1, as well as for cases where the confederate was not considered to have done so in condition 2.

The items investigated by the experiment and their results are listed in Table 13.1. The data to be removed were selected based on the scores of the post-experiment questionnaire. The table lists the questions of the post-experiment questionnaire, the average score and standard deviation in each condition, and a p value for the difference between these conditions. For eye contact, we calculated the average score for question (p), that is, "Did you feel that the interviewee established eye contact with the IA recorder?" We removed the data if the score for this question was too far from the average score of each condition, i.e., greater than $M + 2 \times SD$ or less than $M - 2 \times SD$. As a result, one case was judged as an outlier in each condition. We conducted a two-tailed unpaired t test and found a significant difference between conditions ($t(26) = 3.65 \times 10^{-8}$, $p < 0.01$), while the assumption of equivalent variances was not violated ($F(13, 13) = 1.38$, $p > 0.05$). The score in condition 1 was significantly higher than that in condition 2; therefore, we can regard the gaze movements of the confederate (i.e., interviewee) as sufficiently distinguishable for the subjects included in the analyzed data.

*Impression toward the interviewee.* In this experiment, for simplicity, we assumed that subjects were led to regard their relationships with the confederate as being positive. Therefore, the scores on question (e), i.e., "Did you have a good impression toward the interviewee?," were analyzed. Both average scores (5.79 for condition 1 and 5.21 for condition 2) were greater than 4, which corresponds to a neutral feeling in the scale used in this experiment. Note that a two-tailed unpaired t test showed no significant difference between conditions ($t(26) = 0.12$, $p > 0.10$), while the assumption of equivalent variances was not violated ($F(13, 13) = 2.57$, $p = 0.10$). Therefore, the instruction and the dummy task in which subjects had to evaluate the

**Table 13.1** List of questions on the post-interaction questionnaire (translated from Japanese) and the statistics of their scores in condition 1 and condition 2. Main questions are (a), (c), (e), and (p): (a) asking about the impression toward the android, (c) asking about the impression of the relationship between the interviewee and the android, (e) asking about the impression of the interviewee, and (p) asking about the perception of eye contact (n.s. in the table denotes a non-significant difference.)

| | | Condition 1 | | Condition 2 | | p value |
|---|---|---|---|---|---|---|
| | | Mean | SD | Mean | SD | |
| (a) | Did you have a positive impression toward the IA recorder? | 5.50 | 0.94 | 4.14 | 1.51 | $p < 0.01$ |
| (b) | Did the interviewee look at the IA recorder? | 6.50 | 0.65 | 2.71 | 0.61 | $p < 0.01$ |
| (c) | Did the interviewee seem to have a positive impression toward the IA recorder? | 5.29 | 1.07 | 3.71 | 0.61 | $p < 0.01$ |
| (d) | Did you feel that the IA recorder had a good impression toward the interviewee? | 5.43 | 1.50 | 4.50 | 1.56 | n.s. |
| (e) | Did you have a good impression toward the interviewee? | 5.79 | 0.70 | 5.21 | 1.12 | n.s. |
| (f) | Was the answer given by the interviewee suitable? | 6.07 | 0.73 | 5.36 | 1.28 | $p < 0.10$ |
| (g) | Was the attitude of the interviewee good? | 6.36 | 0.74 | 5.79 | 0.58 | $p < 0.05$ |
| (h) | Were you irritated/attracted by the IA recorder's behavior while the interviewee answered questions? | 5.07 | 1.27 | 5.07 | 1.69 | n.s. |
| (i) | Did the interviewee seem to be respectful toward the IA recorder? | 5.93 | 1.27 | 3.50 | 1.29 | $p < 0.01$ |
| (j) | Did the interviewee seem to be respectful toward you? | 5.86 | 0.95 | 5.43 | 0.65 | n.s. |
| (k) | Did the IA recorder seem to nod when you spoke? | 5.14 | 1.66 | 5.36 | 0.63 | n.s. |
| (l) | Did the IA recorder seem to nod when the interviewee spoke? | 6.29 | 1.07 | 5.71 | 0.83 | n.s. |
| (m) | Was the job interview made easier by the IA recorder? | 4.64 | 1.22 | 3.71 | 0.83 | $p < 0.05$ |
| (n) | Could you understand the interviewee well? | 5.29 | 1.14 | 5.00 | 1.24 | n.s. |
| (o) | Was the performance of the IA recorder good? | 6.29 | 0.61 | 5.54 | 0.97 | $p < 0.05$ |
| (p) | Did the interviewee seem to establish eye contact with the IA recorder? | 6.36 | 1.08 | 2.93 | 1.27 | $p < 0.01$ |

interviewee positively seemed to work. Thus, we regarded the subjects as feeling a positive relationship with the interviewee.

*Impression toward the relationship between the interviewee and the android.* For simplicity, we also assumed that the subjects were controlled to regard the relationship between the confederate and the android as positive or negative, depending on the condition. We compared the average scores for question (c), that is, "Did you feel that the interviewee had a good impression toward the IA recorder?," between the different conditions, and found a significant difference ($t(26) = 6.09 \times 10^{-5}$, $p < 0.01$). The score in condition 1 is significantly higher than that in condition 2; therefore, we can conclude that the subjects' feelings about the relationship between the interviewee and the android was successfully affected by having both the android and the confederate establish or not establish eye contact.

*Evaluation of impression toward the android.* We compared the average scores for question (a), that is, "Did you have a good impression toward the IA recorder?" A two-tailed unpaired t-test revealed a significant difference between conditions ($t(26) = 0.0008$, $p < 0.01$), while the assumption of equivalent variances was not violated ($F(13, 13) = 2.58$, $p = 0.09$). The score in condition 1 is significantly higher than that in condition 2; therefore, it seems that subjects who observe eye contact between the interviewee and the android form more positive impressions toward the android than subjects who do not observe eye contact.

## 13.3   The Effect of a Person with Negative Prepossession

In the previous section, we found that the impression toward the third person is influenced by the behavior of the conversational target when the relationship between the conversational target and oneself is positive. In this section, we investigate the case when the relationship between the conversational target and the subject is controlled to be negative by letting the second person exhibit non-communicative gaze behavior toward the subject.

### 13.3.1   Subjects

We hired 30 Japanese adults (ages: mean = 21.2, SD = 2.01 [y]) from a temporary employment agency. Fifteen subjects were assigned to condition 1 (eight males and seven females), and the remaining fifteen subjects were assigned to condition 2 (eight males and seven females). Written informed consent was obtained from all subjects.

### 13.3.2   Apparatus

Apparatus including the android (Repliee-Q2), as well as the android's behavior, the positions of the cameras and chairs, and the size of the experimental room were the same as in experiment 1.

### 13.3.3  Procedure

#### 13.3.3.1  Task and Instruction

There was no change from experiment 1 regarding the task that the subjects conducted; only the initial instructions were different.

The subjects scored the answers given by the interviewee as described in experiment 1. However, it was necessary to induce a negative bias in the relationship between the interviewer and interviewee. Therefore, the instructor told the subject to assign low scores to the answers given by the interviewee in order to negatively bias the attitude of the subject toward the interviewee. Here, we asked the subject to give 1 or 2 points on the seven-point scale for each answer.

### 13.3.4  Stimulus

The stimulus and the android's behavior were also the same as in experiment 1. Furthermore, the behavior of the confederate again varied according to the condition. In condition 3, it was intended that the subject would recognize a positive relationship between the confederate and the android; therefore, the confederate looked at the android to establish eye contact, as in experiment 1. In condition 4, the subject was meant to perceive a negative relationship between the confederate and the android; therefore, the confederate did not look at the android, as in condition 2.

### 13.3.5  Measurement

The same questionnaires were administrated, and video footage was recorded as in experiment 1.

### 13.3.6  Results

#### 13.3.6.1  Observation

Three coders (one of the authors and two volunteers) analyzed the behavioral data captured by the video cameras using the same coding rules as in experiment 1. We calculated the average Cohen's kappa value from eight arbitrary sets of data encoded by two of the observers to investigate the reliability of the observation, as described for experiment 1. As a result, we confirmed that the analysis was reliable ($\kappa = 0.98$). Therefore, in the following tests, we used the data measured by the author.

We conducted two-tailed unpaired t tests and found there was no significant difference between condition 3 and condition 4 regarding the time for which the subjects looked at the android (mean = 11.5 [frame], SD = 25.1 in condition 3, mean = 3.14 [frame], SD = 8.33 in condition 4, t(27) = 1.19, n.s.), the time and the number of times that the subject and the android established eye contact (time: mean = 1.93, SD = 5.13 in condition 3, mean = 1.14, SD = 4.28 in condition 4, t(27) = 0.45, n.s.; number of times: mean = 0.13, SD = 0.35 in condition 3, mean = 0.07, SD = 0.27 in condition 4, t(27) = 0.53, n.s.), or the number of times that the subjects nodded while the android was nodding (mean = 1.13, SD = 1.46 in condition 3, mean = 0.43, SD = 0.65 in condition 4, t(27) = 1.66, n.s.).

There is no significant difference between the conditions with regard to the subject's behavior and the synchronized behavior that occurred between the subjects and the android.

### 13.3.6.2 Checking Conditions for the Experiment

Before analyzing the effect of the interviewee's gaze movement on the subjects' feelings, we should confirm whether the experiments were actually administered as intended: eye contact between the confederate and the android should be observed in order for a positive relationship to be perceived in condition 3, whereas it should not be observed in condition 4. Moreover, we have to confirm that the relationship between the subject and the confederate was negative.

*Perception of eye contact.* As in experiment 1, we wish to analyze the effect of observing synchronized nonverbal behavior between other agents on the observer's feeling about either agent. Therefore, we removed the data in cases where the confederate was not observed to engage in eye contact with the android in condition 3, as well as cases where the confederate was perceived as having made eye contact with the android in condition 4.

The results are presented in Table 13.2, which lists the questions in the post-experiment questionnaire, the average score and standard deviation in each condition, and a p value for the difference between these conditions. For eye contact, we calculated the average score for question (p). We removed the data if the score for this question was too far from the average score of each condition, i.e., greater than $M + 2 \times SD$ or less than $M - 2 \times SD$. As a result, one case in condition 3 and two cases in condition 4 were judged as outliers. We conducted a two-tailed unpaired t test and found a significant difference between conditions ($t(24) = 9.23$, $p < 10^{-8}$), while the assumption of equivalent variances was not violated ($F(13, 11) = 2.56$, $p = 0.13$). The score in condition 3 is significantly higher than that in condition 4; therefore, we regard the gaze movements of the confederate (i.e., interviewee) as sufficiently distinguishable for subjects whose data were analyzed.

*Impression toward the interviewee.* We assumed that subjects were led to regard their relationship with the confederate as negative. Therefore, the scores on question (e) were analyzed. Both average scores (3.73 for condition 3 and 3.86 for

Table 13.2  List of questions in the post-interaction questionnaire (translated from Japanese) and the statistics of their scores in condition 3 and condition 4. Main questions are (a), (c), (e), and (p): (a) asking about the impression toward the android, (c) asking about the impression toward the relationship between the interviewee and the android, (e) asking about the impression toward the interviewee, and (p) asking about the perception of eye contact

|     |                                                                                              | Condition 3 |      | Condition 4 |      | p value |
|-----|----------------------------------------------------------------------------------------------|------|------|------|------|---------|
|     |                                                                                              | Mean | SD   | Mean | SD   |         |
| (a) | Did you have a positive impression toward the IA recorder?                                    | 5.27 | 0.99 | 4.43 | 1.47 | $p < 0.05$ |
| (b) | Did the interviewee look at the IA recorder?                                                  | 5.67 | 1.73 | 2.79 | 1.30 | $p < 10^{-5}$ |
| (c) | Did the interviewee seem to have a positive impression toward the IA recorder?                | 4.27 | 1.20 | 3.21 | 0.83 | $p < 0.05$ |
| (d) | Did you feel that the IA recorder had a good impression toward the interviewee?               | 4.87 | 1.21 | 4.93 | 1.44 | n.s. |
| (e) | Did you have a good impression toward the interviewee?                                        | 3.73 | 1.64 | 3.86 | 1.34 | n.s. |
| (f) | Were the interviewee's answers suitable?                                                      | 4.20 | 1.65 | 4.79 | 1.34 | n.s. |
| (g) | Was the attitude of the interviewee good?                                                     | 4.20 | 1.70 | 4.07 | 1.60 | n.s. |
| (h) | Were you irritated by/satisfied with the IA recorder's behavior while the interviewee answered questions? | 4.53 | 1.63 | 4.86 | 1.90 | n.s. |
| (i) | Did the interviewee seem to be respectful toward the IA recorder?                             | 4.60 | 1.74 | 3.07 | 1.64 | $p < 0.05$ |
| (j) | Did the interviewee seem to be respectful toward you?                                         | 4.93 | 1.41 | 4.79 | 1.67 | n.s. |
| (k) | Did the IA recorder seem to nod when you spoke?                                               | 4.87 | 1.61 | 3.79 | 2.05 | $p < 0.10$ |
| (l) | Did the IA recorder seem to nod when the interviewee spoke?                                   | 5.73 | 0.92 | 5.64 | 1.24 | n.s. |
| (m) | Was the job interview made easier by the IA recorder?                                         | 4.53 | 1.29 | 2.79 | 1.62 | $p < 0.05$ |
| (n) | Could you understand the interviewee well?                                                    | 4.07 | 1.58 | 4.57 | 1.31 | n.s. |
| (o) | Was the performance of the IA recorder good?                                                  | 5.47 | 0.85 | 4.93 | 1.41 | n.s. |
| (p) | Did the interviewee seem to establish eye contact with the IA recorder?                       | 5.40 | 1.07 | 2.93 | 0.67 | $p < 10^{-8}$ |

condition 4) are less than 4, which corresponds to a neutral feeling. Note that a two-tailed unpaired t test showed no significant difference between conditions ($t(24) = 0.20$, $p > 0.10$), while the assumption of equivalent variances was not violated ($F(13, 11) = 1.50$, $p > 0.10$). Therefore, the instruction and the dummy task where subjects had to evaluate the interviewee as negative seemed to work, and thus we could regard subjects having a negative perception of the interviewee.

*Impression toward the relationship between the interviewee and the android.* We also assumed that subjects were led to regard the relationship between the confederate and the android as positive or negative depending on the condition. We compared the average scores for question (c) between condition 3 and condition 4. We found a significant difference ($t(24) = 2.71$, $p < 0.05$) between conditions, while the assumption of equivalent variances was not violated ($F(13, 11) = 2.08$, $p > 0.10$). The score in condition 3 is significantly higher than that in condition 4; therefore, we can conclude that the subjects' perception of the relationship between the confederate and the android had been successfully affected by having both the android and the confederate establish or not establish eye contact.

*Evaluation of impression toward the android.* We compared the average scores for question (a). A two-tailed unpaired t test revealed a significant difference between conditions ($t(24) = 2.306$, $p < 0.05$), while the assumption of equivalent variances was not violated ($F(11, 13) = 2.17$, $p > 0.10$). The score in condition 3 is significantly higher than that in condition 4; therefore, it appears that subjects who observed eye contact between the interviewee and the android formed a more positive impression of the android than those who did not observe eye contact.

## 13.4 Discussion

Based on the results relating to the impression toward the android (question (a)), regardless of whether subjects were directed to have a positive (experiment 1) or negative (experiment 2) impression toward the interviewee, any positive impression toward the android is enhanced when eye contact is observed between the interviewee and the android. This is considered to support the hypothesis that one's impression toward a robot can be influenced by whether another person appears to be engaging it in nonverbal communication.

It is not clear how we can lead people to establish eye contact with a robot. However, the experimental results indicate that, even if the robot can establish eye contact only with a limited number of people or looks like it is doing so, other people's positive impressions toward it will be strengthened. These changes lead people to look at the robot more frequently and, in turn, affect yet more people's impressions in the same way. As a result, positive impressions and looking behavior toward the robot are reinforced in such a way that the introduction of robots into human society could be accelerated and facilitated.

Moreover, we believe that the difference between experiments 1 and 2, that is, the difference in the subject's impression toward the interviewee, might appear as a difference in effect size from observing the eye contact between the interviewee and the android. The effect size of question (a) is 1.08 in experiment 1 and 0.67 in experiment 2. According to Cohen's criteria, the effect size in experiment 1 is large, whereas that in experiment 2 is mid-range. Therefore, the strength of the effect of nonverbal behavior between agents might be influenced by the impression toward either agent.

In addition, we found significant differences in the scores for question (m), "Was the job interview made easier by the IA recorder?," in both experiments 1 and 2. These scores seem to indicate the subjects' evaluation of the android's mediation abilities. Therefore, we might assume that observing the android's eye contact with another person led subjects to evaluate the android's mediation ability highly (i.e., its ability to contribute toward establishing an atmosphere conducive to communication). This resulted in positive feelings toward the android, as indicated in the score for question (a). However, we cannot ignore an alternative interpretation, whereby a generally positive impression toward the android induced a positive evaluation of its mediation ability.

Here, we examine whether the current results are consistent with Heider's balance theory. We interpret the relationship between the subject and the confederate as having been made to appear positive by the experimental instructions in conditions 1 and 2, but negative in conditions 3 and 4. The relationship between the confederate and the android is directed to be positive by controlling the use of eye contact in conditions 1 and 3, but negative in conditions 2 and 4. If Heider's balance theory applies in the case where the relationship between another person and a robot is controlled by a nonverbal channel, then the relations between the subject and the android would become positive in conditions 1 and 4, but negative in conditions 2 and 3. As the relationship between the subject and the android in condition 1 is better than that in condition 2, Heider's balance theory is supported by experiment 1. However, it is not supported by experiment 2, because the relationship between the subject and the android in condition 3 is better than that in condition 4. For these reasons, the results are only partially consistent with Heider's balance theory.

Taken as a whole, the current results do not appear to support Heider's theory. However, we are cautious in concluding this, as there are other possible interpretations:

(a) *The relationship between the subject and the confederate is insufficient in experiment 2.* The effect size on the score for question (a) in experiment 1 is smaller than in experiment 2. This might imply that the instruction to bias the subjects' impression toward the interviewee worked only to reduce the positive feeling toward him, instead of making it negative as intended. Therefore, if the positive impression toward the interviewee could be further reduced, the effect size might become small. Furthermore, if it reached a negative level, the inverse effect predicted by Heider's theory might appear. To investigate this possibility, one possible direction would be to examine this effect in some real situations where a robot mediates between persons in a "hostile" relationship, such as a debate scenario.

(b) *Heider's balance theory is satisfied in a limited layer of communication.* If we focus only on whether eye contact occurred between agents and regard this to determine a positive or negative relationship, the relationships between the subjects and the interviewee were all positive in the current experiments. The android was evaluated as having better mediation ability in the condition where it established eye contact with the interviewee. We could regard these scores as being reflective of the subjects' evaluation of the potential relationship at the level of their eye contact with the android, which might be imagined through its mediation ability. If this were true, the relationship between subjects and the android could be regarded as positive in conditions 1 and 4, but negative in conditions 2 and 3, which would appear to satisfy Heider's theory at the level of eye contact. To examine this hypothesis, it would be interesting to control only the occurrence of eye contact among three agents.

In the experiments, it was assumed that subjects could recognize the gaze of both the confederate and the robot. Therefore, we used an android whose humanlike appearance allowed subjects to easily recognize when eye contact had been established with the confederate. However, as we believe such a humanlike appearance is not a necessary requirement for the balancing effect, experiments using robots with mechanical appearances should utilize this effect in designing more acceptable robots. In addition, only one type of eye contact was considered in the current experiment, which was performed by training the confederate to look at the android for a certain period while speaking. However, different forms, timing, or durations of eye contact might enhance the impression of this effect, and should therefore be investigated. Furthermore, factors like culture and gender also exert an influence on one's gaze pattern. Therefore, subjects with different cultural backgrounds or of different genders would have different impressions of gaze and this effect. To make our hypothesis more general and useful, various such factors should be investigated.

## 13.5  Conclusion

Experiments involving subject–confederate–android interaction were performed. The gaze of the confederate was directed so as to allow the subject to observe or not observe eye contact between the android and the confederate; the findings support the hypothesis that one's impression toward a robot can be influenced by whether another person looks at the robot and engages in nonverbal communication.

In the experiments, the relationship between the subject and the conversational target is biased by the instructions and the subject's appraisal of the interviewee's answers. The results indicate that if the conversational target has a positive relationship with the subject and establishes eye contact with the third person (i.e., if the confederate establishes eye contact with the robot), the subject will accept the robot more readily. Alternatively, if a robot moves to a position where it can establish eye contact with someone (a conversational target), the impression of another person (the subject) toward the robot will be enhanced. Of course, in this situation, we have to consider how the robot's movement influences the impression it makes, and how the robot should behave in order to establish eye contact.

Although the current results are not completely consistent with Heider's theory, it seems as if the current evidence is insufficient to make this conclusion, as other interpretations are possible. To gain deeper insight into this issue, we need to conduct other experiments and investigate the effects of controlling not just nonverbal aspects, but also other factors of the triadic interaction.

**Acknowledgements** We developed the android, Repliee Q2, in collaboration with Kokoro Company, Ltd.

# References

1. Shimada, Michihiro, Yuichiro, Yoshikawa, Mana, Asada, Naoki, Saiwaki, and Hiroshi Ishiguro. 2010. Effects of observing eye contact between a robot and another person. *International Journal of Social Robotics* 143–154.
2. Kanda, Takayuki, Hiroshi Ishiguro, Michita Imai, and Tetsuo Ono. 2004. Development and evaluation of interactive humanoid robots. *Proceedings of IEEE* 92: 1839–1850.
3. Takano, Eri, Yoshio, Matsumoto, Yutaka, Nakamura, Hiroshi, Ishiguro, and Kazuomi Sugamoto. 2008. The psychological effects of attendance of an android on communication. In *11th International Symposium on Experimental Robotics*.
4. Robins, Ben, Paul Dickerson, Penny Stribling, and Kerstin Dautenhahn. 2004. Robot-mediated joint attention in children with autism. *Interaction Studies* 5: 161–198.
5. Melzoff, Andrew N. 1995. Understanding the intentions of others: Re-enactment of intended acts by 18-month-old children. *Developmental Psychology* 31: 838–850.
6. Goetz, Jennifer, Sara, Kiseler and Aaron, Powers. 2003. In *The 12th IEEE International Workshop on Robot and Human Interactive Communication*.
7. Shimada, Michihiro, Takashi, Minato, Shoji, Itakura, and Hiroshi, Ishiguro. 2006 *Proceedings of the 6th IEEE-RAS International Conference on Humanoid*.
8. Minato, T., M., Shimada, H., Ishiguro, S., Itakura. 2004. Development of an android robot for studying human-robot interaction. 17th International Conference on Industrial and Engineering Applications of Artificial Intelligence and Expert Systems, 424–434 (2004).
9. Bailenson, Jeremy N., and Nick Yee. 2005. Digital Chameleons: Automatic assimilation of nonverbal gestures in immersive virtual environments. *Psychological Science* 16: 814–819.
10. Yamaoka, Fumitaka, Takayuki, Kanda, Hiroshi, Ishiguro, and Norihiro, Hagita. 2006. In *Proceedings of ACM 1st Annual Conference on Human-Robot Interaction*, 313–320.
11. Yoshikawa, Yuichiro, Kazuhiko, Shinozawa, Hiroshi, Ishiguro, Norihiro, Hagita, and Takanori, Miyamoto. 2006. In *Proceedings of Robotics: Science and Systems II*.
12. Yoshikawa, Yuichiro, Kazuhiko, Shinozawa, and Hiroshi, Ishiguro. 2007. In *Proceedings of the 29th Annual Conference of the Cognitive Science Society*, 725–730.
13. Hayashi, K., D., Sakamoto, T., Kanda, M., Shiomi, S., Koizumi, H., Ishiguro T., Ogasawara, and N., Hagita. 2007. Humanoid robots as passive-social medium-a field experiment at train station. In *ACM/IEEE 2nd Annual Conference on Human-Robot Interaction*, 137–144.
14. Sakamoto, Daisuke, and Tetsuo Ono. 2006. Sociability of robots: Effect evaluation of robots on the impression formation between humans (in Japanese). *The Journal of Human Interface Society* 8: 61–70.
15. Heider, F. 1958. *The psychology of interpersonal relations*. Singapore: Wiley.
16. Bailenson, J.N., and N. Yee. 2005. Digital chameleons—automatic assimilation of nonverbal gestures in immersive virtual environments. *Psychological Science* 16: 814–819.

17. Yuichiro, Yoshikawa, Shunsuke, Yamamoto, Suminoka, Hidenobu, Ishiguro, Hiroshi, and Asada, Minoru. 2008. In *Proceedings of the ACM/IEEE International Conference on Human-Robot Interaction*, 319–326
18. McCarthy, A., K., Lee and D., Muir. 2001. Eye gaze displays that index knowing, thinking, and guessing. In *The Annual Conference of the American Psychological Society*.
19. Siegman, A.W. 1976. Do noncontingent interviewer mm-hmms facilitate interviewee productivity? *Journal of Consulting and Clinical Psychology* 44: 171–182.
20. Breed, G.R. 1972. The effect of intimacy : Reciprocity or rates. *British journal of social and Clinical Psychology* 11: 17–46.

# Chapter 14
# Can an Android Persuade You?

**Kohei Ogawa, Christoph Bartneck, Daisuke Sakamoto,
Takayuki Kanda, Tetsuo Ono and Hiroshi Ishiguro**

**Abstract** The first robotic copies of real humans have become available. They enable their users to be physically present in multiple locations simultaneously. This study investigates the influence that the embodiment of an agent has on its persuasiveness and its perceived personality. Is a robotic copy as persuasive as its human counterpart? Does it have the same personality? We performed an experiment in which the embodiment of the agent was the independent variable and the persuasiveness and perceived personality were the dependent measurements. The persuasive agent advertised a Bluetooth headset. The results show that an android is perceived as being as persuasive as a real human or a video recording of a real human. The personality of the participant had a considerable influence on the measurements. Participants who were more open to new experiences rated the persuasive agent lower on agreeableness and extroversion. They were also more willing to spend money on the advertised product.

**Keywords** HRI · Android · Persuasion

---

This chapter is a modified version of a previously published paper [1], edited to be comprehensive and fit with the context of this book.

---

K. Ogawa (✉) · D. Sakamoto · T. Kanda · H. Ishiguro
ATR Intelligent Robotics and Communication Laboratories, 2-2-2 Hikaridai,
Seika-cho Soraku-gun, Kyoto 619-0288, Japan
e-mail: ogawa@irl.sys.es.osaka-u.ac.jp

C. Bartneck
Department of Industrial Design, Eindhoven University of Technology, Den Dolech 2,
5600MB Eindhoven, The Netherlands

K. Ogawa · T. Ono
Department of Media Architecture, Future University-hakodate School of Systems
Information Science, 116-2 Kamedanakano-cho, Hokkaido, Hakodate 041-8655, Japan

D. Sakamoto
Graduate School of Information, Science and Technology, The University of Tokyo,
7-3-1 Hongo, Bunkyo-ku, Tokyo 113-0033, Japan

© Springer Nature Singapore Pte Ltd. 2018
H. Ishiguro and F. Dalla Libera (eds.), *Geminoid Studies*,
https://doi.org/10.1007/978-981-10-8702-8_14

## 14.1 Introduction

A great advantage of having a robotic copy of yourself is that it allows you to be physically present in two locations simultaneously. In particular, politicians may appreciate the ability to give two speeches at the same time during an election campaign. However, while the physical appearances of androids have become almost indistinguishable from their human originals (see Fig. 14.1), it is not clear to what degree androids are able to convey the same personality and persuasive power as their human originals. Moreover, androids need to show a significant advantage over screen characters to justify the extra costs. A simple video transmission is currently easier and cheaper than using a robotic copy, but some situations require a representation that is truly 3D. Human doppelgangers, for example, are frequently used to confuse paparazzi and terrorists. A robotic doppelganger could take its owner's place and ease some of the ethical difficulties associated with this dangerous business. It would be of considerably less consequence if a robotic doppelganger took a bullet than if a human doppelganger did. In this study, we did not want to focus on the pure appearance of a robotic doppelganger, but on the persuasive power and personality that androids may have. After all, it is desirable that your robotic copy possesses the same persuasiveness and personality as yourself. Persuasion can be defined as a social influence. It occurs when one person attempts to induce change in the beliefs, attitudes, or behavior of another person or group of people [2]. Previous studies showed that the success of persuasion depends on the

**Fig. 14.1** Geminoid HI-1 and Hiroshi Ishiguro

source of the message [3, 4], the strength of the argument [5], and the person being influenced [3, 6].

The persuasiveness of technology has become an important research field [7], and many robots are used in contexts where their main or primary purpose is to change the attitude, behavior, or opinions of humans [8]. The first studies on the persuasiveness of virtual characters and robots show promising results. Zanbaka et al. [2] compared the persuasiveness of virtual characters with that of real humans by communicating the benefits of comprehensive examinations to college students. They concluded that virtual characters are perceived as being as persuasive as real humans and that the realism of the character had no effect on its persuasive power. Shinozawa et al. used either a screen character or a robot to give recommendations to users. Their results showed that a robot's recommendation was more effective than that of a screen character [9]. Powers et al. compared people's responses to a screen agent and a robot in a health interview [10]. Their results showed only a few behavioral differences, but considerable differences in attitude. The participants spent more time with the co-located robots and had a more positive attitude. Kidd and Breazeal studied users' perceptions of a robot in the same room as compared with that of a robot shown on a screen [11]. They hypothesized that when the robot is physically present, it will be viewed as more persuasive than when it is telepresent. Their results showed that a robot is more engaging than an animated character and is perceived as more credible and informative, as well as providing a more enjoyable interaction.

However, it is not clear to what degree androids may compare to their human originals in terms of persuasiveness and personality, in particular the influence of the embodiment of a persuasive agent on its persuasiveness and perceived personality. It has been shown that when the personality of a computer voice matches the users' personality, (a) participants regarded the computer voice as more attractive, credible, and informative and (b) participants were more likely to buy a product from the computer [12]. It is therefore necessary to measure not only the perceived personality of the persuasive agent, but also the personality of the participants. In summary, we are interested in the following research questions:

1. What influence does the embodiment of an agent have on its persuasiveness and perceived personality?
2. To what degree does the personality of the users influence their perception of the persuasiveness and personality of a persuasive agent?

## 14.2    Method

We performed a between-participant experiment in which three conditions were applied to the persuasive agent. In the human condition, Hiroshi Ishiguro presented a persuasive message, in the video condition a recording of Ishiguro's persuasive message was presented, and in the android condition Geminoid HI-1 persuaded the audience. The appearance of all three persuasive agents was very similar, which allowed us to focus on the embodiment of the agent, instead of its visual attractiveness.

## 14.2.1 Measurements

The participants' perceptions of the persuasive argument and message were assessed through a semantic differential questionnaire developed by Zanbaka, Goolkasian, and Hodges [2], which is based on the previous work of Mullennix et al. [13].

Items related to the perception of the argument and the perception of the message were measured on a Likert-type scale. The items for each were as follows: perception of the argument (bad–good; foolish–wise; negative–positive; beneficial–harmful; effective–ineffective; convincing–unconvincing); perception of the message (stimulating–boring, vague–specific, unsupported–supported, complex–simple, convincing–unconvincing, uninteresting–interesting). Zanbaka et al. performed a principle components analysis of both perceptions. The results of the factor analysis of the items related to the argument showed only one factor with a high reliability (Cronbach's alpha = 0.90). The factor analysis of items related to the message resulted in two factors. The interesting factor (stimulating, specific, supported, convincing, and interesting) accounted for 39% of the variance, and the conservative factor explained 19% of the variance. The Cronbach's alpha for the interesting factor was 0.76. We translated all the items into Japanese using the back-translation method.

In addition, we evaluated the persuasiveness of the speaker by asking the participants before and after the persuasive speech how much they would be willing to pay for the product. This repeated measure allowed us to compensate for individual differences. A certain participant, for example, might simply not like a given product. We calculated the variable price by subtracting the participant's evaluation before seeing the agent from that after seeing the agent. Hereafter, we refer to this collection of questionnaires as the "persuasion questionnaire."

Several models and measurement tools have been proposed for evaluating personality, including the acknowledged Big Five Model [14], a brief version of the Big Five Model [15], Mowen's Personality Scale [16], and the established Myers–Briggs Type Indicator [17]. Many of these instruments consist of more than 100 items, and their completion can require up to one hour. Since we intended to use several measurement instruments, it seemed unreasonable to dedicate that much attention to only one tool. We therefore used the NEO Five-Factor Inventory (NEO-FFI) that contains only 60 items, which is designed to take only 15 min to fill and is available in the Japanese language. This questionnaire is a short version of the NEO PI-R instrument of the same author [18]. Despite its brevity, the validity and reliability of this tool have been demonstrated. Ishiguro and the participants filled this questionnaire about themselves.

The five factors in this personality questionnaire are neuroticism, extraversion, openness, agreeableness, and conscientiousness, each of which is measured on a 0–48 scale. A person with a high neuroticism score can be described as "sensitive, emotional, and prone to experience feelings that are upsetting." A person with a low neuroticism score is "secure, hardy, and generally relaxed even under stressful conditions." A high extrovert score describes a person as "extrovert, outgoing, active, and high-spirited, and preferring to be around people most of the time," while a low score refers to a person who is "introverted, reserved, and serious and prefers to be alone or with a few close friends." "Open to new experiences and having broad interests and very imaginative" describes a person with a high openness score, and "down-to-earth, practical, traditional, and pretty much set in his/her ways" describes a person with a low openness score. A high agreeable score refers to a person who is "compassionate, good-natured, and eager to cooperate and avoid conflict," and a low score to a person who is "hardheaded, skeptical, proud, and competitive and tends to express anger directly." People with a high conscientiousness score are described as "conscientious and well-organized. They have high standards and always strive to achieve their goals," while people with a low score can be described as "easygoing, not very well-organized, and sometimes careless. They prefer not to make plans."

Unfortunately, the NEO-FFI version for rating another person has not yet been translated into Japanese. We therefore used the Japanese Property-Based Adjective Measurement questionnaire [19]. Its three components are highly correlated with the extraversion, openness, and agreeableness components of the NEO-FFI (Hayashi 1978).

The Geminoid HI-1 android has received a considerable amount of media attention, and hence, it is possible that the participants had previously seen or interacted with it. We therefore asked the participants whether they had previously seen (e.g., on television) the android or Ishiguro (seen-agent), whether they had met them (met-agent), or whether they knew them personally (know-agent). This allowed us to take a possible bias into account in the statistical analysis.

In summary, we measured the persuasiveness of the presentation by its components argument, interesting and conservative. In addition, we calculated the change in the price estimation of the headset by subtracting the value before the product presentation from the value after the presentation (price). We measured the personality of the participants and of Hiroshi Ishiguro using the NEO-FFI questionnaire. We measured the perceived personality of the persuasive agent using the Japanese Property-Based Adjective Measurement questionnaire. Finally, we measured the participants pre-knowledge of the android and Ishiguro.

## 14.2.2 Setup

We used the Geminoid HI-1 android for this experiment, since it allowed a direct comparison with its human equivalent, Hiroshi Ishiguro. The android's movement

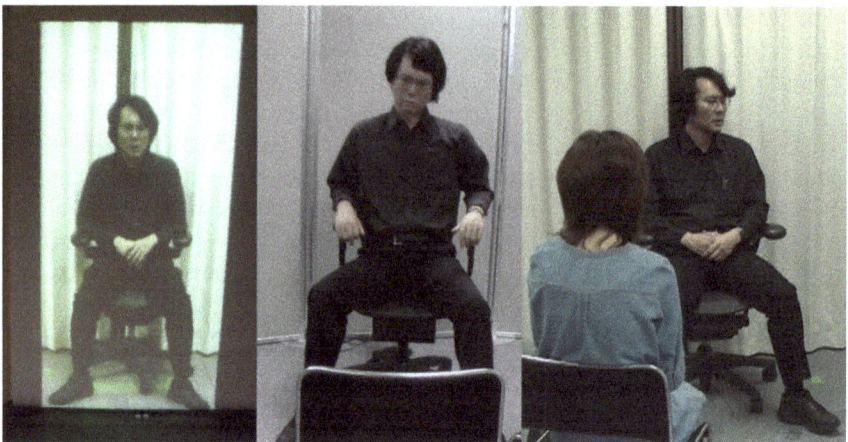

**Fig. 14.2** Three experimental conditions: video (left), android (middle), and human (right)

was based on motion data captured from Ishiguro performing the persuasive speech. The recording also included Ishiguro's voice, so that the lip movement of the android matched the speech signal.

One limitation of the android is that it cannot grip and hold products or press small buttons reliably. We therefore decided to advertise a Bluetooth headset, since it can be demonstrated without the android being required to handle it. Moreover, it may be assumed that a robot may be perceived as being more knowledgeable about electronic products than, for example, food products. The expertise of a speaker does have a considerable influence on his/her persuasiveness, which also holds true for the persuasiveness of machines [20, 21]. The headset was placed over the ears of the android and Ishiguro during the presentation.

A recording of Ishiguro performing the persuasive message was used in the video condition. For the recording, we placed a large television behind the camera that displayed the script of the persuasive message so that Ishiguro could more easily remember it. The same screen was placed behind the participants in the human condition. This procedure allowed Ishiguro to minimize the variations between his presentations. The video was projected onto a 110 by 175 cm screen, which approximates the actual size of Ishiguro and the android. The resolution of the video was 720 by 480 pixels. Figure 14.2 shows the experimental setup for the three conditions.

The advertised headset did not contain any label or brand icon, so that the participants were not able to identify the headset. It was therefore impossible for the participants to simply know the price of the product.

## *14.2.3 Procedure*

The participants took part in the experiment in small groups. After welcoming the participants in room A, the experimenter asked them to fill and sign an informed consent form. Next, the experimenter asked the participants to fill a questionnaire that contained demographic questions and the NEO-FFI personality questionnaire. The participants were then asked how much they would pay for 30 products that were presented to them in a custom-made catalog. The products included furniture, electronic devices, and accessories (see Fig. 14.3).

The experimenter then guided the participants into room B, where the persuasive agent (android, human, or the television screen) was located. The participants were seated on chairs that were arranged in a circle, 1 m away from the persuasive agent (see Fig. 14.2).

The experimenter left the room, and the persuasive agent presented a Bluetooth headset. After the presentation was completed, the experimenter guided the participants back into room A where they filled a questionnaire that contained the question asking how much the participant would pay for the Bluetooth headset that had just been presented, the persuasion questionnaire, and the Japanese Property-Based Adjective Measurement questionnaire. In parallel to the experiment, we asked Ishiguro to fill the personality questionnaire.

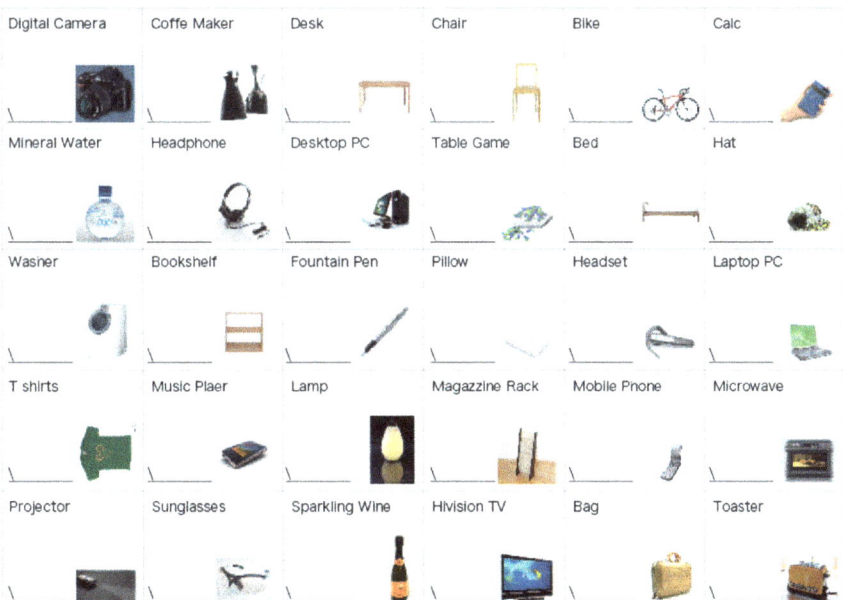

**Fig. 14.3** Product catalog

## *14.2.4 Participants*

Twenty male and 12 female subjects participated in the study. They were between 19 and 25 years old (mean 21.1), and they received 3000 yen for their effort. The participants were recruited from a temporary work placement company called Arbeit Network, which specializes in the work placement of students. All of the participants were students from a wide range of fields, including history, information science, and psychology. Fifty-six percent of the participants had never seen Ishiguro or the Geminoid HI-1 android (e.g., on television), 78% had never met them, and 91% did not know them personally.

## 14.3 Results

A reliability analysis across the six arguments items resulted in a Cronbach's alpha of 0.684, which is below the value of 0.90 reported in Zanbaka, Goolkasian, and Hodges' original paper [2]. The reliability of the interesting factor was 0.861, which is above Zanbaka's value of 0.76. The Cronbach's alpha for the three components of the Property-Based Adjective Measurement was 0.57 for openness, 0.86 for agreeableness, and 0.716 for extraversion. The reliability and validity estimations for the NEO-FFI are available in McCrae and Costa's paper [22].

We performed analysis of covariance (ANCOVA) in which the persuasive agent (human, video, android) was the independent variable and seen-agent and gender were the covariants. Price, argument, interesting, and conservative were the dependent variables. Neither covariant had a significant influence on the measurements. The persuasive agent also did not have a significant influence on the measurements (see Table 14.1).

We performed a second ANCOVA in which the persuasive agent was the independent variable and the seen-agent and the personality of the participant were the covariants. The perceived extraversion, openness, and agreeableness of the persuasive agent were the dependent variables. It should be noted that the Japanese Property-Based Adjective Measurement questionnaire does not have scales for the measurement of neuroticism or conscientiousness and therefore they do not appear in the further analysis. Figure 14.4 shows the mean personality scores for all three conditions.

**Table 14.1** F and P values of the ANCOVA on price, argument, conservative, and interesting

|              | $F_{(2, 27)}$ | P     |
|--------------|---------------|-------|
| Price        | 0.259         | 0.774 |
| Argument     | 0.266         | 0.768 |
| Conservative | 0.040         | 0.961 |
| Interesting  | 0.179         | 0.837 |

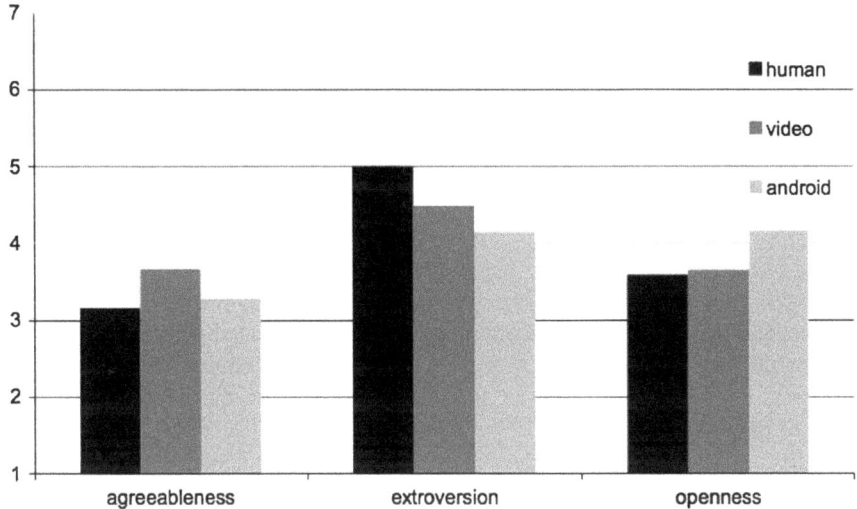

**Fig. 14.4** Mean personality scores for the persuasive agents

The persuasive agent did not have a significant influence on any of the measurements, although a significant level for openness was approached (F(2, 22) = 2.567, p = 0.100). Post hoc t-tests with Bonferroni-corrected alpha showed that the human agent was rated almost significantly (p = 0.153) less open (3.590) than the android agent (4.169). The covariant seen-agent and gender had no significant influence on the measurements. The personality of the participant also significantly influenced the measurements. The openness of the participants had a significant influence on their rating on the extraversion of the agent (F(1, 22) = 8.700, p = 0.07).

We performed a linear regression analysis to explore the relationship between the openness of the participants and their personality ratings for the agent. The openness of the participants was significantly correlated with the ratings for the agent on neuroticism, extraversion, and openness (see Table 14.2). However, the personality ratings for the agent accounted for only 23.2% of the variance in the openness of the participant. Scatter plots revealed that the agreeableness and extraversion ratings for the agent decreased with the rising openness of the participant.

**Table 14.2** Pearson correlation between the openness of the participant and the personality ratings for the agent (Italics indicate significant correlations at P < 0.005)

|  | Openness participant | Agreeableness agent | Openness agent |
|---|---|---|---|
| Agreeableness agent | −0.381 |  |  |
| Extraversion agent | −0.389 | 0.323 |  |
| Openness agent | −0.116 | 0.202 | *−0.089* |

**Table 14.3** Mean scores for Ishiguro's self evaluation, mean score for agent, T and P values for extraversion, openness, and agreeableness

|  | Mean self score | Mean agent score | T | P |
|---|---|---|---|---|
| Extraversion | 3.64 | 4.523 | 4.877 | 0.001 |
| Openness | 6.41 | 3.828 | −26.194 | 0.001 |
| Agreeableness | 3.50 | 3.367 | −0.945 | 0.352 |

Next, we performed a linear regression analysis between the participants' openness and the factors in the persuasive questionnaire (argument, interest, conservative, and price). Only the price was significantly correlated with the openness of the participant (r = 3.57, p = 0.022). A box plot revealed that the more open participants are to new experiences, the more they increase the amount they are willing to pay for the headset.

Finally, we were interested in the degree to which the participants' evaluation of the agent's personality matches the score that Ishiguro gave himself. We divided the scores from the 48-point scale of NEO-FFI questionnaire by 48/7 = 6.85 to allow us to compare the scores with those of the 7-point scale of the Property-Based Adjective Measurement. Table 14.3 shows the mean scores of Ishiguro and the participants. We then performed three one-sample t-tests against the corresponding value from Ishiguro's questionnaire. The ratings for extraversion and openness were significantly different, and the mean scores from the Japanese Property-Based Adjective Measurement questionnaire hovered closely around the center of the scale.

## 14.4 Discussion and Conclusions

The focus of this study was on the effects of the embodiment of the persuasive agent. Embodiment refers to the visual and haptic representation of the agent and not to the agents' voice. The same human voice was used in all conditions. Other studies explicitly focused on the influence of the agents' voice [13, 23].

Zanbaka et al. [2] had previously shown that college students found a virtual character as persuasive as a real human being. Their results were in line with those of other studies that showed that virtual characters are often treated similarly to real humans [24, 25]. We extend their results by concluding that a robotic copy of a real human is perceived as being as persuasive as its human original. Androids can therefore be considered an alternative for presenting persuasive messages.

We also observed that the embodiment of the agent may influence its perceived openness. The android was perceived as more open than its human or robotic counterpart. This may give the android a slight advantage over the video agent and justify the extra expense. We hope that if the number of participants was increased, the effect would become significant.

Despite the considerable media attention that Ishiguro and his android have received, it did not seem to have influenced the participants. Seeing a report on television may still be a different experience from standing in front of the "real McCoy." Overall, the results suggest that the openness of the participants may play an important role in the participants' perception of the personality of the agent. The openness rating was negatively correlated with the agreeableness and extroversion ratings for the agent. Participants who were open also showed an increased willingness to spend money on the advertised headset.

However, the personality ratings that the agent received do not completely match the rating that Ishiguro gave himself. The short interaction time with the agent may not have been sufficient for the participants to understand the agent better. Ishiguro's great openness to new experiences may not be communicated in the context of an advertisement. To gradually get to know people and androids remains a pleasant necessity.

In addition to the doppelganger scenario described in the introduction, we can also envision another application domain for persuasive androids: advertisements. The androids could be used as sales agents in supermarkets and many other stores. Today, audio and video messages are already being used to persuade customers to purchase certain goods, and the first studies on the effectiveness of virtual agents are becoming available [26].

### 14.4.1   Limitations and Future Work

The results of our study are limited to the android used in this study and may not be generalizable to other robots. Further research is necessary to determine in more detail the aspects of the embodiment that contribute to the persuasiveness and personality of an android. We were also limited by the physical limitation of the Geminoid HI-1 android. It cannot move as smoothly as humans and is not yet able to grasp objects. Future androids may have much better abilities and therefore become even more persuasive.

Another drawback of this study was the limited number and diversity of the participants. To achieve more generalizable results, this study should be extended with a more diverse sample, in particular with more participants who are not university students. All the participants were Japanese, and it has been shown that the cultural background of the users influences their perception of a robot [27, 28]. It would therefore be interesting to repeat this experiment with users from other cultures.

**Acknowledgements**   This work was partially supported by Grant-in-Aid for Scientific Research (S), KAKENHI (20220002).

# References

1. Ogawa, K., C. Bartneck, D. Sakamoto, T. Kanda, T. Ono, and H. Ishiguro. 2009. Can an android persuade you? In *18th IEEE international symposium on robot and human interactive communication (RO-MAN 2009)*, 516–521.
2. Zanbaka, C., P. Goolkasian, and L. Hodges. 2006. Can a virtual cat persuade you? The role of gender and realism in speaker persuasiveness. In *SIGCHI conference on human factors in computing systems*, Montreal, Quebec, Canada.
3. Carli, L.L. 1990. Gender, language, and influence. *Journal of Personality and Social Psychology* 59: 941–951.
4. Priester, J.R., and R.E. Petty. 1995. Source attributions and persuasion: Perceived honesty as a determinant of message scrutiny. *Personality and Social Psychology Bulletin* 21: 637–654.
5. Petty, R.E., and J.T. Cacioppo. 1981. *Attitudes and persuasion—classic and contemporary approaches*. Dubuque, Iowa: W.C. Brown Co. Publishers.
6. Haugtvedt, C.P., and R.E. Petty. 1992. Personality and persuasion: Need for cognition moderates the persistence and resistance of attitude changes. *Journal of Personality and Social Psychology* 63: 308–319.
7. Fogg, B.J. 2003. *Persuasivetechnology: Using computers to change what we think and do*. Amsterdam, Boston: Morgan Kaufmann Publishers.
8. Dautenhahn, K. 2003. Roles and functions of robots in human society: Implications from research in autism therapy. *Robotica* 21: 443–452.
9. Shinozawa, K., F. Naya, J. Yamato, and K. Kogure. 2005. Differences in effect of robot and screen agent recommendations on human decision-making. *International Journal of Human-Computer Studies* 62: 267–279.
10. Powers, A., S. Kiesler, S. Fussell, and C. Torrey. 2007. Comparing a computer agent with a humanoid robot. In *Proceedings of the ACM/IEEE international conference on human-robot interaction*, 145–152, Arlington, Virginia, USA.
11. Kidd, C.D., and C. Breazeal. 2004. Effect of a robot on user perceptions. In *Proceedings 2004 IEEE/RSJ international conference on intelligent robots and systems (IROS 2004)*, vol. 4, 3559–3564.
12. Nass, C., and K.M. Lee. 2000. Does computer-generated speech manifest personality? An experimental test of similarity-attraction. In *SIGCHI conference on Human factors in computing systems*, 329–336, The Hague, The Netherlands.
13. Mullennix, J.W., S.E. Stern, S.J. Wilson, and C.l. Dyson. 2003. Social perception of male and female computer synthesized speech. *Computers in Human Behavior* 19: 407–424.
14. Goldberg, L.R. 1992. The development of markers for the Big-Five factor structure. *Psychological Assessment* 4: 26–42.
15. Saucier, G. 1994. Mini-markers: A brief version of Goldberg's unipolar Big-Five markers. *Journal of Personality Assessment* 63: 506–516.
16. Mowen, J.C. 2000. *The 3 m model of motivation and personality: Theory and empirical applications to consumer behavior*. Boston: Kluwer Academic Publisher.
17. Jung, C.G. 1923. *Psychological types: Or the psychology of individuation*, ed. C.G. Jung, trans. H. Godwin Baynes. London: Routledge and K. Paul.
18. Costa, P.T., and R.R. McCrae. 1992. Normal personality assessment in clinical practice: The NEO personality inventory. *Psychological Assessment* 4: 5–13.
19. Hayashi, F. 1978. The fundamental dimensions of interpersonal cognitive structure. *Nagoya University, School of Education, Department of Education Psychology Bulletin* 25: 233–247.
20. Fogg, B.J., and T. Hsiang. 1999. The elements of computer credibility. In *SIGCHI conference on Human factors in computing systems*, 80–87, Pittsburgh.
21. Bartneck, C. 2001. How convincing is Mr. Data's smile: Affective expressions of machines. *User Modeling and User-Adapted Interaction* 11: 279–295.
22. McCrae, R.R., and P.T. Costa. 1987. Validation of a five-factor model or personality across instruments and observers. *Journal of Personality and Social Psychology* 52: 81–90.

23. Stern, S.E., J.W. Mullennix, C. L. Dyson, and S.J. Wilson. 1999. The persuasiveness of synthetic speech versus human speech. *Human Factors: The Journal of the Human Factors and Ergonomics Society* 41: 588–595.

24. Slater, M., and M. Usoh. 1994. Body centered interaction in immersive virtual environments. In *Artificial life and virtual reality*, ed. N. Magnenat-Thalmann, and D. Thalmann, 125–148. New York: Wiley.

25. Pertaub, D.-P., M. Slater, and C. Barker. 2002. An experiment on public speaking anxiety in response to three different types of virtual audience. *Presence: Teleoperators & Virtual Environments* 11: 68–78.

26. Suzuki, S.V., and S. Yamada. 2004. Persuasion through overheard communication by life-like agents. In *IEEE/WIC/ACM international conference on intelligent agent technology (IAT 2004)*, 225–231, Beijing.

27. Bartneck, C., T. Suzuki, T. Kanda, and T. Nomura. 2007. The influence of people's culture and prior experiences with Aibo on their attitude towards robots. *AI & Society—The Journal of Human-Centred Systems* 21: 217–230.

28. Bartneck, C. 2008. Who like androids more: Japanese or US Americans? In *17th IEEE international symposium on robot and human interactive communication, RO-MAN 2008*, 553–557, München.

# Chapter 15
# Attitude Change Induced by Different Appearances of Interaction Agents

**Shuichi Nishio and Hiroshi Ishiguro**

**Abstract** Human–robot interaction studies have thus far been limited to simple tasks, such as route guidance or playing simple games. However, with the advance in robotic technologies, the stage has been reached where the requirements for highly complicated tasks, such as conducting humanlike conversations, should be explored. When robots start to play advanced roles in our lives, such as in health care, attributes such as the trust and reliance of the person and the persuasiveness of the robot also become important. In this study, we examined the effect of the appearance of robots on people's attitudes toward them. Past studies have shown that the appearance of robots is one of the elements that influence people's behavior. However, the effect of the robots appearance on a person when conducting serious conversations that require high-level activity remains unknown. Participants were asked to hold a discussion with teleoperated robots having various appearances, such as an android with a high similarity to a human and a humanoid robot having humanlike body parts. Through the discussion, the teleoperator attempted to persuade the participants. We examined the effect of the robots appearance on their persuasiveness, as well as on people's behavior and their impression of the robots. A possible contribution to machine consciousness research is also discussed.

**Keywords** Appearance · Interaction · Humanoid robot · Android robot
Teleoperation

---

This chapter is a modified version of a previously published paper [1], edited to be comprehensive and fit with the context of this book.

---

S. Nishio (✉) · H. Ishiguro
Advanced Telecommunications Research Institute International,
Keihanna Science City, Kyoto, Japan
e-mail: nishio@botransfer.org

H. Ishiguro
Department of Systems Innovation, Graduate School of Engineering Science,
Osaka University, Osaka, Japan

© Springer Nature Singapore Pte Ltd. 2018
H. Ishiguro and F. Dalla Libera (eds.), *Geminoid Studies*,
https://doi.org/10.1007/978-981-10-8702-8_15

249

## 15.1   Introduction

Recently, researchers have been pursuing the key elements for helping people feel more comfortable with computers and creating an easier and more intuitive interface for various information devices. This pursuit has also begun spreading in the field of robotics, and human–robot interaction (HRI) is now one of the most actively studied fields. A number of studies have investigated peoples responses to robot behaviors and the manner in which robots should behave in order to improve interactions [2].

However, most studies on HRI were limited to simple tasks, such as guidance or playing simple games. This is mainly because artificial intelligence is still imma-ture. Despite active studies, using current technology it is impossible to build a robot that behaves and talks like a human being. This issue prevents researchers from using robots for effective examinations of the characteristics of human nature that can only be seen through intelligent interactions or conversations. One such task is persuasion. Persuasion is an attempt to induce a change in the beliefs, attitudes, or behavior of another person or a group of people. Persuasion can be considered one of the most humanlike activities and has long been studied in the field of social psychology [3]. Recently, with the spread of information systems and agent technologies in peoples everyday life, the persuasiveness of various forms of agents is becoming one of the research areas that receives the most attention [4]. Thus far, several studies have examined the manner in which virtual agents or robots can persuade people. In the study in [5], the effect of a virtual screen agent and a real robot making recommen-dations to people was compared, and the results showed that the robot displayed a stronger influence. Other studies also showed that people were considerably more persuaded or influenced by robots in the real world than by agents in the virtual world [6, 7]. On the other hand, in [8] the authors showed that virtual agents can be as persuasive as real humans. They also showed that the realism of the agent had no effect on its degree of persuasiveness.

In these studies, teleoperation, often referred to as the *Wizard-of-Oz* method, was used to control various agents. Teleoperation is one possible solution for overcoming the limitation of artificial intelligence, which is necessary for robots to acquire prac-tical roles in the real world. In particular, semi-teleoperation systems will be useful in simple situations where robots behave autonomously, and manual teleoperation will be performed only when the interaction enters complex phases that machine logic cannot handle. With the development of efficient teleoperation technology, robots will be able to conduct interactions with people that are similar to those between people.

Moreover, the studies on persuasion lead to the question of the manner in which people are affected by the difference in the appearance of various agents or robots. In human–human interaction, appearance or attractiveness is known to influence peo-ple's behaviors [9–11]. In addition, the results of recent neuroscience studies showed evidence that there are primitive differences in our perception of features having different appearances [12–14]. However, thus far, not many studies have investigated the manner in which the appearance of robots affects people's behaviors or impressions. In the study presented in [15], the authors compared the interaction of children with autism with robotic dolls with different facial masks. In the study in [16], the influence of variations in robotic head designs on people's acceptance of and cooperation with robots was examined. In the study report in [17], the authors compared peoples perceptions of a mechanical robot with and without gloves and a head cover and the results showed that the people perceived the robot's personality differently. These studies concerned changes only in small parts of the robotic body, and the interaction between people and robots was small. Another study using a different approach was reported in [18], where the attitudes of children were compared while they conversed with a teleoperated robot and with a person. In this study, the robot was given the same features as the person in the comparison. That is, the appearance of the robot and the person was almost the same. The results showed that the children's attitude toward the robot gradually became closer to their attitude toward the person in the comparison. This study, however, was a case study involving only two children, one of whom was the daughter of the comparison person.

In this study, we investigated the effect of changing the appearance of an agent by comparing peoples interaction with teleoperated robots and a human. Participants held discussions with various agents, including a human, all of which in fact were the same person in different guises, that tried to persuade the participants. Our main interest here was not only in the effectiveness of the teleoperated robots persuasion, but also in how participants' impressions were affected by the different appearances. In the following sections, we first describe our experimental procedure and the results, and then discuss the findings and refer to the possible contribution of the results of the study to consciousness research.

## 15.2  Method

### 15.2.1  Participants

Fourteen Japanese participants were recruited from nearby universities (eight males and six females). Each participant in the experiment was paid 3,000 yen. The experimenter first gave the participants a briefing on the aim and procedures of the experiment. Following the briefing, the participants signed an informed consent form and filled the pre-experiment questionnaire.

## 15.2.2  Procedure

The experiment consisted of five sessions in total: one initial habituation session and four subsequent discussion sessions, each with a different conversation agent.

First, each participant joined a habituation session. In this session, participants conversed with the *Human* agent for three minutes. The *Human* agent was instructed to ask participants about their hobbies and daily life, in a friendly manner. This habituation session was aimed at accustoming the participants to the experiment room and familiarizing them with the person who performed as the *Human* agent. This person also filled the teleoperator role for the other agents. By conversing with the person behind the agents at the beginning, participants were expected to form an initial impression of the teleoperator. This was important, as our aim was to observe the changes in the participants attitude according to the different appearances of the agents after forming a social relationship with the person.

After entering the experiment room, participants sat on a chair in front of each agent (Fig. 15.1). The distance between the chair and the conversation agent was set at 2.2 m. Then, through the agent, the teleoperator explained that the person behind the agent was the same person as in previous sessions and announced the discussion topic for the session, as follows: "Hello again. This time I have a different appearance, but please don't worry about that. Now, let's discuss a new topic. The topic is ..." The duration of each session was 8 min. When the time limit was reached, the experimenter entered the room, stopped the discussion, and led the participant to another room. The participant then filled a post-session questionnaire. The combination of discussion topic and agent was determined randomly to achieve a counterbalance. Every participant took part in discussions on all four topics, each with a different agent.

**Fig. 15.1**  Scene from the experiment (*Android* condition)

### 15.2.3  Conversation Agents

We compared four conversation agents, each with a different appearance: *Object*, *Humanoid*, *Android*, and *Human* (Fig. 15.2). All the agents, except *Human* agent, were teleoperated by a person who acted as the *Human* agent.

#### 15.2.3.1  Object

We created the *Object* agent by placing two trash cans on a chair (Fig. 15.2a). A speaker was positioned behind the agent to transmit the utterances of the teleoperator. This was intended to be the most primitive type of robot; it has a physical body but not an anthropomorphic appearance. The trash cans were stacked so that their height was closer to that of the *Android* agent and the *Human* agent. In addition, the trash cans were covered with a piece of black cloth so that participants would not become preoccupied with them. The black color of the cloth was chosen so that the *Object* agent would appear similar to the other agents.

#### 15.2.3.2  Humanoid

We used "Robovie R2" as the *Humanoid* agent (Fig. 15.2b). Robovie R2 is a humanoid robot manufactured by V-Stone Corporation, Japan. It is equipped with body parts similar to those of humans, has 17 degrees of freedom in the whole body, and is driven by electric motors. In this experiment, however, we used no robotic motions; the robot stayed as motionless as the *Object* agent. This was to allow comparison with the *Object* agent, and was also necessary to avoid motor gear noise, which would disturb the conversation. Moreover, we removed the cartoonlike head cover with which the robot was equipped and exposed the naked mechanics of the

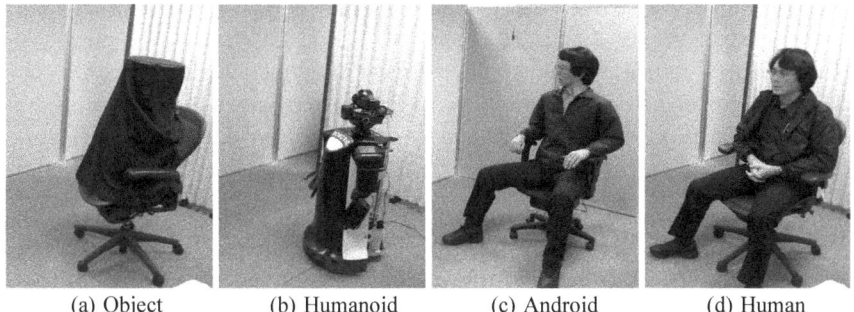

(a) Object          (b) Humanoid          (c) Android          (d) Human

**Fig. 15.2**  Conversation agents

robotic head. This was intended to eliminate any disturbing effect while still retaining the anthropomorphic features, such as eye- and mouthlike parts.

### 15.2.3.3  Android

As the *Android* agent, we used "Geminoid HI-1" (Fig. 15.2c). Geminoid HI-1 is an android robot developed by ATR Intelligent Robotics and Communication Laboratories [19]. It resembles a human being and is modeled on a real person. It is covered by silicone skin molded using a cast taken from the source person, and the frame structure is designed based on MRI scans of the source model. HI-1 has fifty pneumatic actuators inside its head and body that enable it to generate smooth and quiet motions. Further, when not receiving commands from the operator, HI-1 is designed such that it continuously shows subtle, "unconscious" motions. Without these motions, HI-1 appears dead [20]. This agent was prepared as the robot with the most humanlike appearance.

### 15.2.3.4  Human

The source person of Geminoid HI-1 served as the *Human* agent (Fig. 15.2d). This person (male, age 44) joined the habituation session and also teleoperated the other agents. As a *Human* agent, he was instructed not to make many gestures and to show behavior as similar as possible to that of HI-1.

### 15.2.3.5  Teleoperation System

The *Object*, *Humanoid*, and *Android* agents were teleoperated by the person who played the *Human* agent. For this purpose, a modified version of the Geminoid teleoperation system was used [21]. This system was positioned in a different room sufficiently far from the experiment room to prevent the teleoperators voice from reaching the participants directly.

   Two video cameras and two microphones were installed in the experiment room and the captured signals were transmitted to the teleoperation console. One of the cameras showed the agent in operation and the second showed the participant. Headphones were used to listen to the utterances of the participants. The operator also teleoperated the robotic motion of the *Android* agent. This was achieved by two means:

- A simple graphical user interface was prepared to send commands for controlling the robots motions. The operator was able to select several motions, such as nodding, shaking the head, or turning the head to the left/front/right with a single mouse click.
- A motion capture system (Hawk, MotionAnalysis Corp.) was used to obtain the operators lip/jaw position in real time. Using this measurement, control commands

were issued to generate HI-1's lip/jaw movements synchronously with its utterances.

### 15.2.4 Discussion Topics

In each of the main sessions, participants discussed one of the four discussion topics:

- Should children be allowed to use cellular phones?
- Should elderly peoples ownership of a driving license be restricted?
- Should parents allow children to watch as many TV programs as they want?
- Are voting rights necessary for university students?

The topics were carefully chosen so that the participants could easily understand them and conduct discussions. In the pre-experiment questionnaire, the participants indicated their attitudes on each of the topics on a 7-point Likert scale (disagree = 1, agree = 7). Based on their pre-experiment response, the teleoperator was instructed to persuade the participants to change their opinion on the topic, that is, when they were for something, to persuade participants to be against it, and vice versa. When participants' answers were neutral, the teleoperator was instructed to persuade them into holding a strong opinion.

### 15.2.5 Questionnaire

In the pre-experiment and post-session questionnaires, the participants rated their opinions on the discussion topics on a seven-point Likert scale. Additionally, in the post-session questionnaire, the participants evaluated their impressions of the discussion in their responses to 14 questions, based on a 7-point Likert scale. In both scales, "7" denoted strong agreement (or a positive opinion) and "1" denoted strong disagreement (or a negative opinion). The original questionnaires were in Japanese. After the experiment, we interviewed the participants about their impressions of the agents.

## 15.3 Results

### 15.3.1 Attitude Change by Persuasion

We first examined the change in the participants attitudes after persuasion by different agents. The measure for attitude change is defined as

$$change\ score = \begin{cases} if(pre - exp.response) > 0: \\ \qquad (pre - exp.response) - (post - exp.response) \\ if(pre - exp.response) == 0: \\ \qquad absolute(post - exp.response) \\ if(pre - exp.response) < 0: \\ \qquad (post - exp.response) - (pre - exp.response) \end{cases}$$

For example, when the pre-experiment response was "2" (negative) and the post-experiment response was "5" (positive), persuasion was deemed effective (*change score* = +3) as there was a change in the participant's opinion. However, if the post-experiment response was "1," it meant that the participant still held a negative opinion, and that the persuasion was not effective (*change score* = −1).

Although the teleoperator believed that he was successful in persuading most of the participants, the shift in opinion was determined to be quite small ($M = 0.68, SD = 1.19$) and a Kruskal–Wallis test showed no significant difference between agents ($\chi^2 = 2.80, df = 3, p = 0.42$).

### 15.3.2 Impressions Toward Different Agents

In the case of the post-session questionnaire, we first checked for multicollinearity by examining the correlation matrix of the questionnaire results and removed items with high correlations. The removed items were Q1: *Do you like the agent?* ($r = 0.86$ with Q2) and Q12: *Can you touch the top of the agent?* ($r = 0.99$ with Q8). With the removal of these two items, the determinant recovered to 0.02. We then performed factor analysis (principal component method accompanied by Promax rotation) and obtained three major factors. Table 15.1 shows these factors and their loadings (*fl*). These factors explained 66.2% of the variances in total before rotation (F1 = 44.50%, F2 = 11.21%, F3 = 10.48%).

Next, we obtained the factor scores and analyzed the variances using a Kruskal–Wallis test. The results showed that the difference between the agents was significant in all three scores (F1: $\chi^2 = 37.73, df = 3, p < 0.01$, F2: $\chi^2 = 20.59, df = 3, p < 0.01$, F3: $\chi^2 = 14.39, df = 3, p < 0.01$). Following these results, we performed multiple comparisons using the Steel-Dwass method; the results showed significant differences as follows: *Object, Humanoid < Android < Human* for F1, *Object, Humanoid, Android < Human* for F2 and *Humanoid, Android < Human* for F3 (Fig. 15.3).

In F1, we can see the factor loadings of "presence" (Q14, *fl* = 0.88) and "eye contact" (Q9, *fl* = 0.87) show high values (Table 15.1). We can also see the items related to personality (Q13: dignity, *fl* = 0.75; Q3: ability, *fl* = 0.41) and those related to human likeness (Q11: thing, *fl* = −0.74; Q10: human, *fl* = 0.60) hold a high weight on this factor. Of the two major items, "presence" seems to be related to the former

**Table 15.1** Results of factorial analysis of the post-session questionnaire results. Q1 and Q12 were removed to avoid multicollinearity. The values in columns headed F1 to F3 denote loadings for the three major factors. Factor loadings smaller than 0.40 in absolute are omitted from the table

| Question | | F1 | F2 | F3 |
|---|---|---|---|---|
| Q2 | Can you trust the agent? | | 0.84 | |
| Q3 | Do you think the agent was an able man? | 0.41 | 0.55 | |
| Q4 | Was it easy to talk with the agent? | | 0.89 | |
| Q5 | Was the agent calm? | | | 0.80 |
| Q6 | Was the agent speaking naturally? | | | 0.67 |
| Q7 | Was the agent moving naturally? | 0.60 | | |
| Q8 | Was the agent weird? | | −0.79 | |
| Q9 | Did you have eye contacts with the agent? | 0.87 | | |
| Q10 | Did the agent look like a human being? | 0.60 | | |
| Q11 | Did the agent look like a thing? | −0.74 | | |
| Q13 | Did the agent show dignity? | 0.75 | | |
| Q14 | Did the agent show presence? | 0.88 | | |

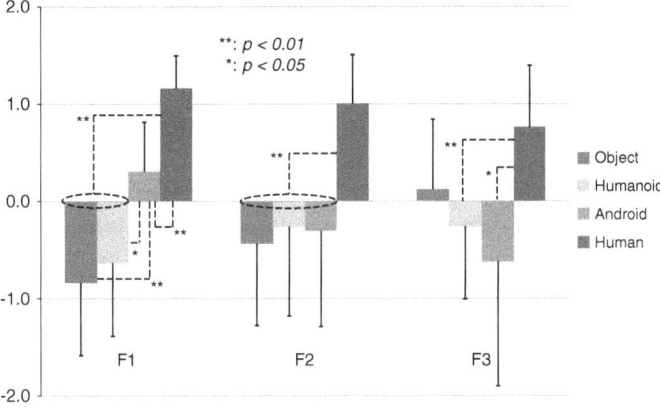

**Fig. 15.3** Means and standard deviations of factor scores together with results of multiple comparisons

group of items addressing personality, and "eye contact" to the latter group of items addressing human likeness. Thus, we can see that this factor shows the extent to which the agents expressed the presence of their personality and human likeness. The result of the variance analysis showed that the synthesized score rises as the agents appearance becomes closer to human. Thus, we can easily understand that this factor is related to the "human likeness" of the agent. It is of interest here that personality presence is also correlated with this likeness. That is, as agents appearance becomes closer to human, the agent begins to show more personality, perhaps better reflecting the original personality of the teleoperator. However, we can state that if the appearance of the agent is far from human, the agent is unlikely to show a humanlike presence.

However, F2 shows a different trend. Although the factor is composed by items related to personality such as "trust" (Q2, $fl = 0.84$), "ability" (Q3, $fl = 0.55$), and "easy to talk with" (Q4, $fl = 0.89$), the factor score showed a clear difference between human and artificial agents. In the post-experiment questionnaire, we also asked the participants about their impression of the four agents in free-style questions. Some of the participants reported that listening to a voice coming out of a "black object" (*Object* agent) or from a "machine" (*Humanoid* agent) was a strange experience. [22] showed the importance of matching appearance and voice. In their experiment, the combination of neither a computer appearance with a human voice nor a human appearance with a mechanical voice was successful. In our experiment, the mismatch of agent appearance and human voice may have caused the weird feeling toward the artificial agents, which resulted in the ease with which they could conduct a conversation with them being low.

Another reason for the difference between an artificial agent and a human may be the teleoperation. Every artificial agent was teleoperated, but only the *Human* agent was facing the participant directly. This difference may have affected both the participant side and the operator side. However, the fact that the "presence" (Q14, $fl < 0.40$) item did not contribute strongly to this second factor seems not to support this hypothesis. Further examination is required of the difference between face-to-face conversation and teleoperation.

The last factor, F3, shows dependency on "calmness" (Q5, $fl = 0.80$) and "speaking naturally" (Q6, $fl = 0.67$). This is difficult to interpret, but some participants noticed that the *Android* agent was shaking, which might have distracted them, resulting in their impression that its speech was unnatural. It was determined that the shaking of the geminoid was caused by small motions of the embedded air actuators. F3 also shows a significant difference between *Humanoid* agent and *Human* agent. As some participants reported that the "small" motions of the *Humanoid* agent looked strange, the lack of motion in the *Humanoid* agent may have caused this difference.

On the basis of these considerations, we can conclude that interlocutors tend to form different impressions of agents having different appearances, even if the agents are teleoperated by a single person. The results also showed that some aspects of impression gradually change together with the agents likeness to humans, such as that induced by the teleoperator's presence, while others depend merely on whether the agent is artificial or not. The latter finding may simply mean that researchers in this field still have to wait for advances in robotics; that is, missing features remain to be implemented. The results also showed that the effect of persuasion does not differ among the different types of agents, including *Human* agent. We cannot, however, determine on the basis of these experimental results whether this can be elaborated upon or not.

## 15.4 Contribution to Machine Consciousness

Embodiment is a key topic when considering consciousness [23]. Can consciousness arise without a body? Does the shape of consciousness change when the body changes? Teleoperated robots, including geminoids, can serve as tools for studying how consciousness is deformed when the normally unseparated binding between the mind and body is broken and the mind is reconnected to the body of another form, as in this study. In particular, when the robotic body quite closely resembles a human body, such as in the case of geminoids, it would be much easier for people to obtain the sense of a remote presence, the feeling of "being out there." Recent studies have begun to show that this feeling or even transfer of body ownership does occur for objects that do not belong to us. While the effects of visio-tactile coordination have been studied through the "rubber-hand illusion" [24], recent studies have shown that teleoperation of robots without tactile feedback, that is, mere visio-motor coordination, is sufficient for body ownership transfer and may even lead to the transfer of tactile sensation [25]. This phenomenon is known to strongly depend on the outer appearance of the target object and would be less likely to occur when the target does not appear as a human body [26].

The acquisition of such a sense of presence remotely has long been studied in the field of virtual reality. According to [27], *presence occurs when what is said about consciousness occurs within the domain of a virtual reality.* The authors of this study proposed that virtual reality can be utilized to study the internal mechanism of consciousness, and with the future progress in technologies, virtual reality systems will be an effective and controllable tool for forming a strong immersive effect.

However, one should recall that one important factor that forms human beings is missing here: interaction. Interaction with others, in particular with caregivers, is a crucial factor for the human developmental process, and thus, may also be an essential factor in forming our consciousness. An even more radical idea may be that consciousness is a dynamic transition that is continuously updated through interaction with others that appear to own consciousness or through simulations of interaction with others. As interaction is a dynamic and continuous change of state between multiple entities, not only a single side but all the sides of the interaction need to be considered and examined as a single system with mutual feedback. That is, when using teleoperated robots or virtual reality systems, the manner in which the operated entity affects the interaction partner and the whole interaction should be considered. In this context, teleoperated robots are much effective in the sense that they can directly influence people in the physical world; not only does the operator feel the sense of presence remotely, but the interlocutor also feels the presence of the operator remotely. By examining interactions or conversations made through the medium of teleoperated robots, it can be seen, at the same time, the manner in which the operators' consciousness is affected by a remote presence, and the interlocutor or the whole interaction system is affected by the presence of the teleoperator. Our study in this paper showed that this presence and even the impression of the teleoperator are easily affected by the outer appearance of the teleoperated entity. If presence is what people feel about others' consciousness, the results here suggest that the feeling of others' consciousness may be affected by their appearance. The reason people do not feel consciousness in a thermostat, for example, may not only be because it has no consciousness; it is also because it looks like a thermostat.

**Acknowledgements** This work was partially supported by Grant-in Aid for Scientific Research (S), KAKENHI (20220002).

# References

1. Nishio Shuichi, and Hiroshi Ishiguro. 2011. Attitude change induced by different appearances of interaction agents. *International Journal of Machine Consciousness* 3 (01): 115–126.
2. Fong, T., I. Nourbakhsh, and K. Dautenhahn. 2003. A survey of socially interactive robots. *Robotics and Autonomous Systems* 42: 143–166.
3. Petty Richard E. and Duane T. Wegener. 1998. *Attitude change: Multiple roles for persuasion variables*, 4th edn, 323–389. New York: McGraw-Hill.
4. Fogg, B.J., and T. Hsiang. 1999. The elements of computer credibility. In *Proceedings of the SIGCHI conference on human factors in computing systems*, 80–87. Pittsburg.
5. Shinozawa, K., F. Naya, J. Yamato, and K. Kogure. 2005. Differences in effect of robot and screen agent recommendations of human decision-making. *International Journal on Human-Computer Studies* 62 (2).
6. Kidd, C.D., and C. Breazeal. 2004. Effect of a robot on user perceptions. In *Proceedings of 2004 IEEE/RSJ international conference on intelligent robots and systems*, vol. 4, 9, 3559–3564.
7. Powers Aaron, Sara Kiesler, Susan Fussell, and Cristen Torrey. 2007. Comparing a computer agent with a humanoid robot. In *Proceedings of the human-robot interaction*, 145–152.
8. Catherine Zanbaka, Paula Goolkasian, and Larry Hodges. 2006. Can a virtual cat persuade you?: The role of gender and realism in speaker persuasiveness. In *Proceedings of the SIGCHI conference on human factors in computing systems*, 1153–1162. New York, NY, USA: ACM.
9. Bassili, J.N. 1981. The attractiveness stereotype: Goodness or glamour? *Basic and Applied Social Psychology* 2 (4): 235–252.
10. Grammer, K. 1993. *Signale der Liebe*. Hamburg: Hoffmann und Campe.
11. Solnick, S.J., and M.E. Schweitzer. 1999. The influence of physical attractiveness and gender on ultimatum game decisions. *Organizational Behavior and Human Decision Processes* 79 (3): 199–215.
12. Perani, D., F. Fazio, N.A. Borghese, M. Tettamanti, S. Ferrari, J. Decety, and M.C. Gilardi. 2001. Different brain correlates for watching real and virtual hand actions. *Neuroimage*, 14 (3): 749–758.
13. Kilner, J.M., Y. Paulignan, and S.J. Blakemore. 2003. An interference effect of observed biological movement on action. *Current Biology*, 13 (6): 522–525.
14. Chaminade, T., J. Hodgins, and M. Kawato. 2007. Anthropomorphism influences perception of computer-animated characters' actions. *Social Cognitive and Affective Neuroscience* 2 (3): 206–216.
15. Robins, B., K. Dautenhahn, R. te Boerkhorst, and A. Billard. 2004. Robots as assistive technology—does appearance matter. In *Proceedings of the 13th IEEE international workshop on robot and human interactive communication (RO-MAN 2004)*, 277–282.
16. Goetz, J., S. Kiesler, and A. Powers. 2003. Matching robot appearance and behavior to tasks to improve human-robot cooperation. In *Proceedings of the 12th IEEE international workshop on robot and human interactive communication (RO-MAN 2003)*, 55–60.

17. Syrdal, D.S., K. Dautenhahn, S. Woods, M. Walters, and K.L. Koay. 2007. Looking good? appearance preferences and robot personality inferences at zero acquaintance. In *Proceedings of AAAI summer symposium on multidisciplinary collaboration for socially assistive robotics*, 86–92. AAAI press.

18. Nishio, S., H. Ishiguro, and N. Hagita. 2007. Can a teleoperated android represent personal presence?—A case study with children. *Psychologia* 50 (4): 330–343.

19. Ishiguro, H., and S. Nishio. 2007. Building artificial humans to understand humans. *Journal of Artificial Organs* 10 (3): 133–42.

20. Minato Takashi, Michihiro Shimada, Hiroshi Ishiguro, and Shoji Itakura. 2004. Development of an android robot for studying human-robot interaction. In *Proceedings of the 17th international conference on innovations in applied artificial intelligence*, 424–434. Springer.

21. Nishio, S., H. Ishiguro, and N. Hagita. 2007. Geminoid: Teleoperated android of an existing person. In *Humanoid robots: New developments*, A.C. de Pina Filho, ed. Vienna, Austria: I-Tech Education and Publishing.

22. Nass, C., and S. Brave. 2005. *Wired for speech: How voice activates and advances the human-computer relationship*. Cambridge, MA: MIT Press.

23. Clowes, R.W., S. Torrance, and R. Chrisley. 2007. Machine consciousness: Embodiment and imagination. *Journal of Consciousness Studies* 14 (7): 7–14.

24. Botvinick, M. 1998. Rubber hands 'feel' touch that eyes see. *Nature* 391 (6669): 756.

25. Watanabe, T., S. Nishio, K. Ogawa, and H. Ishiguro. 2011. Body ownership transfer to android robot induced by teleoperation (in Japanese). *Transaction on IEICE* J94-D(1).

26. Armel, K.C., and V.S. Ramachandran. 2003. Projecting sensations to external objects: Evidence from skin conductance response. *Proceedings of the Royal Society of London B: Biological Sciences* 270 (1523): 1499–1506.

27. Sanchez-Vives, M.V., and M. Slater. 2005. From presence to consciousness through virtual reality. *Nature Reviews Neuroscience* 6: 332–339.

# Chapter 16
# Do Robot Appearance and Speech Affect People's Attitude? Evaluation Through the Ultimatum Game

**Shuichi Nishio, Kohei Ogawa, Yasuhiro Kanakogi, Shoji Itakura and Hiroshi Ishiguro**

**Abstract** In this study, we examined the factors that influence humans recognition of robots as social beings. Participants took part in sessions of the Ultimatum Game, a procedure frequently used in the fields of both economics and social psychology for examining peoples' attitudes toward others. In the experiment, several agents having different appearances were used, and speech stimuli, which were expected to induce a mentalizing effect on the participants' attitude toward the agents, were utilized. The results show that, while appearance per se did not elicit a significant difference in the participants attitudes, the mentalizing stimuli affected their attitudes in different fashions, depending on the robots' appearance. These results showed that elements such as a simple conversation with an agent and the agents' appearance are important factors that cause people to treat robots in a more humanlike fashion and as social beings.

**Keywords** Appearance · Speech · Ultimatum game · Attitude change

---

This chapter is a modified version of a previously published paper [1], edited to be comprehensive and fit with the context of this book.

---

S. Nishio (✉) · H. Ishiguro
Advanced Telecommunications Research Institute International,
Keihanna Science City, Kyoto, Japan
e-mail: nishio@botransfer.org

K. Ogawa · H. Ishiguro
Department of Systems Innovation, Graduate School of Engineering Science,
Osaka University, Osaka, Japan

Y. Kanakogi · S. Itakura
Graduate School of Letters, Kyoto University, Kyoto, Japan

© Springer Nature Singapore Pte Ltd. 2018
H. Ishiguro and F. Dalla Libera (eds.), *Geminoid Studies*,
https://doi.org/10.1007/978-981-10-8702-8_16

## 16.1  Introduction

In contrast to traditional industrial robots used in factory automation, service robots, which are expected to form a new industry and become a next-generation communication medium after the Internet, perform services for people and are being actively studied worldwide. In addition to the traditional research topics of robot navigation and manipulation, for service robots new topics, such as how they will communicate and exist among people, need to be studied. So that robots can closely associate with and exist alongside people, communicate with them smoothly, and be trusted and take responsible roles in society; they should elicit peoples positive impressions; that is, people must feel that robots are reliable and trustworthy. One example can be seen in banks. In Japan, every automated teller machine screen displays an avatar with a humanlike appearance that greets customers. Machines with similar functionalities already exist in daily environments. Are humanlike appearance and speech important for creating a condition that allows humans to respect machines and rely on them? Are these the keys characteristics of artificial agents that elicit trust?

Issues related to the manner in which people treat robots also exist. Robots, in particular humanlike models, have rather fragile mechanical parts and require careful handling. Thus, people must "respect and treat them like people, using social politeness." How can we make people treat robots with the same (or nearly the same) respect or morality as they would afford to humans?

Past research found that a robots' appearance affects people's behavior [2, 3]. This effect also operates at the unconsciousness level and tends to change people's behaviors and perceptions. Kilner compared the influence on a persons' body motion of watching the motions of other people and robots. Although the people and robots performed the same motion, the participants were affected only by the human motion [4]. Komatsu showed that different appearances of virtual agents change people's perception of sounds [5]. These findings suggest that the appearance of robots is one of the factors that affects people's behavior.

A further element that affects people's behavior is *mentalizing* [6]. Mentalizing, which is often referred to as a "theory of mind", means attributing mental states to other entities. Normally, this word describes the developmental process of recognizing minds in other people. However, this idea is also applicable to non-human agents. People tend to anthropomorphize humanlike objects, and they tend to anthropomorphize even objects having appearances that are far from humanlike and treat them as if they have mental states [7]. This process of mentalizing non-human objects occurs as a result of simple triggers. For example, Gallagher measured participants' brain activity while playing a simple computer rock, scissors, and paper game and their results showed that their activation patterns differed according to whether they were told that their counterpart was a computer or a person [8]. In this case, it was merely the description of the counterpart that changed the participants' responses. Itakura showed that people mentalized a robot simply by watching it make eye contact and perform a simple task [9]. Would such induced mentalizing change people's behavior toward robots? If so, how?

In this study, we used the *Ultimatum Game* (UG) for measuring behavior changes toward robots. The UG has been used to measure behavior in the fields of economics and sociology to determine how people accept others' behaviors [10]. In this study, we used this game to compare how people behave toward different non-human artificial agents, such as humanoid or android robots, having different appearances. At the same time, to examine how mentalizing clues affect people's behavior, we operated the agents such that they showed simple stimuli that cause mentalizing.

## 16.2   Ultimatum Game

The UG is a bargaining game that is frequently used in the fields of both experimental economics and social psychology as a tool to quantitatively measure peoples' attitudes, such as fairness and their behavior when they encounter unfairness [10].

The UG is a simple game played by two players: a *proponent* and a *respondent*. The two players interact to decide how to divide a given amount of money. The proponent decides how to distribute it between him/herself and the respondent and proposes his/her decision to the respondent. If the respondent accepts the proposal, the rewards are distributed based on the proposal. If the respondent rejects the proposal, all the money is forfeited; neither receives any reward. When both participants make reasonable judgments, the respondents' optimal strategy is always to accept any proposal having a distribution amount larger than zero. The optimal strategy for the proponent is to propose the smallest possible nonzero amount to the respondent.

However, when playing the UG, people tend to behave in a non-optimal manner. Proponents tend to make fair offers such as a 60/40 division, and when respondents receive offers of less than 20%, they tend to reject them [11]. Proponents tend to play fairly, and respondents receiving an unfair offer tend to show punishment behavior, even at the cost of forfeiting their rewards. This tendency becomes particularly clear when the respondent knows that this unfair offer was intentionally made by the proponent and then his/her rejection rates increase significantly [11]. When the respondent learns that the unfair offer was deliberate, he/she frequently feels anger toward the proponent and is satisfied with the decision to reject the proposal. However, if he/she learns that the unfair offer was made unintentionally, e.g., randomly, he/she tends to feel guilty [12]. When the opponent is not a social entity, but only a machine without intention, we assume that proponents will be less likely to make fair proposals and respondents will be unlikely to reject them, even when they are unfair.

The UG can be used as a powerful tool for studying people's attitudes toward others and how they accept them. By examining the scores of the UG sessions between people and artificial agents, we can observe whether people treat the agent as an artifact having no intentions or as an existence that does have intentions. In this study, we used the UG to examine how people's attitudes are affected by changing the appearance and the behavior of agents. We used the *truncated Ultimatum Game* (tUG), a variation of the UG [13]. In the tUG, participants cannot freely decide the division of the amount; instead, they choose from a set of predefined division rates. We prepared the following three choices, following Falk [11]:

- 80/20 (proponent takes 80%),
- 50/50 (both take 50%),
- 20/80 (respondent takes 80%).

Because the possible rates are limited, the proponent has to decide whether to divide fairly or unfairly, and the respondent can clearly see the proposer's intention.

In the UG, normally each participant encounters the same opponent only once to eliminate revenge feelings. However, we pitted each participant against the same opponent multiple times and each participant played the UG with a limited number of agents to reflect the nature of our study. This UG variation is called the *multiperiod Ultimatum Game* (mUG) [14]. In the mUG, the amount to be divided is usually increased or decreased in each game to maintain the equilibrium point. However, we fixed the total amount in each game to simplify the rules and because we were not focusing on the equilibrium itself. As in the normal UG, the mUG proponents tend to show fair behavior [14, 15]. Slembeck reported that, in a study similar to ours addressing a case where the total distributed amount did not change [16], the behavior of the proponents showed the same tendency as in the normal UG, while the number of rejections by respondents gradually decreased with the number of trials. This trend suggests that the respondents gradually learned their opponents' strategies and shows that they acted in a coordinated manner while continuing to punish unfair proposals.

In the experiment in this study, participants played games with multiple types of artificial agents that had different appearances. During the game, the agents also had a short conversation with the participants to change the "human likeness of the agent." Our hypotheses were as follows:

H1  When the agent resembles a human, the fairness of the proposals and the refusal rate of the participants will become closer to those that a human would apply to another human

H2  After the conversation, the fairness of the proposals and the refusal rate of the participants will become closer to those that a human would apply to another human.

## 16.3  Methods

### 16.3.1  Participants

Twenty-one university students participated in the experiment (12 males, 9 females) whose average age was 21.2 years (S.D. = 2.56). All participants received an explanation about the procedure, the measurement items, and the equipment. They gave their written consent. To raise their incentive, participants were told that their compensation would be based on the game's outcome. The actual amount of compensation was in fact fixed for every participant and explained after all the experimental procedures were completed.

### 16.3.2   Overview of Experiment Procedure

Participants first participated in a practice session, where an experimenter other than the person for the Human condition (described below) took the role of the opponent. During the session, the steps of the UG were described and 10 games were played. The experimenter as an opponent made the same predetermined proposals and responses for all the participants.

After the practice session, participants completed the following steps for each agent:

Session 1 (S1)   Performed the first session,
Question 1 (Q1)   Moved to a different room and took a short break,
[S2]   Returned to the experiment room and performed the second session,
[Q2]   Moved again to a different room.

The participants repeated these steps for the four artificial agents in random order. After they finished the sessions with all the agents, they filled post-experiment questionnaires.

## 16.3.3   Equipment and Agents

In the experiment room, the participants sat facing their opponent and a judge sat beside the desk (Fig. 16.1). In front of the participant and the opponent, three cards were placed that indicated the possible choices. Between the participant and the opponent, a partition was positioned that allowed the players to see each other while hiding their selection from the other player (Fig. 16.2).

**Fig. 16.1**  Equipment alignment

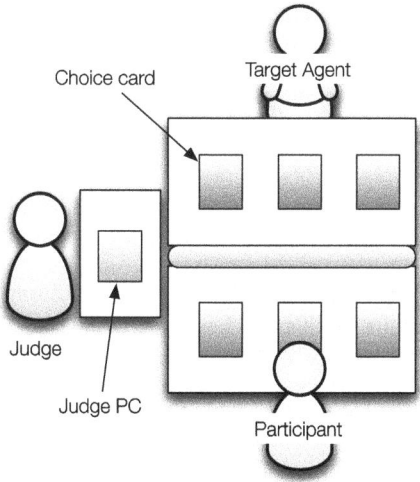

**Fig. 16.2** Experimental
scene

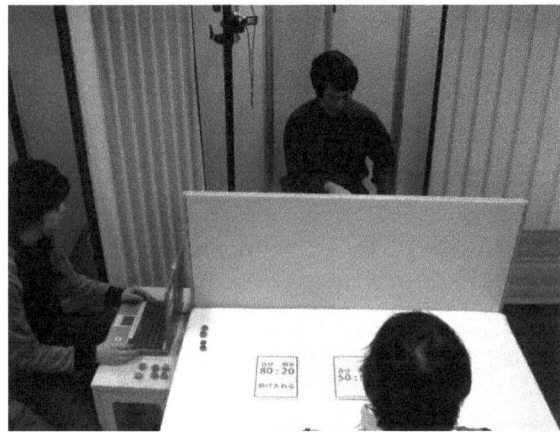

**Fig. 16.3** Computer
condition

As the opponents, we used the following four artificial agents with different
degrees of similarity to humans:

- Computer terminal (Computer condition),
- Humanoid robot (Humanoid condition),
- Android robot (Android condition),
- Human (Human condition).

We used a laptop computer in the Computer condition (Fig. 16.3). Unlike in the
other conditions, we placed the computer on a box on the desk where the opponent

**Fig. 16.4** Humanoid
condition

was seated to allow the participants on the opponent side to see it; however, the participants could not see the display on the screen. This allowed the setting in the Computer condition to resemble the other conditions more closely. In the Humanoid condition (Fig. 16.4), we used a humanoid robot named "Robovie R2," which has a face and arms and 17 degrees of freedom driven by electric motors. In the Android condition, we used "Geminoid HI-1" (Fig. 16.5), which is a teleoperated android, the appearance of which resembles that of a living man [17]. It has 50 degrees of freedom driven by pneumatic actuators. Finally, in the Human condition (Fig. 16.6), an experimenter (male graduate student) took the role of the opponent.

Before starting S1 with each agent, the participants received a brief description of the agent. This description included such concise hardware specifications as the agents' name and degrees of freedom. No information was mentioned about whether the agent was moving autonomously or was teleoperated. The judge also warned the participants not to talk to the opponent agent.

## 16.3.4   Mentalizing Stimulus

In S2, when the participants returned to the experiment room, the agent started a short conversation. The agent lifted its face (except in the Computer condition), looked at the participant, and said "Hello." The agent continued to make small talk by randomly choosing one of the following four short sentences: "How are things

**Fig. 16.5** Android
condition

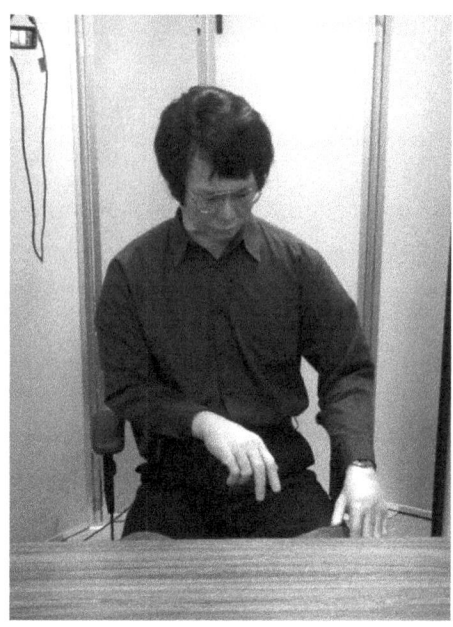

**Fig. 16.6** Human condition

going?"; "Nice weather, isn't it?"; "You look tired?"; and "It's cold today, isn't it?" When the participant responded, the agent gave a short answer. All the subsequent utterances of the participants were ignored. These utterances were chosen and activated by a teleoperator in another room. All the agent utterances had been prerecorded in his own voice by the same person who performed the Human condition.

### 16.3.5   Details of Each Session

In both sessions (S1 and S2), participants played 10 games. In the first game of each session, the agent always played the role of the proponent and the participant played the respondent. After each game, they exchanged roles; thus, the participants played each of the two roles five times.

At the beginning of each game, the judge looked at the proponent (either the participant or the agent) and said, "You are the proponent. Please select a distribution rate." After the proponent made the selection, the judge typed the result into his laptop PC and announced the result by pointing at each player: "You have xx yen and you have xx yen." Next, the judge looked at the respondent: "Please accept or reject." Then, the judge typed the respondent's choice into his PC, nodded to the proponent, and announced whether it was rejected or accepted. When the proposal was accepted, the judge placed the allocated amount of 10-yen coins on the desk of each player.

The proponents and respondents indicated their decisions by pointing at the three cards that were placed in front of each player, on which the three proposal choices (80/20, 50/50, 20/80) and two responses (accept or reject) were written (Fig. 16.7). If the agents had spoken their choices aloud, this utterance might have invoked a mentalizing effect in the participants. To avoid this effect, the judge looked at the card to which the proponent or respondent pointed and announced the selection results.

When making a selection, except in the Computer condition, the agents moved their right arm and pointed at the card of their choice. At the same time, the agents nodded and looked at the card. After the judge had input the result into his PC, the agents returned to their initial pose. These motions were controlled by the teleoperator in the other room.

**Fig. 16.7**  Proposal/response choice cards

Each selection of the agent was determined by the following algorithm. When the agent played the proponent role, 80/20 (agent benefits) and 50/50 (fair) were chosen with a probability of 6:4 to observe the participants' response when the agent made an unfair offer. We supposed that, if we created the impression that the agents behaved selfishly, the participants' behavior would show significant differences. When the agents took the role of respondent, they always accepted a 20/80 division (agent benefits) and accepted the other offers (50/50 and 80/20) at a 50% chance level to create the impression that they were neither responding randomly nor using the optimal strategy.

While playing each game, the agents always kept their heads turned downward to avoid eye contact with the participants and any mentalizing effect, which occurs when a robot makes eye contact and performs a meaningful motion [9]. The agents raised their heads and looked at the participants only when talking at the beginning of S2.

However, in the Computer condition, the agent could not point at the card like the other agents. Therefore, the judge looked at the computer screen. In fact, nothing was shown on the screen; the proposals and responses for the computer agent were listed on the judge's laptop computer, and the judge made his announcements based on this predefined list. To validate this action of looking at his laptop, in all conditions, the judge performed the act of inputting the results to his laptop before the announcements.

Note that another important role for the judge is to allocate the actual money. This allocation is important when playing the UG, as noted in personal communication with Prof. Güth. In this experiment, we fixed the amount of money at 100-yen per trial and the allocated money consisted of ten 10-yen coins. This maintained the incentives of the participants to obtain good results. Except in the Human condition, since the agents did not have the ability to hold coins, the judge distributed them to both players.

## 16.4  Results

Next, we show the results of the games and questionnaire responses. Since we found no gender differences in the results, we omitted them in the following.

### 16.4.1  UG

We examined the following two measures: (a) the number of rejections of agent proposals and (b) the number of fair proposals made to agents. In (a), we examined both the fair and unfair proposals made by the agents. As described in the introduction, these values can be used as indicators of whether the participants accepted the agents as social entities.

**Fig. 16.8**  Number of
rejections of unfair proposals

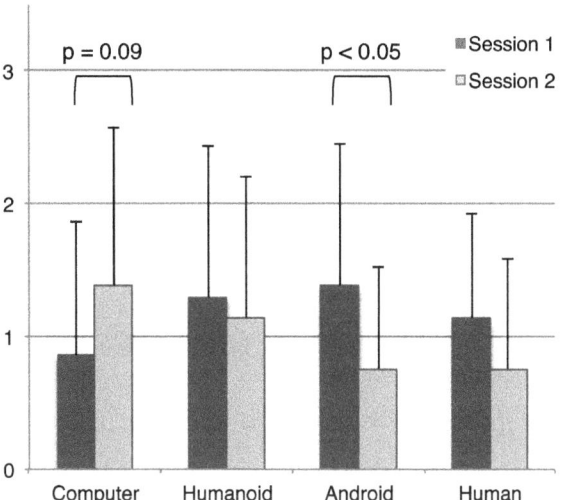

**Fig. 16.9**  Number of fair
proposals to agents

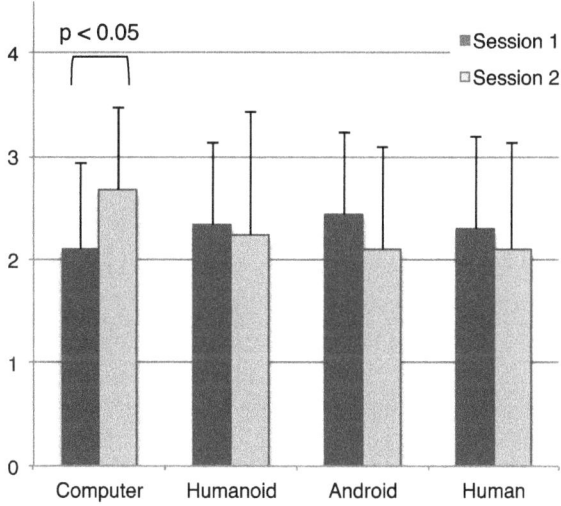

We performed a two-factor (agent, session) ANOVA test, where the sessions were
treated as a repeated factor. For the number of rejections, the results showed the only
significance was in the number of rejections of unfair proposals in the interaction
between agents and sessions $(F(3, 80) = 2.55, MSe = 0.41, p < 0.01$, Fig. 16.8). A
simple main effect test showed significance in the Android condition $(F(1, 160) =
4.04, MSe = 1.00, p < 0.05)$ and marginal significance in the Computer condition
$(F(1, 160) = 2.90, p = 0.091)$.

Only a marginal significance for the number of fair proposals was found in
the interaction between agent and sessions $(F(3, 80) = 2.63, MSe = 0.64, p = 0.056$,
Fig. 16.9). A simple main effect test showed significance in the Computer condition
$(F(1, 160) = 3, 98, MSe = 0.86, p < 0.05)$.

### 16.4.2   Post-experiment Questionnaires

After the experiment, the participants answered the following questions:

QE1   Did you feel as if you were playing a person?
QE2   Was your impression the same before and after the break?
QE3   Was the opponent like a human?
QE4   Was the opponent like a living creature?
QE5   Was the opponent like a machine?

The participants gave their responses on a seven-point Likert scale ($-3$: Disagree to $+3$: Agree). A single-factor (agent) ANOVA test showed significance for these questions, except for QE2.

QE1   $F(3, 80) = 41.85, MSe = 1.57, p < 0.01$
QE2   $F(3, 80) = 0.61, MSe = 2.74, p = 0.61$
QE3   $F(3, 80) = 36.79, MSe = 1.38, p < 0.01$
QE4   $F(3, 80) = 32.21, MSe = 1.72, p < 0.01$
QE5   $F(3, 80) = 38.55, MSe = 1.70, p < 0.01$

The results of multiple comparisons using the Bonferroni method are shown in Fig. 16.10.

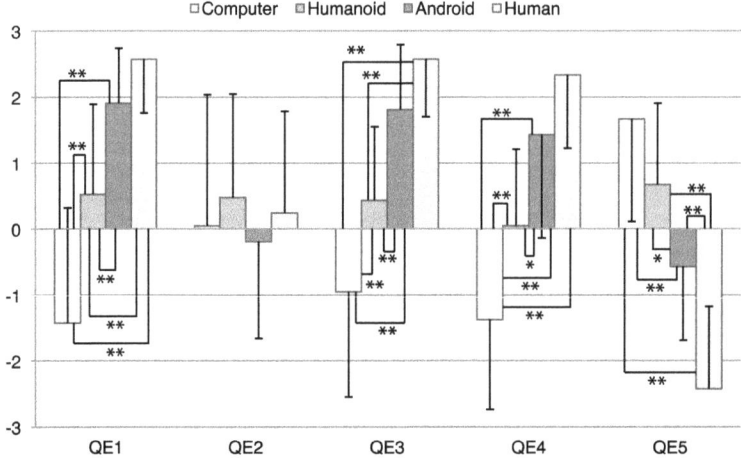

**Fig. 16.10** Results of post-experiment questionnaires: multiple comparison by agent (mean with standard deviation, *: $p < 0.05$, **: $p < 0.01$)

## 16.5  Discussion

First, according to the results for the post-experiment questionnaire (QE3), the participants perceived the degree of human likeness of the agents in the following order: Computer condition < Humanoid condition < Android condition. The results for the android agent were particularly close to those for a human. At the same time, their responses to QE5 showed that participants correctly recognized that the android agent was a machine.

Next, we examined whether our hypotheses were supported. The participants' attitudes toward each agent in the first session validated H1, "When the agent looks like a human, the fairness of the proposals and the refusal rate of the participants will become closer to those that a human would apply toward a human." However, no differences between these and the UG results could be discerned. Therefore, the first hypothesis is not supported by the results. This corresponds with the "Media Equation" theory [7]. Since this may be a result of the nature of the game and our experimental procedure, further investigation is required.

In the case of H2, "After the conversation, the fairness of the proposals and the refusal rate of the participants will becomes closer to those applied to a human," the results for the interaction between sessions and agents were interesting and significant. Before and after conversations with the agents, the participants showed different types of attitude changes that depended on the agents' appearance. In particular, the number of rejections in the second session for the Android condition decreased and became closer to those for the Human condition. Thus, the number of rejections in the second session corresponded to the human likeness of the agents. Fewer rejections in the second session may reflect the repetition effect of UG [16]. However, the results showed significant changes only in the Computer and Android conditions. Therefore, we conclude that, in these two conditions, the mentalizing effect caused by the utterances exceeded the repetition effect. In other words, since the appearance of the android agent was not sufficiently close to that of a human, the number of rejections in the first session was rather high. However, through the conversations between the sessions, the participants received a strong mentalizing effect, and thus, the number of rejections decreased and became closer to those for the human agent. The same tendency was seen for the humanoid agent as for the android agent in the first session. However, because of the androids' mechanical appearance, the mentalizing effect was weak and no significant change occurred between the two sessions.

A different interpretation regarding the participants' attitude to the android agent is also possible. Although the android agents appeared to resemble people, through the conversations the participants may have strongly felt the differences between them and the real human, which decreased their sense of intentionality in the android agent. That is, the utterances strengthened the machine likeness of the android, which strengthened the participants' attitudes toward "unintended proposals." Although this interpretation matches the "uncanny valley effect," in the post-experiment questionnaires, the responses of most participants to QE1 (it felt like playing with a person), QE3 (opponent was like a human), and QE4 (opponent was like a living creature) for the Android condition were similar to those for the Human condition. At the same time, the responses to these questionnaires for the Android condition were significantly different from those for the Computer and Humanoid conditions. Therefore, this interpretation is not supported.

Why did the participant attitudes change toward the computer agent? In the first session, the number of rejections was low, and after the conversations, it became higher. In general, two attitudes seemed to prevail in the UG: reciprocity, a tendency to be fair and to punish intentional unfairness, and generosity toward unintended unfairness. In the first session, participants may have had the impression that the computer agent responded according to an automatic program. Thus, even when the proposal was unfair, participants may have thought that it was "unintended, and so they forgave it." The conversations, however, strongly affected them and helped them to recognize the computer agent's human likeness. Thus, in proportion to the gap in human likeness, the tendency toward reciprocity operated strongly, leading to a large increase in the number of rejections. Through conversations with the agent, participants attempted to correct their incorrect impressions about the agent and increased their rejections, as well as their fair offers.

## 16.6   Conclusion

Using the UG, we examined how appearance affects the behavior of people and how the effect of mentalizing stimuli changes depending on an agent's appearance. Although our results showed no differences in behavior when we investigated only the agents' appearances, they did show that the effect of mentalizing stimuli depends on the appearance of agents. Such simple stimuli as having a short conversation changed people's behavior toward artificial agents, and it then more closely resembled that toward others. This insight is useful for designing both a robots' appearance and its functionality. However, since we failed to identify the reasons for this change, further investigation is required.

These results raise the concern that utilizing humanlike factors in computers, such as conversation, may lead to a strong mentalizing effect that causes people to think of machines as humanlike and "selfish and thus decreases their trust in them." Therefore, humanlike avatars and utterances may not constitute good designs for equipment and robots where precise and accurate operation is expected, such as automated teller machines. We need to further consider the appropriateness of human likeness and how it depends on social roles.

**Acknowledgements** The authors thank Professor Werner Güth of Max Planck Institute of Economics and Dr. Christian Becker-Asano of University of Freiburg for their valuable advice on the Ultimatum Game.

# References

1. Nishio, S., K. Ogawa, Y. Kanakogi, S. Itakura, and H. Ishiguro. 2012. Do robot appearance and speech affect people's attitude? Evaluation through the ultimatum game. In *21st IEEE International symposium on robot and human interactive communication (RO-MAN 2012)*, 809–814.
2. DiSalvo, C.F., F. Gemperle, J. Forlizzi, and S. Kiesler. 2002. All robots are not created equal: the design and perception of humanoid robot heads. In *Proceedings of the 4th conference on Designing interactive systems: processes, practices, methods, and techniques*, DIS '02, 321–326.
3. Yee, N., J.N. Bailenson, and K. Rickertsen. 2007. A meta-analysis of the impact of the inclusion and realism of human-like faces on user experiences in interfaces. In *Proceedings of the CHI*, 1–10.
4. Kilner, J.M., Y. Paulignan, and S.J. Blakemore. 2003. An interference effect of observed biological movement on action. *Current Biology* 13 (6): 522–525.
5. Komatsu, T., and S. Yamada. 2007. How do robotic agents' appearances affect people's interpretations of the agents' attitudes? In *Proceedings of the CHI*, 2519–2524.
6. Frith, U., and C.D. Frith. 2003. Development and neurophysiology of mentalizing. *Philosophical Transactions of the Royal Society of London Series B* 358 (1431): 459–473.
7. Reeves, Byron, and Clifford Nass. 1996. *The media equation: How people treat computers, television, and new media like real people and places*. New York, NY, USA: Cambridge University Press.
8. Gallagher, H.L., A.I. Jack, A. Roepstorff, and C.D. Frith. 2002. Imaging the intentional stance in a competitive game. *NeuroImage*, 16 (3, Part A): 814–821.
9. Itakura, S. 2008. Development of mentalizing and communication: From viewpoint of developmental cybernetics and developmental cognitive neuroscience. *IEICE Transactions on Communications* 91 (7): 2109–2117
10. Güth, Werner, Rolf Schmittberger, and Bernd Schwarze. 1982. An experimental analysis of ultimatum bargaining. *Journal of Economic Behavior & Organization* 3 (4): 367–388.
11. Falk, Armin, Ernst Fehr, and Urs Fischbacher. 2003. On the nature of fair behavior. *Economic Inquiry* 41 (1): 20–26.
12. Yamagishi, T., Y. Horita, H. Takagishi, M. Shinada, S. Tanida, and K.S. Cook. 2009. The private rejection of unfair offers and emotional commitment. *Proceedings of the National Academy of Sciences of the United States* 106 (28): 11520–11523.
13. Ohmura, Y., and T. Yamagishi. 2005. Why do people reject unintended inequity? Responders' rejection in a truncated ultimatum game. *Psychological Reports* 96 (2): 533–541.
14. Güth, Werner. 1995. On ultimatum bargaining experiments—A personal review. *Journal of Economic Behavior & Organization* 27 (3): 329–344.
15. Roth, E. 1989. An experimental study of sequential bargaining. *American Economic Review* 79 (3): 355–84.
16. Slembeck, T. 1999. Reputations and fairness in bargaining - experimental evidence from a repeated ultimatum game with fixed opponents. Technical report 9904, Department of Economics, University of St.Gallen, May 1999.
17. Nishio, S., H. Ishiguro, and N. Hagita. June 2007. Geminoid: Teleoperated android of an existing person. In *Humanoid Robots: New Developments*, Armando Carlos de Pina Filho ed, 343–352. Vienna, Austria: I-Tech Education and Publishing.

# Chapter 17
# Isolation of Physical Traits and Conversational Content for Personality Design

Hidenobu Sumioka, Shuichi Nishio, Erina Okamoto and Hiroshi Ishiguro

**Abstract** In this paper, we propose the "Doppel teleoperation system," which isolates several physical traits of a speaker, for investigating the manner in which personal information is conveyed to others during conversation. An underlying problem related to designing a personality for a social robot is that the manner in which humans judge the personalities of conversation partners remains unclear. The Doppel system allows each of the communication channels to be transferred, so that the channel can be chosen in its original form or in that generated by the system. For example, voice and body motions can be replaced by the Doppel system, while the speech content is preserved. This allows the individual effects of the physical traits of the speaker and the content of the speaker's speech on the identification of personality to be identified. This selectivity of personal traits provides a useful approach for investigating which information conveys our personality through conversation. To show the potential of our system, we experimentally tested the extent to which the conversation content conveys the personality of speakers to interlocutors when their physical traits are eliminated. The preliminary results show that, although interlocutors experience difficulty identifying speakers using only conversational content, when their acquaintances are the speakers, they can recognize them. We indicate several potential physical traits that convey personality.

**Keywords** Personality · Conversation · Android robot · Teleoperation

This chapter is a modified version of a previously published paper [1], edited to be comprehensive and fit with the context of this book.

H. Sumioka · S. Nishio (✉) · E. Okamoto · H. Ishiguro
Advanced Telecommunications Research Institute International,
Keihanna Science City, Kyoto, Japan
e-mail: nishio@botransfer.org

E. Okamoto · H. Ishiguro
Department of Systems Innovation, Graduate School of Engineering Science,
Osaka University, Osaka, Japan

H. Ishiguro and F. Dalla Libera (eds.), *Geminoid Studies*,
https://doi.org/10.1007/978-981-10-8702-8_17

## 17.1  Introduction

An individual's personality traits are fundamental information for deciding other people's behavior when communicating with her/him. Recent studies on human–robot communication have revealed that people attribute human characteristics to a robot, and their emotions and behavior are affected by the robot's personality traits, including its appearance [2, 3]. It has also been reported that a robot's personality traits can negatively or positively affect task performance [4]. Therefore, the design of a robot's personality is a crucial issue for enabling it to smoothly assist human people.

To design a robot personality that is task-appropriate, it is reasonable to draw inspiration from the manner in which humans judge the personality traits of others. Such individual judgments have long been studied in cognitive science and psychology. Recent progress in studies on personality has provided the dimensions of personality called the "Big Five," which can be used to describe human personalities [5]. This model allows personality to be measured not only according to the targets' self-ratings but also according to informants' ratings of them in order to clarify what information conveys the personality traits of the targets. Many studies now suggest that there exists a strong relationship between physical traits and personality. For example, studies that used criterion measures based on self and peer ratings showed that a person's appearance, including facial expressions [6–8] and clothing style [9], enables others to accurately judge his/her personality. While these studies were based on photographs of the face or the full body, the authors of other studies have argued that body movements [10] and voice [11, 12] also provide useful information for judging personality traits, in particular extraversion.

Although the results of these studies indicated several communication channels in which personal traits are presented, the experimental setting was limited to a case where an evaluator observed a person; that is, there was no conversation between them, although it is likely that the content of a conversation would provide the most information for judging her/him. A crucial difficulty in examining the relationship between personality and the physical traits that provide the personality during a conversation is to isolate the physical traits of an individual person from the conversation and to control their effects. Such isolation and control would allow the investigation not only of the independent effects of physical traits and personal thoughts but also of the mutual interaction among them as they pertain to the identification of personality. This investigation would provide new insight for designing the personality of social robots that communicate with people through multiple channels.

Interactive artificial agents may help us overcome the difficulty mentioned above, since they have been utilized as controllable "humans" to facilitate the understanding of the cognitive mechanism of human adults or infants [13, 14]. In this context, in their studies researchers have addressed the problems of the behavior and appearance of agents by using a robot called an android that has a very human-like appearance and have thus contributed to both cognitive science and robotics [15]. While typical androids are controlled as stand-alone agents, a teleoperated android called a "geminoid," which closely resembles a living individual [16, 17], has been developed as a telecommunication medium to address several telepresence and self-representation issues [18, 19]. It has been shown that it can convey personality traits, as well the presence of an operator, which a virtual agent may not provide in the real world. This system enables an operator to conduct nonverbal and physical interactions, including body touches, gestures, and facial expressions, as well as verbal interactions with others, by remotely operating an android having an appearance that may differ from that of the operator.

Although the geminoid system provides a means of isolating physical appearance from personality traits, it still transfers not only conversational content but also many of its operator's physical traits, such as body movements, facial expressions, and speech features. We solved this problem by assuming a speaker who provides the content of the conversation and an operator who acts as a "mediator," which might distort the speech features of the speaker, as well as controlling the geminoid's movement. The assumption of a mediator allows the separation of personal information in the conversation into physical traits (appearance, body movements, and paralanguage) and speech content (personal thought). The isolation of physical traits allows the speakers' physical traits to be selectively replaced with ones belonging to the geminoid and its operator, while the content of the speakers utterances and the physical traits of the mediator and the geminoid are conveyed to the interlocutors.

In this paper, we propose a teleoperation system called "Doppel," which isolates several physical traits from the content of conversations. By controlling the physical traits conveyed to an interlocutor, this system allows the researcher to analyze the individual effects of a speaker's physical traits and the content of the speaker's utterances on the identification of the personality. To show the potential of our system for investigating the manner in which the personalities of speakers are conveyed to interlocutors, we examined the interlocutors' identification of the speakers during conversations.

In the rest of this paper, we first describe our proposed system. Next, we report an experiment for evaluating the extent to which conversation content provides the personalities of speakers to their interlocutors. Preliminary results show that, although interlocutors experience difficulty identifying speakers using only conversational content, they can recognize whether they are talking with strangers or with their acquaintances. Finally, we discuss what information may provide the personalities of speakers to their interlocutors during conversations.

## 17.2  Doppel Teleoperation System

Figure 17.1 shows an overview of our proposed "Doppel Teleoperation System," which is based on a telecommunication system for a teleoperated android and uses a "geminoid," which resembles a living individual [16, 17]. In the existing system, an operator communicates with people at a remote location. Unlike a video conference system, where only visual and vocal information is provided, our system conveys

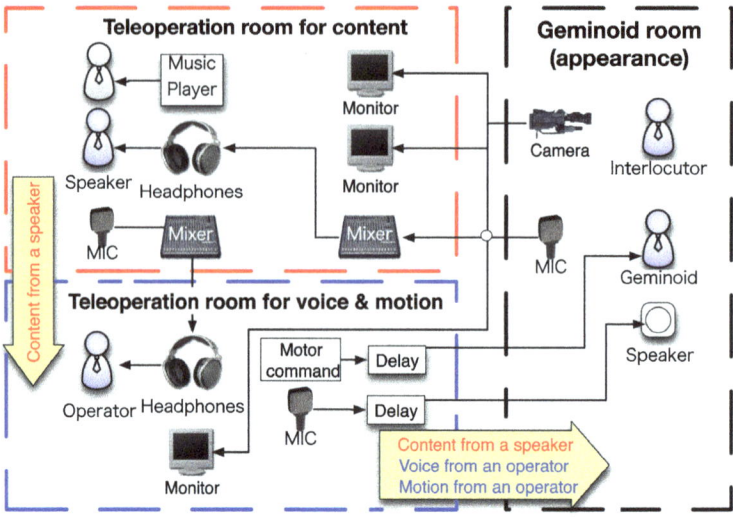

**Fig. 17.1** Overview of Doppel teleoperation system. Large arrows show communication channels to be conveyed to the operator or interlocutor and their sources

the operator's presence. We extended this system such that it can isolate individual communication channels by separating the teleoperation system into two subsystems: one that allows a speaker to conduct a conversation with an interlocutor and a second that allows an operator to control the geminoid's voice and motions. In the proposed system, the speaker communicates with the interlocutor through the geminoid, listens to the interlocutor, and talks into a microphone in another room. The operator listens to the speaker and repeats the speaker's words in her/his speaking manner in another teleoperation room. Therefore, the system conveys the conversation content from a speaker to an interlocutor during conversations and replaces the speaker's physical traits, which indicate the speaker's appearance, vocal information including paralanguage, and body movement, with those belonging to the geminoid or its operator: The appearance is that of the geminoid, and the vocal information and motion information are those of the operator. In the following section, we describe more detailed information about the system.

## 17.2.1 Appearance: Geminoid

A speaker's appearance is replaced with that of another physical entity that has a human-like appearance. The influence of appearance on personality identification should be investigated not only when human characteristics are modulated but also when they are eliminated. In the study, we used a geminoid (Fig. 17.2) that resembles an actual individual. One can also use a telenoid [20] to remove the influence

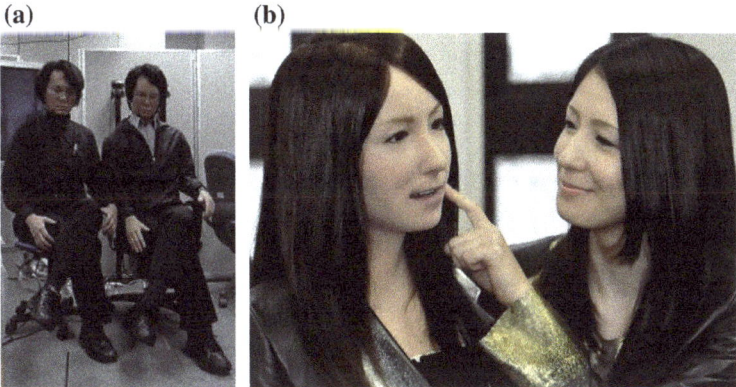

**Fig. 17.2** Geminoids: **a** Geminoid HI-1 (right) and model (left); **b** Geminoid F (left) and model (right)

of individual characteristics on the identification. Geminoid HI-1 resembles a living male (Fig. 17.2a). It has 50 degrees of freedom (DoFs) including 13 for facial expressions. The appearance of Geminoid F is similar to that of a living female (Fig. 17.2b). Most of its 12 DoFs are used for facial expressions.

Both geminoids have two different controllers: a conscious behavior controller and an unconscious one [16]. While the conscious behavior controller drives the geminoid's behavior under an operator's direction, such subtle expressed motions as breathing, blinking, and trembling are added by the unconscious behavior controller to maintain natural behavior. In addition to this semi-automatic control, the geminoid's lip movements are synchronized with those of its operator by applying facial feature tracking software that uses a camera placed in front of the operator.

### 17.2.2 Content of Conversation: Speaker

A speaker decides the geminoid utterances by monitoring the conversation between the geminoid and the interlocutor. The speaker's words to the interlocutor are conveyed—not to the interlocutor—to the operator, who hears them in another operation room through headphones.

### 17.2.3 Voice and Motion: Operator

The operator controls the geminoid to convey verbal and nonverbal information about the speaker and the operator him/herself to an interlocutor. Ideally, the verbal and nonverbal information presented by the geminoid should be generated by systems for speech information processing and motion generation instead of by a human operator. Because of the limitations of current technology, we decided to use a human operator.

The utterances of the speaker are repeated by the operator in his/her manner of speaking in front of a microphone, which is connected to the sound system behind the geminoid, following instructions from the experimenter. In addition to conveying the speaker's words, the operator controls the number of the speaker's physical traits that she or he provides to the interlocutor in paralanguage and movements. For example, if the operator repeated the speaker's words and mimicked the speaker's manner of speaking, the speaker's speech features would help the interlocutor identify the speaker. Thus, the geminoid's voice and movements presented to an interlocutor are a mixture of verbal and nonverbal information from a speaker and an operator.

## 17.3 Experiment: Personal Identification Based on Conversational Content

Our proposed system allows us to isolate the communication channels of a speaker and to design a new experimental setting that existing methodologies cannot easily achieve. As a first step toward examining the manner in which people identify the personality traits of others during conversation, we investigated whether people can identify a person using only conversational content and the extent to which physical traits affect the identification.

In the following experiments, we used Geminoid F. So that the speaker would be identified through conversational content, not through personal information, an operator replaced the speaker's dialect and specific words that might identify the speaker (e.g., the speaker's nickname) with standard dialect and general words (e.g., you), although the content of the speaker's utterances was preserved.

Since ordinary people may have difficulty making such replacements and controlling their behavior, we assigned an actor to be the operator. A female actor was assigned in coordination to the Geminoid F's appearance. Geminoid F's lip and head movements were synchronized with those of the operator, and other body movements and facial expressions were ignored, except for blinking, which was realized by the unconscious controller. The communication channels and their sources are summarized in Table 17.1.

### *17.3.1 Working Hypothesis and Prediction*

Although physical appearance, motions, and voice reflect personality traits, the conversation content should also provide considerable information for identifying personality traits, because it includes a person's thoughts, opinions, and feelings. It

**Table 17.1**  Sources of communication channels during conversation in experiments

| Channel | Source |
|---|---|
| Appearance | Geminoid F |
| Lip motion | Operator |
| Voice sound | Operator |
| Speaking speed | Operator |
| Accent | Operator |
| Conversation content | Speaker |

conveys more personal information if the speakers are acquaintances. Therefore, our objective was to verify the following hypotheses.

Hypothesis 1 (H1)  People can identify a speaker using only conversation content.

Hypothesis 2 (H2)  People can more correctly identify speakers using only conversation content when the speaker is an acquaintance than when the speaker is a stranger.

We conducted experiments with two different conditions to verify these hypotheses: the *no-acquaintance* condition (*NAc*), where the speaker and the interlocutor did not know each other, and the *acquaintance* condition (*Ac*), where the speaker was a friend of the interlocutor and they knew each other well. We examined H1 by evaluating the accuracy with which the interlocutors guessed the identity of the speakers from among four candidates. H2 was tested by comparing the accuracy of the identification in the *NAc* and *Ac* conditions.

### 17.3.2  Participants

Since Geminoid F has a female physical appearance, only female participants were recruited in order to eliminate the possibility that gender differences simplify the identification of the actual speaker. Seventy-six Japanese females participated in the experiment. We formed 19 pairs of participants who did not know each other for the *NAc* condition and 19 pairs of friends for the *Ac* condition. We assigned one of each pair to be the speaker and the other to be the interlocutor. The average age of the participants was 25.3 years (SD = 6.7).

### 17.3.3  Procedure

The subjects playing the role of interlocutors chatted with the speakers about given topics in order to identify the speaker from among four possible candidates:

the subject assigned to be a speaker ($S_s$), the model of the geminoid F ($S_m$), the geminoids operator ($S_o$), and the experimenter's assistant ($S_e$). The last three persons remained the same throughout the experiments, and we confirmed that the interlocutor did not know them. The model and operator of Geminoid F were never selected as actual speakers in the experiment. Therefore, the selection of $S_m$ or $S_o$ by the interlocutor is assumed to be caused not by the conversational content but by other physical traits of Geminoid F and the operator. It is implied that, while the interlocutor's identification was based on the appearance of Geminoid F if she selected $S_m$, it was based on the operator's movements and voice if she selected $S_o$.

Each experiment consisted of six 3-min sessions. An experimenter selected a discussion topic and an actual speaker from $S_s$ and $S_e$ before each session. The topic was chosen from two different types of topic: common and controversial (see Table 17.2 for examples). The common topics were those that people frequently introduce in conversation. Some topics were related to personal history items, such as a memory of a Christmas gift in childhood or a favorite type of man. The controversial topics addressed matters that people rarely introduce. The selected speaker was told to discuss the given topic through the geminoid with the interlocutor while the person who was not selected listened to music on headphones so that he/she would not hear the conversation between the actual speaker and the interlocutor. Three consecutive selections of the same speaker were avoided so that the interlocutor would not recognize the speaker because of the conversation length.

Before each experiment, each possible speaker was asked to talk about two different topics provided by the experimenter. Each talk was videotaped for 2 min. The interlocutors watched the videos of all the talks to discern the personalities of all the speakers. Then, they rated their personalities on the Japanese Property-Based Adjective Measurement questionnaire [21], which consists of 20 items, the responses to which are given on a seven-point scale for pairs of antonyms. The questionnaire has

**Table 17.2** Examples of conversation topics

| Common topics |
| --- |
| Have you ever dreamed in color? |
| What do you plan to do after you retire? |
| Do you remember any special Christmas gifts in your childhood? |
| Tell each other about a time you were really sick |
| What type of man do you like? |

| Controversial topics |
| --- |
| Should all college students have the right to vote? |
| Should we revoke the driving licenses of senior citizens to reduce traffic accidents? |
| What do you think about the whaling issue? |
| How can we prevent child abuse? |
| What do you think about World War II? |

high correlation with the extraversion, openness, and agreeableness components of the Big Five Model [5].

After rating the personalities of all the speakers, each interlocutor in turn was led to the experiment room where Geminoid F was located. The operator ($S_o$) and speakers ($S_s$ and $S_e$) were located in different rooms. The operator was told to replace the speaker's dialect and specific words that might identify the speaker with the standard dialect and general words. She was also asked to use her own tone and rephrase the speaker's words, while preserving the speaker's content. The speakers were informed that their words might be changed by the geminoid's operator. After briefly explaining the specifications of Geminoid F and the number of sessions and their duration to the interlocutor, the experimenter informed her that the actual speaker might change for each session. The participants were also told that the geminoid was controlled by the operator ($S_o$) whom they saw in the video and that they would talk with this operator, expressing their own thoughts or according to what one of the other speakers was saying. During a session, the actual speaker and the interlocutor asked each other questions about a given topic and responded to each other. After each session, the interlocutor guessed the identity of the speaker and explained her guess. She also rated the speaker's personality on the questionnaire [21]. After all the sessions were completed, the interlocutor was debriefed about the experiment.

### 17.3.4 Evaluation

The interlocutor's performance was evaluated by the accuracy rate that indicates the frequency with which she correctly identified the actual speaker. To examine the effect of the differences in the personality of the possible speakers on the interlocutors' guesses, we also calculated the difference between the scores of the Japanese Property-Based Adjective Measurement questionnaire on which each interlocutor rated the speakers before the experiment.

## 17.4 Results

The average accuracy rate of identifying the actual speaker for the *NAc* and *Ac* conditions and the total average across subjects were 0.28, 0.31, and 0.29, respectively. No significant difference was found between conditions ($p = 0.13 > .05$ by binomial test). This result indicates that the participants experienced difficulty identifying the speaker from conventional content, refuting our first hypothesis. We also tested the second hypothesis by comparing the average accuracy rates of the two conditions. However, the results showed no significant difference between them ($p = 0.85 > .05$ by Wilcoxon test). This suggests that the difference between the two conditions does not support our second hypothesis.

We calculated the accuracy rates per type of topic to determine whether the participant's identification of the actual speaker depended on the type of topic. The average accuracy rates for the common and the controversial topic were 0.32 and 0.30 in the *Ac* condition and 0.26 and 0.30 in the *NAc* condition, respectively. There was no significant difference between different types of topic ($p = 0.8$ in the *Ac* condition and $p = 0.62$ in the *NAc* condition according to a Wilcoxon signed rank test), suggesting that the identification performance of the interlocutors was not biased by the type of topic.

One might think that the difficulty in identifying the actual speakers was caused by similar personality traits among the possible speakers. However, as shown in Table 17.3, we found no positive correlation between the accuracy rate and the distances of the personality scores of the possible speakers, which were measured prior to the experiments. The results showed rather weak negative tendencies. Therefore, our results do not support the possibility that similarity in personality traits complicated the judgment of the speakers.

In the *Ac* condition, the interlocutor talked not only with her friend but also with a stranger $S_e$. If she more frequently identified the friend correctly than the stranger, the second hypothesis was supported within the *Ac* condition, where the interlocutors were aware that friends existed among the possible speakers. Therefore, we compared the performances of identifying the actual speaker for the condition where the friend was the speaker with that for the condition where a stranger was the speaker in the *Ac* condition.

Figure 17.3 shows the rate of identifying each speaker across the interlocutors in the *Ac* condition. As can be seen, the interlocutors seem to frequently identify the actual speaker as a candidate other than their friends. This may have been caused by the strong conservative bias of the interlocutors when identifying the actual speaker as a friend. To distinguish the accuracy from the bias effects, we therefore computed the $A'$ and $B''$ scores [22] of the hit rate (i.e., the rate of correctly identifying that a friend was the actual speaker) and the false alarm rate (i.e., the rate of incorrectly identifying that a friend was the speaker). The scores showed that the interlocutors were sensitive to friends ($A' = 0.80$), although they showed a strong conservative bias against identifying them as the speaker ($B'' = 0.70$). Both scores were higher than those when identifying a stranger ($S_e$) as the actual speaker ($A' = 0.54$, $B'' = 0.036$). In fact, when the accuracy rate of the interlocutors identifying the speakers as their friends is calculated, it can be seen that this rate is significant. Figure 17.4 shows the accuracy rates of the interlocutors identifying the actual speakers in the

**Table 17.3** Coefficients of correlation between accuracy rate and differences in personality measures among possible speakers

| Condition | $S_m$ versus $S_e$ | $S_m$ versus $S_s$ | $S_o$ versus $S_e$ | $S_o$ versus $S_s$ | $S_e$ versus $S_s$ |
|---|---|---|---|---|---|
| No-acquaintance | −0.30 | −0.08 | −0.35 | −0.55 | −0.45 |
| Acquaintance | −0.37 | −0.39 | −0.38 | −0.21 | −0.24 |

**Fig. 17.3** Rate of each possible speaker identified by interlocutors in *acquaintance* condition (*Ac*)

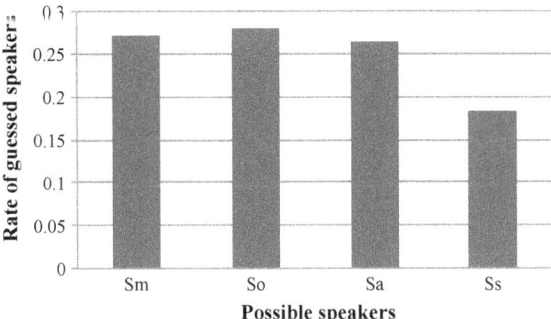

**Fig. 17.4** Accuracy rates of identifying different actual speakers in the *acquaintance* (*Ac*) condition

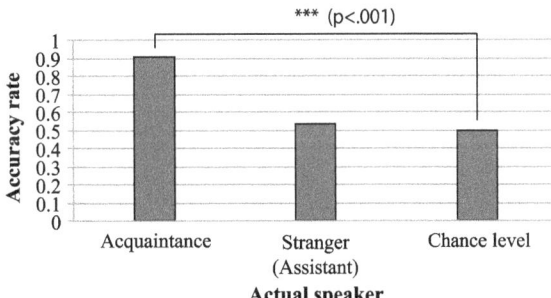

*Ac* condition. The rate at which the interlocutors accurately identified their friends as the actual speakers is considerably higher than the rate that would be expected by chance. In fact, the difference between the rates is significantly different ($p = 0.00 < 0.001$ by binomial test). A difference such as this is not seen in the accuracy rates for identifying the assistant as the actual speaker (see the middle bar in Fig. 17.4). These results indicate that the interlocutors correctly identified their friends as the actual speakers, supporting the second hypothesis in the *Ac* condition.

We also calculated the probabilities of identifying each possible speaker as the actual speaker in order to examine the extent to which physical traits affect the identification of speakers (Table 17.4). Although the probability of selecting the geminoid model or the operator is slightly higher than that of selecting the participants as the speaker, we found no significant difference among them. This implies that no specific physical trait is stronger than any other when several traits are presented.

## 17.5   Discussion

Our results revealed that it is difficult for people to identify a person without physical traits, such as physical appearance, body movements, and speech features. In fact, after the experiments, some interlocutors reported that they felt that the geminoid

**Table 17.4** Selection probability of possible speakers

| Speakers | Selection probability |
|---|---|
| Model of Geminoid F | 0.285 |
| Operator | 0.241 |
| Assistant | 0.263 |
| Subject as a speaker | 0.206 |

had another new personality, and not that of one of the possible speakers. Although some results did not support our hypotheses, they did provide fruitful insights.

The accuracy and response bias scores suggested that the interlocutors low accuracy rate of identifying their friends may have been caused by their conservative bias. This finding was supported by the high accuracy rate when the interlocutors identified their friends as the actual speakers (Fig. 17.4). Since our second hypothesis was partially supported, not between conditions but in the *Ac* condition, further verification is needed.

The accuracy rates shown in Table 17.4 suggest that no physical traits exist that have significant effects on personal identification. This implies that the identification of personality during conversation results from mutual interaction between the people conversing. Although transmission of personality traits has been studied with a focus on one aspect of conversation (see [23] as an example about linguistic style), the mutual interaction during conversation should also be investigated to design the personality of social robots that communicate with people through multiple channels, because mismatches between physical traits may lead to people misunderstanding a robots behavior and then failing to communicate with it. However, this investigation seems difficult in the case of existing approaches that need to extract a single modality from all the modalities and exclude the others. Our system allows not only the single effect of the physical traits but also the mutual interaction among them to be examined by selectively controlling the presented information. In future studies, we will conduct experiments with different combinations of physical traits to investigate how physical traits and conversational content interact with other traits.

One concern about this system is the influence of the geminoid on the interlocutors judgments. Previous studies reported that people respond to an android as they respond to a human when it shows human-like behavior [3]. Since we designed the geminoid to resemble a human in appearance and movement, we expected the subjects to consider it "human." However, the well-known uncanny valley [24] theory argues that even small flaws in human likeness affect humans perception of androids. Therefore, the manner in which the human likeness of an android affects the human judgmental process must be addressed in the future.

We should note that the interlocutors still extracted some physical traits of their speakers through conversations, although we attempted to eliminate this possibility. More precisely, they used some speech features to guess the actual speakers: timing and duration of speech, and feedback expressions to the interlocutors' comments such as "Really?," "Exactly," and "No way!" In addition, it was determined that the interlocutors may have used the speed of speech and the accent to guess the actual speakers in the preparatory experiments. A detailed investigation of the physical traits including these speech features will also be valuable future work.

## 17.6   Conclusion

In this paper, we proposed the Doppel teleoperation system, which isolates several physical traits from conversations, to investigate how personal information is conveyed to others during conversation. The Doppel system allows the user to choose to transfer each communication channel either in its original form or in the form generated by the system. This allows the user to analyze the individual effects of the physical traits of the speaker and the content in the speaker's speech on the identification of personality. We examined the extent to which the conversation content conveyed the personality of the speakers to the interlocutors, when all the physical traits of the speakers were eliminated. The results showed that, although the interlocutors experienced difficulty identifying the speakers using only conversational content, when they were talking with their friends they were able to identify them. The independent effects of physical traits and their interaction will be explored in more detail. We believe that this system will help researchers understand the cognitive mechanism of human personalities and establish a design basis for robot personality.

## References

1.  Sumioka, Hidenobu, Shuichi Nishio, Erina Okamoto, and Hiroshi Ishiguro. 2012. Isolation of physical traits and conversational content for personality design. In *21st IEEE International Symposium on Robot and Human interactive Communication (RO-MAN 2012)*, 596–601.
2.  Fong, T., I. Nourbakhsh, and K. Dautenhahn. 2003. A survey of socially interactive robots. *Robotics and autonomous systems* 42 (3): 143–166.
3.  Shimada, M., and H. Ishiguro. 2008. Motion behavior and its influence on human-likeness in an android robot. In *Proceedings of the annual Conference of the Cognitive Science Society*, 2468–2473.
4.  Goetz, J., and S. Kiesler. 2002. Cooperation with a robotic assistant. In *ACM SIGCHI conference on human factors in computing systems*, 578–579. ACM.
5.  McCrae, R., A. Zonderman, P. Costa, M. Bond, and S. Paunonen. 1996. Evaluating replicability of factors in the revised neo personality inventory: Confirmatory factor analysis versus procrustes rotation. *Journal of Personality and Social Psychology* 70 (3): 552.
6.  Berry, D.S. 1990. Taking people at face value: Evidence for the kernel of truth hypothesis. *Social Cognition* 8: 343–361.
7.  Berry, D.S. 1991. Accuracy in social perception: Contributions of facial and vocal information. *Journal of Personality and Social Psychology* 61: 298–307.
8.  Little, A.C., and D.I. Perrett. 2006. Using composite images to assess accuracy in personality attribution to faces. *British Journal of Psychology* 98: 111–126.
9.  Naumann, L. 2009. Personality judgments based on physical appearance. *Personality and Social Psychology Bulletin* 35 (12): 1661–1671.

10. Kenny, D.A., C. Horner, D.A. Kashy, and L. Chu. 1992. Consensus at zero-acquaintance: Replication, behavioral cues, and stability. *Journal of Personality and Social Psychology* 62: 88–97.
11. Scherer, K.R., and U. Scherer. 1981. Speech behavior and personality. In *Speech evaluation in psychiatry*, 115–135.
12. Borkenau, P., and A. Liebler. 1992. Trait inferences: Sources of validity at zero-acquaintance. *Journal of Personality and Social Psychology* 62: 645–657.
13. Itakura, S. 2008. Development of mentalizing and communication: From viewpoint of developmental cybernetics and developmental cognitive neuroscience. *IEICE Transaction communications E series B* 91 (7): 2109.
14. Yoshikawa, Y., K. Shinozawa, and H. Ishiguro. 2007. Social reflex hypothesis on blinking interaction. In *Proceedings of the annual conference of the cognitive science society*, 725–730.
15. Ishiguro, H. 2007. Android science-toward a new cross-interdisciplinary framework. *Robotics Research* 28: 118–127.
16. Sakamoto, D., T. Kanda, T. Ono, H. Ishiguro, and N. Hagita. 2007. Android as a telecommunication medium with a human-like presence. In *Proceedings of the international conference on Human-robot interaction*, 193–200.
17. Nishio, S., H. Ishiguro, and N. Hagita. 2007. Geminoid: Teleoperated android of an existing person. In *I-Tech: Humanoid robots-new developments*.
18. Nishio, S., H. Ishiguro, and N. Hagita. 2007. Can a teleoperated android represent personal presence?—A case study with children. *Psychologia* 50 (4): 330–342.
19. Straub, I., S. Nishio, and H. Ishiguro. 2010. Incorporated identity in interaction with a teleoperated android robot: A case study. In *Proceedings of the International Symposium on Robot and Human International Communication*, 119–124.
20. Ogawa, K., S. Nishio, K. Koda, G. Balistreri, T. Watanabe, and H. Ishiguro. 2011. Exploring the natural reaction of young and aged person with telenoid in a real world. *Journal of Advanced Computational Intelligence and Intelligent Informatics* 15 (5): 592–597.
21. Hayashi, F. 1978. The fundamental dimensions of interpersonal cognitive structure. *Bulletin of the Faculty of Education of Nagoya University* 25: 233–247. in Japanese.
22. Grier, J.B. 1971. Nonparametric indexes for sensitivity and bias: Computing formulas. *Psychological Bulletin* 75 (6): 424.
23. Mairesse, F., and M.A. Walker. 2011. Controlling user perceptions of linguistic style: Trainable generation of personality traits. *Computational Linguistics* 37 (3): 455–488.
24. Mori, M., K.F. MacDorman, and N. Kageki. 2012. The uncanny valley. *IEEE Robotics and Automation Magazine* 19 (2): 98–100.

# Chapter 18
# Body Ownership Transfer
# to a Teleoperated Android

**Shuichi Nishio, Tetsuya Watanabe, Kohei Ogawa and Hiroshi Ishiguro**

**Abstract**  Teleoperators of android robots occasionally feel as if the robotic bodies are extensions of their own bodies. When others touch the android that they are teleoperating, even without tactile feedback, some operators feel as if they themselves have been touched. In the past, a similar phenomenon named the "Rubber Hand Illusion" was studied because it reflects the three-way interaction between vision, touch, and proprioception. In this study, we examined whether a similar three-way interaction occurs when the tactile sensation is replaced with android robot teleoperation. The results show that when the operator and the android motions are synchronized, operators feel as if their sense of body ownership is transferred to the android robot.

**Keywords**  Body ownership · Teleoperation · Android robot

## 18.1  Introduction

Recently, we developed a teleoperated android robot named "Geminoid" [2], the appearance of which is designed such that it is similar to that of a real person (Fig. 18.1). Geminoid was made as a research tool to examine how the appearance and behavior of a robot affect people's communication. During various studies, we found that conversations conducted through the geminoid affect not only the people in front of it but also its operators. Soon after starting to operate the geminoid, the operators tend to adjust their body movements to the movements of the geminoid.

This chapter is a modified version of a previously published paper [1], edited to be comprehensive and fit with the context of this book.

S. Nishio (✉) · T. Watanabe · H. Ishiguro
Advanced Telecommunications Research Institute International,
Keihanna Science City, Kyoto, Japan
e-mail: nishio@botransfer.org

T. Watanabe · K. Ogawa · H. Ishiguro
Department of Systems Innovation, Graduate School of Engineering Science,
Osaka University, Osaka, Japan

© Springer Nature Singapore Pte Ltd. 2018
H. Ishiguro and F. Dalla Libera (eds.), *Geminoid Studies*,
https://doi.org/10.1007/978-981-10-8702-8_18

293

**Fig. 18.1** Geminoid HI-1
(left) with its source person
(right)

For example, they talk slowly in order to synchronize their utterances with the geminoid's lip motion and move slightly when the robot does. Some operators even feel as
if they themselves have been touched when others touch the teleoperated android [2].
For example, when someone pokes the geminoid's cheek, operators feel as if their
own cheek is being poked, despite the lack of tactile feedback. This illusion occurs
even when a person who is not the source model of the geminoid is operating it. However, this illusion does not always occur, and it is difficult to cause it intentionally.

A similar illusion, named the "Rubber Hand Illusion" (RHI) [3], exists. In the RHI
procedure, an experimenter repeatedly strokes a participant's hand and a rubber hand
at the same time. The participant can see only the rubber hand and not his/her own
hand. After performing this procedure for a while, participants begin to experience
an illusion; that is, they feel as if the rubber hand is their own hand. Although only
the rubber hand is stroked, the participants feel as if their own hand is being stroked.
This illusion, the RHI, is considered to occur as a result of synchronization between
the visual stimulus (watching the rubber hand being stroked) and the tactile stimulus
(feeling that his/her hand is stroked) experienced by the participant [4]. The resulting
illusionary effect is quite similar to that which occurs in the case of our teleoperated
android, the geminoid. However, in the case of the geminoid, the illusion occurs
without any tactile stimulus; the operator only teleoperates the geminoid and watches
it moving and being poked.

Our hypothesis is that this illusion, body ownership transfer toward a teleoperated android, occurs because the operation of the geminoid and the visual feedback
of seeing the geminoid's motion are synchronized. That is, body ownership is transferred by seeing the geminoid moving in synchrony with the operator. If body ownership transfer can be induced merely by operating the robot without haptic feedback,
this could lead to a number of applications, such as realizing a highly immersive
teleoperation interface and developing prosthetic hands/body parts that can be used
as a person's real body parts.

We attempted to verify this by comparing cases where the participants in an experiment watch the robot moving in synchrony and not moving in synchrony with their motion. Thus, our first hypothesis is

**Hypothesis 1** Body ownership transfer toward the geminoid body occurs through synchronized geminoid teleoperation with visual feedback.

The RHI requires the synchronization of visual and tactile senses. However, the geminoid cannot move without delays, because the sensing and the actuators reaction are limited. In fact, the geminoid usually moves with a 200–800 ms delay. We make adjustments during a conversation by delaying the voice such that the geminoid produces the voice in synchronization with its mouth movements and speech [2]. According to related work, these delays during teleoperation reduce the extent of body ownership transfer. For example, Shimada et al. showed that the RHI effect was significantly decreased when the delay between the visual and tactile stimuli was more than 300 ms [5]. If this applies also to the geminoid's body ownership transfer, we cannot explain why body ownership is transferred to the geminoid during its operation. Therefore, the mechanism of the geminoid's body ownership transfer may differ from the mechanism of the RHI's body ownership transfer. Thus, our second hypothesis is

**Hypothesis 2** In geminoid teleoperation, body ownership is transferred also when the geminoid moves with delays.

## 18.2 Methods

Based on the hypotheses in the previous section and knowledge culled from related studies, our objective was to experimentally verify that body ownership is transferred by the geminoid operation and its visual feedback. When the geminoid is operated, during a conversation its mouth and head are synchronized with those of the operator. However, it is difficult to maintain control when conversation is used as a task. For this reason, we used the operation of the geminoid's arm as the task in our study.

The participants consisted of 19 university students, 12 males and 7 females, whose average of age was 21.1 years (standard variation, 1.6 years). All were right-handed. They received explanations about the experiment and signed consent forms.[1]

### 18.2.1 Procedure

First, the participants operated the geminoid's arm and watched the scene for a constant time. During this part of the experiment, the geminoid's arm was synchronized

---

[1]This experiment was approved by the ethical committee of Advanced Telecommunications Research International Institute (No. 08-506-1).

with the operator's arm. This is the same as asking subjects to watch a scene where a rubber hand is being stroked by a brush in an RHI procedure. Next, we applied only painful stimuli to the geminoid's arm. The skin conductance response (SCR) was measured and the participants filled a self-report questionnaire to measure their subjective experience. We predicted that both the participants responses and the SCR measurements would be affected if body ownership was transferred.

The SCR shows significant values when the autonomic nervous system is aroused, such as when people feel pain [6]. Armel et al. verified body ownership transfer by measuring the SCR [7]. Their idea was that if the RHI occurs, the skin conductance of the subject will be effected when the target object (rubber hand) receives a painful stimulus. The participants in the experiment watched a scene in which the rubber hand was bent strongly after the RHI procedure (synchronized/delayed condition). The results confirmed that the SCR value in the synchronized condition is higher than in the delayed condition.

People looking at the geminoid may believe that it is a real human, because it has a very human-like appearance. For this reason, the participants were asked to examine it before the experiment to clarify that the geminoid is in fact an object. Then, they learned how to operate the robot. At this time, they were also asked to examine the camera that was located above the robot and were informed that they would watch the geminoid by means of this camera. However, this camera was not used, as discussed below.

The participants then entered the experiment room. One marker for motion capture system was affixed to their right arm and electrodes were affixed to the left hand. The marker was positioned 19 cm from their elbow to maintain their arm movement radius. The electrodes were affixed to the ball of the hand and the hypothenar. Then, they were instructed to grasp the hand to which the marker was attached and to turn their palm down, and to turn the other palm up, and to avoid moving both their fingers and wrists. An identical procedure was followed in the main trials. After practicing, they were informed about the main trial procedure and the questionnaire. They were also informed about this experiment's purpose and received instructions for operating the geminoid. They were then instructed to wear a head-mounted display (HMD; Vuzix iWear VR920) and to watch two stimulus images: the geminoid's right little finger being bent (Fig. 18.2, center) and an experimenter injecting the top of the geminoid's right hand (Fig. 18.2, right). The participants watched these images several times, as in Armel et al.s study [7]. They, then practiced again using the HMD . After this preparation, the main trials were conducted. After all the trials were completed, the participants filled the questionnaire.

## 18.2.2  Main Trial

In the main trials, we repeated the following procedure six times. First, the participants operated the geminoid's right arm by moving their right arm in a horizontal direction at 3 s intervals. They watched the geminoid's arm movement through the

**Fig. 18.2** Simulated views shown to participants: (left) normal view showing arm movement range; (center) "finger bending stimulus"; (right) "injection stimulus"

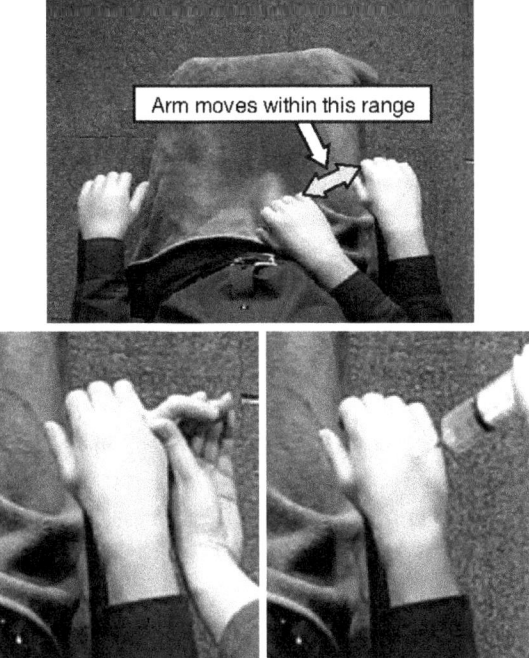

HMD. At this time, they were instructed to look downward so that their posture would be synchronized with that of the geminoid on the HMD, because posture synchronization is important for body ownership transfer. We covered the HMD with a black cloth so that participants could see only the display, because we believed that the extent of body ownership transfer would decrease if participants could see the scene of the experiment room and their own body. Moreover, both the participants and the geminoid wore a blanket to prevent differences in clothing from influencing the body ownership transfer. Figure 18.3 shows the participants during the experiment.

After 1 min of operation, the experimenters signaled to the participants that the test had ended and waited until their SCR returned to normal. The participants were then shown one of the two stimulus images.

## 18.2.3   Experimental System

In this experiment, we required a system for controlling the extent of the delay of the geminoids movement. However, we could not operate the geminoid without delays because of its limitations. We therefore employed a simulation system. That is, we employed a method that allowed us to select and display pictures created by splitting

**Fig. 18.3** Participant setting

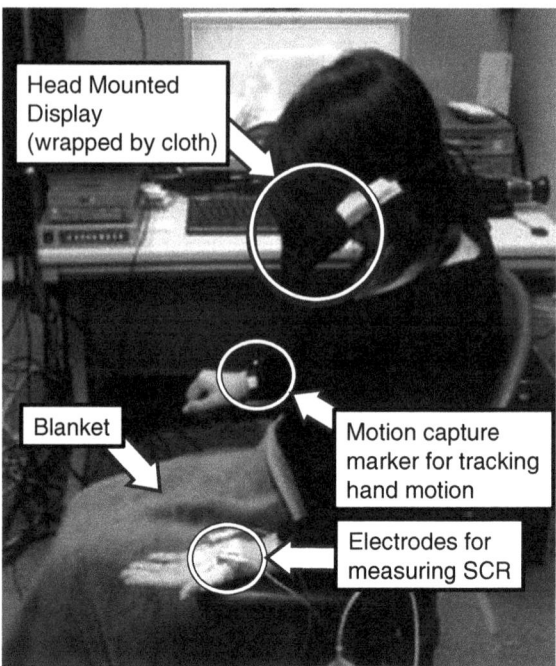

images that had been captured previously, based on the position tracked by a motion capture system (the Motion Analysis HAWK-I camera and control system). Using this system, we implemented a condition where the movement of the geminoid was arbitrarily delayed, as well as a condition where its movements were mostly synchronized with those of the operator. We captured pictures using a high-speed camera (Casio EXILIM EX-F1) at 300 fps and used 5,000 pictures (Fig. 18.2, left). We showed the stimulus images after the operation images, switching to these images from the operation image.

## 18.2.4 Measurement

We collected data using a self-report questionnaire and SCR to evaluate the extent of the participant's body ownership transfer. First, the self-report questionnaire included the following questions:

- (Finger bending) Did it feel as if your finger was being bent?
- (Injection) Did it feel as if your hand was being injected?

Participants filled the questionnaire orally because they were wearing an HMD and therefore could not see to write. They gave their responses on a 7-point Likert scale (1: not strong, 7: very strong) for both questions.

We used a Coulbourn Instruments V71 23 Isolated Skin Conductance Coupler as the biological amplifier; the sensitivity was set to 100 mV/microS, and measurements were taken using a direct current power distribution method. We used a KEYENCE NR-2000 as the A/D converter and set the sampling period to 20 ms (50 Hz).

### 18.2.5 Conditions and Predictions

According to our hypotheses stated in Sect. 18.1, we applied the following three conditions:

*Sync* condition:    The geminoids movement was synchronized with the operators movement without a delay.

*Delay* condition:   The geminoids movement was synchronized with the operators movement with a certain delay.

*Still* condition:    The geminoid remained motionless, although the operator moved.

In the delay condition, the delay was set to 1 s, following Armel et al. study in [7]. On the basis of our hypotheses, we predicted the following.

**Prediction 1:**    The participants would show larger responses in the sync condition than in the still condition.

**Prediction 2:**    The participants would show larger responses in the delay condition than in the still condition.

To verify the above predictions, each participant took part in a total of six trials consisting of the above three conditions for each of the two stimuli, and we examined whether body ownership transfer occurred. The order of the conditions and stimuli was counterbalanced among participants.

## 18.3  Results

In general, skin conductance reacts with a 1–2 s delay to a stimulus [8]. Therefore, usually the maximum value between the point at which the stimulus is applied and the point at which the SCR returns to normal is measured to verify the reaction to the stimulus. However, we confirmed that skin conductance frequently shows a reaction before the stimuli is given. The participants responded in post-trial interviews that they felt as if they received the injection stimuli in their own hands, and it felt unpleasant when the experimenter's hand approached the geminoid's hand after the teleoperation. This suggests that both the reaction to the injection stimulus as well as those before the injection were caused by the body ownership transfer. In these stimulus images, the time between the appearance of the experimenters hand and the application of the stimulus was 3 s. Thus, in this study, we set the starting time of

the range for obtaining the maximum value to 2 s after the stimulus image started
(1 s before the stimulus was applied) and the end time to 5 s after the stimulus was
applied, as in Armel et al. study [7]. Thus, we used a range for obtaining the maximum value from 1 s before the stimulus was applied to 5 s after it was applied.

We now describe the data that were excluded from the evaluations. First, the data
of 2 of the 19 participants were excluded because in their case the electrode for the
SCR failed. Second, we excluded data that were identified as outliers by the Smirnov-
Grubbs test using a significance level of 5%. We excluded one bit in both the delay
and still conditions under the injection stimulus and four bits in the still condition
under the injection stimulus. Finally, we analyzed 17 bits of the sync condition data,
16 bits of the delay condition data, and 16 bits of the still condition data under the
finger bending stimulus, and 17 bits of the sync condition data, 17 bits of the delay
condition data, and 13 bits of the still condition data under the injection stimulus.

The SCR value is normalized by a logarithmic transformation [8]. Therefore, we
logarithmically transformed the SCR value and conducted a parametric test. The
results of a one-way ANOVA confirmed that no significant difference was caused by
the finger bending stimulus ($F(2) = 0.66$, $p = 0.52$), but that a significant difference
was caused by the injection stimulus ($F(2) = 3.36$, $p < 0.05$). The results of a Tukey
HSD multiple comparison test confirmed that a significant difference existed only
between the sync and still conditions under the injection stimulus (sync condition >
still condition, $p < 0.05$). Figure 18.4 shows the average with standard error and the
results of the Tukey HSD multiple comparison test.

We also performed a statistical analysis of the data collected from the responses
to the self-report questionnaire. The results of a one-way ANOVA confirmed that
there was no significant difference under the finger bending stimulus ($F(2) = 2.88$,
$p = 0.06$), but a significant difference under the injection stimulus ($F(2) = 5.25$, $p <
0.01$). The results of the Tukey HSD multiple comparison test confirmed a significant
difference only between the sync and still conditions under the injection stimulus
(sync condition > still condition, $p < 0.01$). Figure 18.5 shows the average with the
standard error and the results of the multiple comparison Tukey HSD test.

**Fig. 18.4** Average with
standard error of SCR
analysis

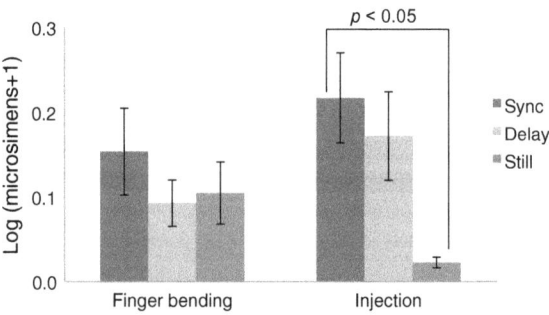

**Fig. 18.5** Average with standard error of questionnaire analysis (left: *did it feel as if your finger was being bent?*, right: *did it feel as if your hand was being injected?*)

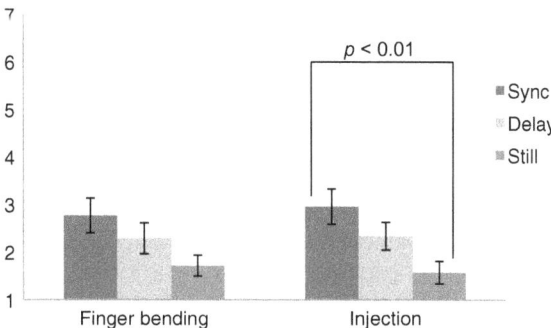

## 18.4   Discussion

According to the results, we now evaluate our predictions and hypotheses. First, the results verified Prediction 1 (participants would show larger responses in the sync condition than in the still condition). In the case of the injection stimulus, the results of both the self-report questionnaire and the SCR showed a significant difference between the sync and still conditions. In both measures, the responses for the sync condition were significantly larger than for the still condition. Thus, Prediction 1 was verified and the corresponding Hypothesis 1 (body ownership is transferred by watching scenes where the geminoid is synchronized with an operator) was supported. That is, we confirmed that geminoid teleoperation induces body ownership transfer toward the geminoid body.

Second, the results did not verify Prediction 2 (participants would show larger responses in the delay condition than in the still condition), since no significant difference was shown between the delay and still conditions. Consequently, we cannot verify Hypothesis 2 (in geminoid teleoperation, body ownership is also transferred when the geminoid moves with delays). However, there was no significant difference between the sync and delay conditions. In past studies on the RHI, the degree of body ownership transfer was reduced by delays between the tactile and visual stimuli [3, 7]. Shimada et al. also showed that when the delay becomes larger than 300 ms, participants reported a significantly low degree of the RHI effect [5]. The result that there was no significant difference between the sync and delay (1 s) conditions in this study may suggest that the mechanism of the geminoid body ownership transfer is different from that of the RHI body ownership transfer. In this experiment, the delay was longer than during a usual teleoperation, because we set the same delay time as in related studies. Since this delay may have influenced our results, further verification is required.

A significant difference was confirmed for the injection stimulus. However, no significant difference was confirmed by the self-report and SCR data for the finger bending stimulus. In our opinion, the reasons for this are twofold, as follows. First, the body part that the participants operated differed from the body part in which they received the stimulus. Because the participants operated the geminoid's arm by moving their own arm, body ownership may have been transferred to the geminoid's hand and arm, but not to its fingers. As a result, the participants did not react to the finger bending stimulus. Second, our finger bending images were defective. Because the arm position in the first image showing finger bending differed from that in the last image, the participants watched scenes where the arm position was shifted instantaneously when the image was switched to the stimulus image. This may have had an unintended influence on the participants. None of the injection images was defective.

The first reason raises two interesting questions: Is body ownership transferred only to the body part operated by the participant? Is body ownership transferred to more widespread body parts? If we could extend body ownership to another body part by operating one part of the body, we could improve operability. Human somatic sensation systems are known to have non-uniform distribution throughout the body; some body parts are more sensitive and others less sensitive. In our opinion, it is difficult to transfer body ownership to a sensitive part. Therefore, we predict that it is difficult to extend body ownership to sensitive parts, whereas it is easy to extend it to insensitive parts. Further examination is required to verify this idea.

In previous RHI research studies, the objective measurements and the self-report data showed a high correlation, as in this study. Moreover, participants indicated high values (5–7 on the 7-point scale) in their self-reports [3]. However, in this study, the self-report average was about 3 on the 7-point Likert scale, which was relatively low. Participants, at least subjectively, did not feel as if part of the geminoid's body had become part of their body. As the participants task in this experiment involved simply moving the arm, participants were very conscious of their own body. This concentrated attention to their own body may have been the reason for the low scores in the self-report evaluation. In their daily life, people are rarely aware of their body movements for fulfilling everyday tasks, such as walking or grasping. It is easier to cause body ownership transfer when the operator waits 5 min after the teleconversation to start adjusting the operation than when the operator starts sooner. We believe that we can increase the extent of body ownership transfer by ensuring that the participants do not experience feelings in their body after they have started concentrating on their task.

The appearance of the object may also influence the extent of the body ownership transfer, as reported by Armel et al. [7] and Pavani [9]. Is it possible to cause body ownership transfer during teleoperation only when the android robots appearance is very humanlike? Will it occur when humanoid robots with robotic appearances or industrial robots are teleoperated? Future studies will answer these questions.

## 18.5  Conclusion

In this study, we examined whether body ownership transfer occurs through the teleoperation of an android robot. We showed, by synchronizing the operation of the robot and its visual feedback, that operators can be caused to feel as if a part of the robot has become a part of their own body. We believe that clarifying the mechanism of body ownership transfer is important for resolving fundamental issues such as how body contributes to organizing our self and how our sense of self-other boundaries are formed and recognized. Future work may investigate further why this body ownership transfer occurs and finds the necessary conditions for causing the sense of body ownership transfer to occur intentionally.

# References

1. Nishio, S., T. Watanabe, K. Ogawa, and H. Ishiguro. 2012. Body ownership transfer to teleoperated android robot. In *International conference on social robotics*, 398–407.
2. Nishio, S., H. Ishiguro, and N. Hagita. 2007. Geminoid: Teleoperated android of an existing person. In *Humanoid robots: New developments*, ed. A. de Pina Filho. Vienna, Austria: I-Tech Education and Publishing.
3. Botvinick, M. 1998. Rubber hands 'feel' touch that eyes see. *Nature* 391 (6669): 756.
4. Tsakiris, M. 2010. My body in the brain: A neurocognitive model of body-ownership. *Neuropsychologia* 48 (3): 703–12.
5. Shimada, S., K. Fukuda, and K. Hiraki. 2009. Rubber hand illusion under delayed visual feedback. *PLoS One* 4 (7): e6185.
6. Lang, P.J. 1993. Looking at pictures: Affective, facial, visceral, and behavioral reactions. *Psychophysiology* 30 (3): 261–273.
7. Armel, K.C., and V.S. Ramachandran. 2003. Projecting sensations to external objects: Evidence from skin conductance response. *Proceedings of Biological Sciences* 270 (1523): 1499–1506.
8. Hori, T., and Y. Niimi. 1985. Electrodermal activity (in Japanese). In *Physiological psychology*, ed. H. Miyata, K. Fujisawa, and S. Kakiki, 98–110. Tokyo: Asakura Publishing.
9. Pavani, F. 2000. Visual capture of touch: Out-of-the-body experiences with rubber gloves. *Psychological Science* 11 (5): 353–359.

# Chapter 19
# Effect of Perspective Change on Body Ownership Transfer

**Kohei Ogawa, Koichi Taura, Shuichi Nishio and Hiroshi Ishiguro**

**Abstract** We previously investigated body ownership transfer to a teleoperated android body caused by motion synchronization between the robot and its operator. Although visual feedback is the only information provided from the robot side, as a result of body ownership transfer, some operators feel as if they are touched when the robot's body is touched. This illusion can help operators transfer their presence to the robotic body during teleoperation. By enhancing this phenomenon, we can improve the communication interface and the quality of the interaction between the operator and interlocutor. In this study, we examined the effect of a change in the operator's perspective on the body ownership transfer during teleoperation. According to the results of past studies on the rubber hand illusion (RHI), we hypothesized that a perspective change would suppress the body owner transfer. Our results, however, showed that under any perspective condition, the participants felt the body ownership transfer. This shows that its generation process differs for teleoperated androids and the RHI.

**Keywords** HRI · Android · Body ownership transfer

This chapter is a modified version of a previously published paper [1], edited to be comprehensive and fit with the context of this book.

K. Ogawa (✉) · K. Taura · S. Nishio · H. Ishiguro
Laboratory ATR, 2-2-2 Hikaridai, Seikacho, Sorakugun, Kyoto 6190288, Japan
e-mail: ogawa@irl.sys.es.osaka-u.ac.jp

H. Ishiguro
Graduate School of Engineering Science, Osaka University, 1-1 Machikaneyama, Toyonaka 5600043, Japan

H. Ishiguro and F. Dalla Libera (eds.), *Geminoid Studies*,
https://doi.org/10.1007/978-981-10-8702-8_19

## 19.1   Introduction

Over the past few years, a very human-like teleoperated android called Geminoid
has been developed. Geminoid resembles the particular person on whom it was
modeled (Fig. 19.1). Research using Geminoids is expected to contribute to the
scientific fields of psychology and cognitive science [2].

Geminoid HI-1, which was used in this study (Fig. 19.1, center) has fifty
pneumatic actuators in its face and body to provide very human-like movements.
A conversation through Geminoid has the potential to improve human telecommunication as compared with such traditional telecommunication media as phones
and TV conference systems [3]. When people converse with a Geminoid, initially
they are usually distracted by the robot's appearance. However, after conversing for
a few minutes, they start interacting with it naturally as if they are talking to the real
person. When they feel as if the real person is in front of them, the operator's
presence has been transferred to Geminoid.

A Geminoid operator uses intuitive applications. Using a face recognition system, Geminoid can mimic the operator's face direction and facial expressions.
Using voice analysis, the operator's utterances and Geminoid's mouth movements
can be synchronized in real time. The operator can also observe scenes through
Geminoid.

After a period of operation during which operators become adapted to the
teleoperation system, they sometimes experience realistic sensations such that they
actually feel in possession of Geminoid's body [4]. When operators participate in a
conversation, some may even feel a poke on the robot's face as if they themselves
had been touched instead of the robot. This situation means that the Geminoid
operator feels as if his/her body is extended to Geminoid in a distant place through
operating and observing it. This phenomenon is called "body ownership transfer"
and occurs when an external object is recognized as part of one's body.

Botvinick and Cohen [4] experimentally investigated the "rubber hand illusion"
(RHI). When a participant's hand was placed out of view and a life-size rubber
hand model was placed in front of a subject, simultaneous paintbrush strokes on

**Fig. 19.1** Three Geminoids and their originals

both the rubber and participant's hands evoked a cross modal perceptual illusion of possession of the rubber hand.

Watanabe et al. [5] investigated whether the body ownership transfer illusion in fact happens during the operation of Geminoid's hands using only visual and movement synchronization. Their results showed that when the hand movements of Geminoid and participants were synchronized, a sense of ownership caused the subject to react when the robot's hand was threatened with a pain-causing stimulus such as a needle.

In Geminoid teleoperation, this illusion causes the operators to feel both kinetic (sense of moving their hands) and visual sensations (video stream taken by video camera). However, the body ownership transfer in Geminoid teleoperation remains unclarified, in particular from the perspective of visual sensations. In this study, we focused on the operator's visual sensation to clarify the relationship between body ownership transfer and visual feedback to operators.

We humans usually use "first-person view" in our daily life because our eyes are located toward the top of our face. In related RHI research, the first-person view was used. However, in Geminoid teleoperation, we have empirically employed the "third-" and "second-person views (views from the other person)," because operators sometimes claimed that it is difficult to conduct natural conversations with an interlocutor through Geminoid using only the first-person view, since they could not observe themselves. However, they felt as if Geminoid was their own body even if they did not use the first-person view.

In this paper, we report an effect of a perspective change on body ownership transfer during Geminoid teleoperation, since such knowledge may help researchers in their development of teleoperated robots. If body ownership transfer can be controlled, it can be applied to a broad range of fields, for example, prosthetic arms or telecommunication media.

## 19.2  Related Studies

Research on the transfer of bodily sensation and ownership has been widely conducted in recent years. Researchers have applied different methods (e.g., out-of-body experiences induced by the visual and multisensory input [6], visual-somatosensory input in virtual reality [7], body swapping experience by manipulating the visual and the multisensory input [8], and body representation in the brain using fMRI [9]) to induce and evaluate the illusion of bodily feeling extension to an object.

Botvinick and Cohen [5] provided the first description of the RHI using paintbrush strokes on both rubber and participants' hands. Using the skin conductance response (SCR), which measures autonomic nervous system arousal in the anticipation of pain, Armel and Ramachandran [10] investigated the transfer of body sensations to a rubber hand. They performed an RHI procedure in two cases used by Botvinic et al. and Armel et al., where they bent a finger of the rubber hand backwards, which obviously appeared painful. The results showed a high SCR value at the time when the finger was bent when the strokes on the two hands were synchronized. Slater et al. [11] conducted similar experiments using a computer-simulated environment. They reported the illusion of ownership displacement through tactile stimulation on a person's hidden real hand synchronized with virtual visual stimulation on a virtual arm projected from the person's shoulder.

In all the above studies and similar rubber hand experiments, the illusion is evoked by multisensory input of synchronous tactile stimulation and visual feedback. However, the transfer of ownership for other perceptual correlations should be confirmed.

Pavani investigated the effect of variance between a participant's hand and a rubber hand [12]. They compared a case where the rubber hand faced the same direction as the participant's hand and another where the rubber hand was rotated 90°. Their results showed that if the direction of the rubber hand is not the same as that of the participant's hand, it interferes with the body ownership transfer. Consistency between the human's own posture and that of the rubber hand is one important factor for body ownership transfer.

## 19.3  Experiment

Through the results of an experiment, we demonstrate an effect of perspective differences for body ownership transfer during Geminoid teleoperation.

### 19.3.1  Hypotheses and Predictions

We conducted the experiment according to the following hypothesis.

Hypothesis: The body ownership transfer is observed not only in the first-person perspective (1PP) condition, which is the highly synchronized condition, but also in other conditions: the second-person perspective (2PP) condition and mirror condition.

### 19.3.2  Conditions

This experiment examines the effect of a perspective change on the body ownership transfer in teleoperation situations using Geminoid. We prepared the following three perspective conditions:

**Fig. 19.2** First-person perspective (1PP)

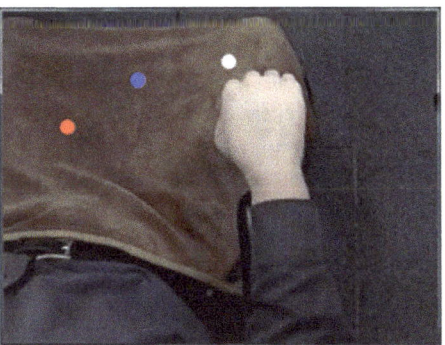

- 1PP: Geminoid's own perspective (Fig. 19.2)
- 2PP: perspective from Geminoid's opposite side (Fig. 19.3)
- Mirror: inverse perspective that resembles a mirror (Fig. 19.4).

We expected that body ownership transfer would occur under the 1PP condition because this has already been investigated in related studies. We found empirically that the 2PP view is sometimes appropriate for teleoperation and that the operator felt the body ownership transfer under the 2PP condition. Therefore, we employed

**Fig. 19.3** Second-person perspective (2PP)

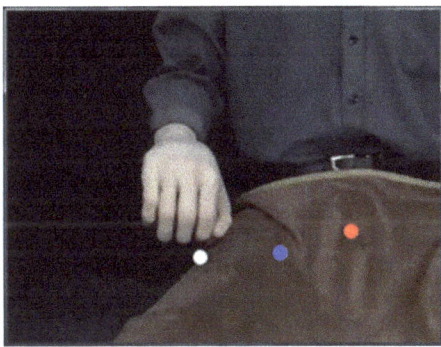

**Fig. 19.4** Inverse perspective like a mirror (mirror)

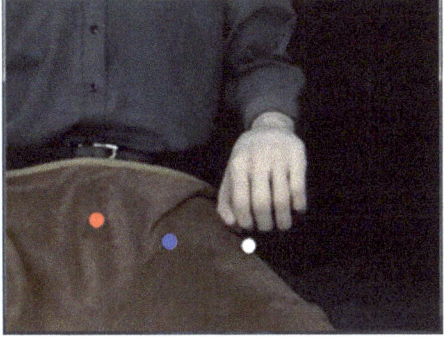

this condition in our experiment. Under 2PP, when a participant moves his/her right hand, the left side hand moves for the participant. This perspective is not usual in daily life. Therefore, we added the mirror condition, which is the inverse perspective of the 2PP condition.

To investigate the body ownership transfer, we observed the differences in movement synchronization of a participant and Geminoid under three conditions. Each participant experienced three different movement synchronizations under one point of view: 1PP, 2PP, or mirror.

The following were the movement synchronization conditions:

- Sync: the participant and Geminoid's hand moved in synchronization.
- Delay: the participant and Geminoid's hand moved in synchronization with a 500-ms delay.
- Still: Geminoid's hand never moved even if the participant moved his/her hand.

After they experienced each trial, the participants controlled Geminoid's hand by using their own hand movements. The stimulus image, which is a movie that shows Geminoid's hand being injected with a needle, was presented after each trial. When the stimulus is presented, some participants may react only to the "threat" of the needle. Therefore, the movement of the needle was stopped 3 s after its appearance on the screen and it was then inserted into Geminoid's hand. To avoid the participant's being unaware of the injection, the needle was injected for 3 s and removed from sight.

### 19.3.3   Procedures

The experiment consisted of two stages: training and testing. Forty-three healthy, right-handed participants (25 men, 18 women) whose ages ranged from 18 to 25 years (M = 20.7, SD = 1.7) were selected. All the participants learned that Geminoid is a robot by watching the real Geminoid in the experiment room before the experiment started. Then, the experimenter explained the experiment. All the participants signed consent forms. This experiment was approved by ethical committee of Advanced Telecommunications Research International Institute (11-506-1).

Participants wore infrared markers on their wrists and electrodes on their palms. The infrared marker was part of the motion capture system for controlling the Geminoid. The electrodes measured the skin conductance response (SCR) values. The participants then wore a head-mounted display (HMD) (Vusic iWear VR920) on which they watched the stimulus movie.

Next, they watched the entire stimulus movie several times to habituate them to the injection. Then, they practiced controlling Geminoid using their own hands for the testing trial. After the testing trial, they answered questionnaires.

### 19.3.4   Geminoid Teleoperation System for the Experiment

In our experiment, it was necessary to control Geminoid's movement delay very precisely. However, such precise control was very difficult, because Geminoid's actuators are driven by pneumatic cylinders. Therefore, we filmed Geminoid's hand movements at 300 frames per second using a high-speed camera and divided the movie into still pictures frame by frame and obtained over 1000 still pictures. To maintain the fixed delay time, the still pictures were displayed in accordance with the participant's hand movements captured by the motion capture system so that smooth visual feedback was achieved with and without a fixed delay.

The participant's hand movements were captured by two infrared markers, one installed on the wrist and the other on the chair's armrest. The armrest marker was positioned 19 cm from the participants' elbow to absorb the size differences of each participant. To avoid biases caused by the experiment room environment and the participants' bodies, the HMD was surrounded by a black cloth to help the participants concentrate on the stimulus movie.

### 19.3.5   Testing Trial

In the test trial, we repeated the following procedures three times. Participants observed the stimulus movie from the perspectives shown in Figs. 19.2, 19.3 and 19.4. Three small, red, blue, and white circles were superimposed on the stimulus movie. The participants controlled Geminoid's hand into the indicated color position for 3 s by sound. To provide consistency between the presented perspective and their posture, the participants bent their heads to see their own arms under the 1PP condition. Under the 2PP and mirror conditions, they looked toward the front. To avoid any clothing bias caused by the difference between the participants and Geminoid's clothing, the laps of both were covered by identical blankets

**Fig. 19.5** Experimental setup

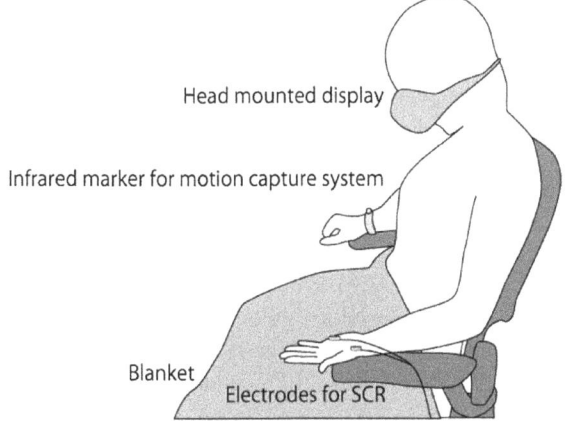

Head mounted display

Infrared marker for motion capture system

Blanket

Electrodes for SCR

(Fig. 19.5). After a 1-min trial, the experimenter signaled that the trial was complete, and the participants took a short break until their SCR values returned to normal. Then, we presented the stimulus movie.

### 19.3.6 Measure

We employed two measures to verify the body ownership transfer. One was SCR, and the other was subjective evaluations of the participants who indicated their sense of body ownership transfer by answering the following questions:

- Question 1 (Q1). Did you feel that your hand was also the robot's hand?
- Question 2 (Q2). Did you feel as if your own hand was receiving an injection?

Participants answered the questionnaires orally because their sight was blocked by the HMD.

The responses to the questions in the questionnaires were given on a seven-point Likert scale, where three denotes the most positive point. To measure the SCR, the main unit we used was a Polymate AP-216 and the amplifier was an AP-U030. Electrodes were affixed to the participants' thenar and hypothenar eminences.

## 19.4 Results

### 19.4.1 Subject Evaluation

The participants' subjective impressions of the body ownership transfer in the movement synchronization for each perspective were evaluated with an ANOVA. The results for Q1, "Did you feel that your hand was also the robot's hand?" showed a significant difference among the three movement synchronization conditions: (1PP condition: $F(2, 28) = 16.40$, $p < 0.001$; 2PP condition: $F(2, 26) = 30.47$, $p < 0.001$; mirror condition: $F(2, 26) = 21.17$, $p < 0.001$). We conducted multiple comparisons of all the movement synchronization conditions for each perspective condition using the Sidak method. The mean values, standard deviations, and significant differences are shown in the left and center graphs of Fig. 19.6. The 1PP condition results showed that the sync condition score exceeded the delay and the still condition scores with significant differences ($p < 0.01$). Hereafter, we write these results as (Sync > Delay**, Sync > Still**). The 2PP results were Sync > Delay**, Sync > Still**, Delay > Still**. The mirror results were Sync > Delay**, Sync > Still**, Delay > Still**.

The results showed a significant difference only in the 2PP conditions for Q2: "Did you feel as if your own hand was receiving an injection?" (1PP condition: $F(2, 28) = 1.60$, $p = 0.22$; 2PP condition: $F(2, 26) = 15.27$, $p < 0.001$; mirror

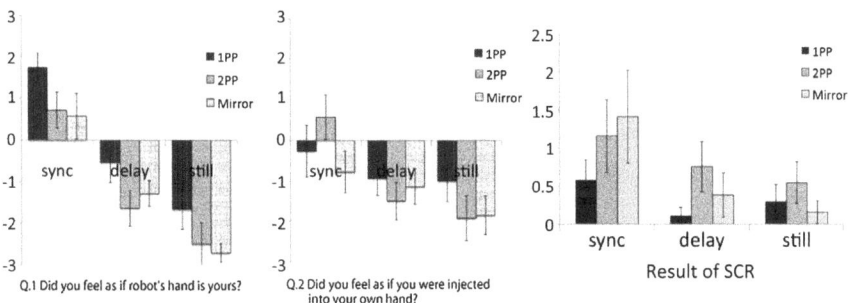

**Fig. 19.6** Results of multiple comparisons among movement synchronization condition for each perspective: left: Q1, center: Q2, right: SCR

condition: $F(2, 24) = 5.16$, $p = 0.001$). We conducted multiple comparisons among all movement synchronization conditions for each perspective condition with the Sidak method. The mean values, standard deviations, and significant differences are shown in the left and center of Fig. 19.6. The 2PP results were Sync > Delay**. We found no significant differences in the other conditions.

### 19.4.1.1   Comparisons Between Perspective Conditions

The participants' subjective impressions of the body ownership transfer under the perspective differences for each movement synchronization condition were evaluated with an ANOVA. The results showed no significant difference for either Q1 (sync condition: $F(2, 40) = 2.11$, $p = 0.13$, delay condition: $F(2, 40) = 1.56$, $p = 0.22$, still condition: $F(2, 40) = 2.00$, $p = 0.15$) or Q2 (sync condition: $(2, 40) = 1.56$, $p = 0.22$, delay condition: $F(2,40) = 0.30$, $p = 0.74$, still condition: $F(2, 40) = 1.18$, $p = 0.32$). The mean values, standard deviations, and significant differences are shown in the left and center of Fig. 19.7.

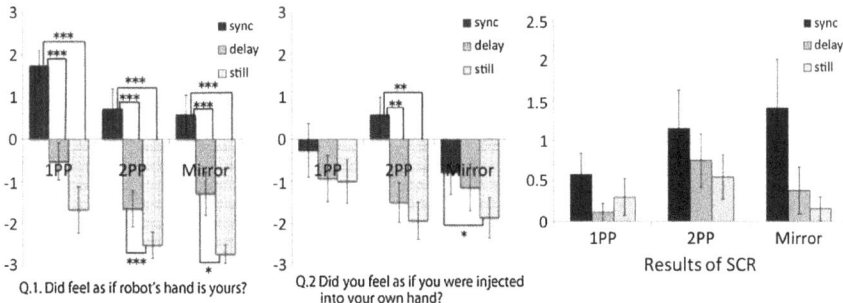

**Fig. 19.7** Results of multiple comparisons among aspects for each movement synchronization: left: Q1, center: Q2, right: SCR. ($*p < 0.05$, $**p < 0.01$, $***p < 0.001$)

## 19.4.2  SCR Evaluation

Before the evaluation, the SCR results of 9 of 43 participants whose SCR showed
no response were eliminated from the evaluation target. SCR data, which are
basically detected 1–2 s after the stimulus is applied, are commonly analyzed by the
peak value from 1 to 2 s after the stimulus to the end of the SCR activation. Thus,
to analyze them, we have to use the data from 1 s after the stimulus. However, the
experimental results showed that the participants' SCR sometimes responded before
the stimulus. In interviews after the experiment, some participants admitted that
they felt anxious as the needle approached Geminoid's hand, because they felt as if
Geminoid's hand was their own. Hence, we treated the pre-stimulus SCR data as
caused by the stimulus.

The time that elapses between the needle's first appearance and its injection into
Geminoid's hand is about 3 s. We selected the peak value in the duration from 2 s
after the needle's appearance until 5 s after its injection into Geminoid's hand
(Fig. 19.8).

The participants' SCR values in the movement synchronization for each perspective condition were evaluated by an ANOVA. The mean values and standard
deviations are shown on the right of Fig. 19.6. The SCR values of 11 participants in
1PP, 11 participants in 2PP, and 10 participants in the mirror condition were
evaluated. The results showed no significant differences between the three move-
ment synchronization conditions in each perspective condition: 1PP condition: $F_{(2, 20)} = 1.30$, $p = 0.30$; 2PP condition: $F_{(2, 20)} = 1.57$, $p = 0.23$; mirror condition:
$F_{(2, 18)} = 3.40$, $p = 0.06$.

The participants' SCR values in the perspective conditions for each movement
synchronization condition were evaluated by an ANOVA. The mean values and the
standard deviations are shown on the right of Fig. 19.7. The number of evaluated
participants was the same as in the previous analysis. The results showed no sig-
nificant differences between the perspective conditions: sync condition: $F_{(2, 29)} = 0.84$, $p = 0.44$; delay condition: $F_{(2, 29)} = 1.60$, $p = 0.22$; still condition: $F_{(2, 29)} = 0.75$, $p = 0.48$.

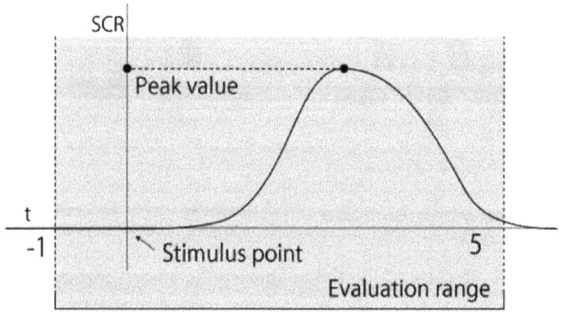

**Fig. 19.8** SCR evaluation range

As a further evaluation, we focused only on the movement synchronization based on the perspective conditions. The value of the three perspective conditions was summed for each movement synchronization condition and evaluated by an ANOVA. The results showed a significant difference between the three movement synchronization conditions: $F(2, 72) = 6.135$, $p < 0.01$. We conducted multiple comparisons of all movement synchronization conditions for each perspective condition with the Sidak method. The results show that the sync condition score exceeded the delay and still condition scores with significant differences ($p < 0.05$).

## 19.5   Discussion

In this paper, we investigated the relationship between the perspective of the visual feedback to Geminoid operators and body ownership transfer. We expected that the conditions not only of the first-person view but also of the third-person view would evoke the body ownership transfer illusion. However, the experimental results did not clearly support our hypothesis.

The results for Q1 indicated that, even if the perspective was changed from 1PP to 2PP or mirror, the participants felt as if Geminoid's hand was their own hand. However, the results for Q2 show that they did not actually feel that their own hand was being injected. The differences between the results for Q1 and Q2 showed that the participants may have felt the body ownership transfer, but they thought the needle was actually being injected, not into their own hand, but into Geminoid's hand. We expected the body ownership transfer to occur, but its intensity was not in fact sufficient to make them feel the injection on their own hand.

The SCR results, however, demonstrated that the participants' SCR measures responded to the stimulus under each perspective condition. Although their subjective impressions were that the needle was in fact being injected, not into their own hand, but into Geminoid, they unconsciously responded to the injection stimulus under all perspective conditions. We believe that the 2PP and mirror perspectives may also arouse the body ownership transfer illusion.

The participants felt the body ownership transfer in the mirror perspective because they recognized their own reflections in the mirror, where people are used to seeing themselves. Gonzalez et al. described a similar phenomenon where body ownership was transferred to agents seeing themselves in mirrors [13]. The participants felt the body transfer ownership in the mirror aspect. The 2PP perspective is less commonly experienced, because people do not get the chance to experience it in their daily life. In this perspective, the participants seemed to experience difficulty controlling Geminoid's hand. However, perhaps this difficulty in the 2PP perspective increased their concentration so that they felt the body ownership transfer. The Q2 results suggest that there was a significant difference only in the 2PP condition (Fig. 19.5, center). In our opinion, the 2PP and mirror perspectives can also arouse the body ownership transfer illusion.

## 19.6    Conclusion

In this paper, we investigated the relationship between the body ownership transfer illusion and alterations in perspective. To demonstrate the relationship, we focused on the illusion that the Geminoid teleoperator sensed as a result of the synchronization of the kinetics and the visual sensation. We described an experiment to clarify the effect of perspective difference, which had not been previously investigated in relation to the body ownership transfer illusion in Geminoid teleoperation. We employed two measures to verify the body ownership transfer: SCR and a subjective evaluation. The results showed that not only the first-person view but also the third-person and the mirror views may cause body ownership transfer.

**Acknowledgements** This research was partially supported by KAKEN-2022000 and by KAKENHI-24650114.

## References

1. Ogawa, K., K. Taura, S. Nishio, and H. Ishiguro. 2012. Effect of perspective change in body ownership transfer to teleoperated android robot. In *Proceedings of the 21st IEEE international symposium on robot and human interactive communication (RO-MAN 2012)*, 1072–1077.
2. Ishiguro, H. 2005. Android science: Toward a new cross-disciplinary framework. In *CogSci 2005*, 1–6.
3. Sakamoto, D., T. Kanda, T. Ono, H. Ishiguro, and N. Hagita. 2007. Android as a telecommunication medium with a human-like presence. In *Proceedings of the ACM/IEEE international conference on human-robot interaction*, 193–200.
4. Botcinick, M. 1998. Rubber hand 'feel' touch that eyes see. *Nature* 391 (6669): 756.
5. Watanabe, T., S. Nishio, K. Ogawa, and H. Ishiguro. 2011. Body ownership transfer to an android by using tele-operation system. *Transaction of IEICE* J94-D (1): 86–93 (in Japanese).
6. Ehrsson, H.H. The experimental induction of out-of-body experiences. *Science* 317: 1048.
7. Lenggenhager, B., T. Tadi, T. Metzinger, and O. Blanke. Video ergo sum: Manipulating bodily self-consciousness. *Science* 317: 1096.
8. Petkova, V.I., and H.H. Ehrsson. If I were you: Perceptual illusion of body swapping. *PLoS ONE* 3 (12): e3832. https://doi.org/10.1371/journal.pone.0003832.
9. Tsakiris, M., G. Prabhu, and P. Haggard. 2006. Having a body versus moving your body: How agency structures body-ownership. *Consciousness and Cognition* 15: 423–432.
10. Armel, K.C., and V.S. Ramachandran. 2003. Projecting sensations to external objects: Evidence from skin conductance response. *Proceedings: Biological Sciences* 270 (1523): 1499–1506.
11. Slater, M., D. Perez-Marcos, H.H. Ehrsson, and M.V. Sanchez-Vies. 2008. Toward a digital body: The virtual arm illusion. *Frontiers* 2. https://doi.org/10.3389/neuro.09.006.
12. Pavani, F. 2000. Visual capture of touch: Out-of-the-body experiences with rubber gloves. *Psychological Science* 11 (5): 353–359.
13. Gonzalez-Franco, M., D. Perez-Marcos, B. Spanlang, and M. Slater. 2010. *The contribution of real-time mirror reflections of motor actions on virtual body ownership in an immersive virtual environment*, 111–114. Virtual Reality: IEEE.

# Chapter 20
# Body Ownership Transfer by Social Interaction

**Shuichi Nishio, Koichi Taura, Hidenobu Sumioka**
**and Hiroshi Ishiguro**

**Abstract** Body ownership transfer (BOT) comprises the illusion that we feel external objects as parts of our own body, which occurs when teleoperating android robots. In past studies, we investigated the conditions under which this illusion occurs. However, these studies were conducted using only simple operation tasks, such as moving only the robot's hand. Does this illusion occur during more complex tasks, such as conducting a conversation? What kind of conversation setting is required to invoke this illusion? In this study, we examined the manner in which factors in social interaction affect the occurrence of BOT. Participants conversed using the teleoperated robot under different conditions and teleoperation settings. The results revealed that BOT does occur during the task of conducting a conversation, and that the conversation partner's presence and appropriate responses are necessary to enhance BOT.

**Keywords** Body ownership · Teleoperation · Android robot
Social interaction

## 20.1 Introduction

When people use a tool, they often feel as if the tool has become a part of their body. When holding a stick, they can "feel" whether the stick is hitting a hard or soft object or which part of the stick has hit an object. When driving a car, people gradually learn

---

This chapter is a modified version of a previously published paper [1], edited to be comprehensive and fit with the context of this book.

---

S. Nishio (✉) · K. Taura · H. Sumioka · H. Ishiguro
Advanced Telecommunications Research Institute International,
Keihanna Science City, Kyoto, Japan
e-mail: nishio@botransfer.org

K. Taura · H. Ishiguro
Department of Systems Innovation, Graduate School of Engineering Science,
Osaka University, Osaka, Japan

© Springer Nature Singapore Pte Ltd. 2018
H. Ishiguro and F. Dalla Libera (eds.), *Geminoid Studies*,
https://doi.org/10.1007/978-981-10-8702-8_20

the dimensions of the car and how the car moves when they steer it. That is, the act of manipulating the object and the visual and/or tactile feedback evokes this illusion. This phenomenon is described as an extension of the body schema; it is known also to occur in monkeys [2] and even in hermit crabs [3]. This phenomenon not only provides clues for investigating the manner in which people recognize their own body, but is also important for obtaining finer tool skills. Good baseball or tennis players who use bats and rackets, respectively, have a strong sense of body schema extension to their tools. Good car drivers can feel the size of the car.

There is one common factor in all of these examples: the user always drives or manipulates the tools and always touches them. That is, the user has control of the tool and always receives tactile feedback, frequently accompanied by visual feedback. This "restriction" can also be understood from the notion of *eccentric projection* [4]. Eccentric projection is a localization of a sensation at the position in space of the stimulus object rather than at the point where the sense organ is stimulated [5]. When people see a red apple, they feel that it is the apple that is red, and do not feel that their retina has been painted red. It is said that visual sensations are projected eccentrically in most cases, but touch sensations are projected only when people have tactile feedback, such as when they are holding the target object [5]. This redirection is possible, as all of the senses that people "feel" are in fact projected sensations based on the activity in the cerebral cortex. However, this redirection does not always operate "correctly," and may result in strange phenomena, such as the phantom limb phenomenon [6]. The body schema extension may be recognized as one of these results.

Another interesting example of this phenomenon is the rubber hand illusion (RHI) [7]. An experimenter repeatedly strokes a participant's hand and a rubber hand simultaneously, while the participant can see only the rubber hand and not his/her own hand. After some time has passed, the participant begins to feel as if the rubber hand is his/her own hand. This illusion, the RHI, is considered to occur as a result of synchronization between the participants visual stimulus (watching the rubber hand being stroked) and tactile stimulus (feeling that his/her hand is being stroked) [8]. In the RHI, the external object onto which the sensation is projected is not touched by the participant; however, the tactile stimulus is provided simultaneously to both the object and the body. Therefore, this can be considered a variation of body schema extension.

Now, is it possible to project a persons tactile sensation to an external object without such real tactile stimuli being applied to his/her body? If such a projection is possible, this may be a strong factor in achieving skillful teleoperation of robots where the target object (robot) is located at a remote location, and therefore, no tactile feedback is possible without an expensive and complicated system. There is an emerging necessity for teleoperated robots in many fields, for example telesurgery and work in dangerous areas such as nuclear plants, or as an intimate communication device. The realization of the projection of sensation to these robotic devices would lead to an improved operation experience and a stronger sense of a remote presence.

The results of our previous study showed that operators of teleoperated android robots (*Geminoids*) felt as if they were being touched when the remotely located robot was touched, even without tactile feedback [9]. This illusion, the

*Body Ownership Transfer (BOT)* illusion, occurs as a result of the synchronization between the act of teleoperating the robot and the visual feedback of the robot's motion [10, 11]. Since our previous study, many studies on BOT have been conducted, and it was also found that BOT has stronger persistence under temporal delays as compared to RHI [11], BOT occurs when various teleoperation interfaces are used [12], and that the will to control the robot (agency) and seeing the robot in motion (visual feedback) are alone sufficient to cause BOT, and thus, the operator's sense of proprioception is not required [13]. However, BOT has been examined only in manipulation tasks where a teleoperator controlled the arm of the robot, either by motion mapping or through a brain–machine interface. Originally, we found that interaction through the geminoid caused BOT. There may be elements in social interaction with others that affect the teleoperator's sense of body ownership of the robot that has not been examined in previous studies of RHI and BOT. The likelihood of this is quite high, as there were no means for performing such investigations before teleoperated androids were created.

In this study, we investigated which elements of social interaction evoke the BOT illusion. Social interaction includes many elements that are not seen in manipulation tasks and is in general believed to require more abstract, higher-level processing than simple tasks. In this study, we initially examined two factors of interaction:

(1) Existence of conversation partner
    Although it is possible for a person to conduct a conversation without seeing his/her partner (as in phone calls), the conversation is enhanced when the partner is in sight. However, the partner can also be considered a "target for the act of interaction. Thus, in this study, we employed the visibility of the partner as one factor of interaction.
(2) Degree of interaction
    The appropriate response reactions of the partner constitute the second crucial factor for enhancing interaction. In the case of a simple manipulation task, the expected response is a physical change in the target object (such as in its shape or location). In interaction, when people speak, they expect others to show some reaction, which can be considered the "feedback for the teleoperation act.

In addition, we included in our investigation whether the participants have control of the robot in a remote location as another factor in the testing. By including this teleoperation factor, we can observe the interaction effect between the above factors and the teleoperation. In addition, our objective was to confirm whether the teleoperation performed in this experiment was sufficient to induce BOT, given that the amount of motion of the robot resulting from teleoperation was much smaller than in previous studies; in previous studies, the robot's arm was moved, which resulted in a rather large amount of motion, whereas in this study, only the head movement was visible, which was a relatively small motion.

## 20.2  Methods

### 20.2.1  *Apparatus*

In this experiment, we used a teleoperated android named "Geminoid F." Geminoids are a series of android robots with an appearance that resembles existing persons [9]. Geminoid F (Fig. 20.1, left) is a geminoid that was created to resemble a female person. It has twelve pneumatic actuators, installed mainly in its face for generating humanlike facial expressions. In our previous studies of BOT, we used Geminoid HI-1, a geminoid created to resemble a male person, as more actuators are installed in the whole body and it is suitable for manipulation tasks. In this experiment, however, we used Geminoid F, as this geminoid is better suited for the conversation task because of its ability to show more natural facial expressions.

Geminoids are teleoperated robots, and they do not move autonomously. A teleoperator controls the geminoid through its teleoperation interface from a remote room (Fig. 20.1 right, and Fig. 20.2). The voices on both sides are transmitted to each side and the teleoperator can see video images of the remote location. The operator can see both the interlocutor (near first-person view) and the face of the geminoid (second-person view). This dual display allows the operator to perform the conversation task and simultaneously recognize the operation results by seeing the robot's motions.

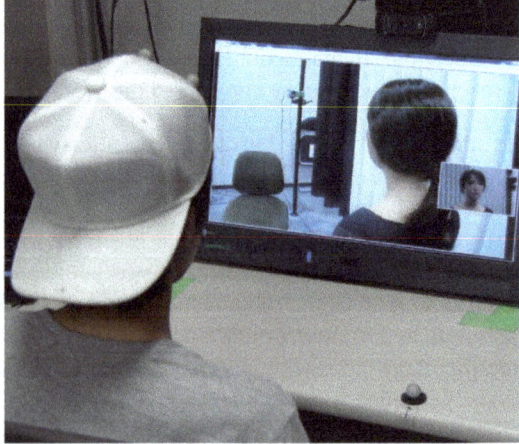

**Fig. 20.1**  Teleoperated android robot Geminoid F (left) and its teleoperation interface (right). In the teleoperation monitor, the operator can see both the interlocutor (near first-person view, main screen) and the face of the geminoid (second-person view, subscreen)

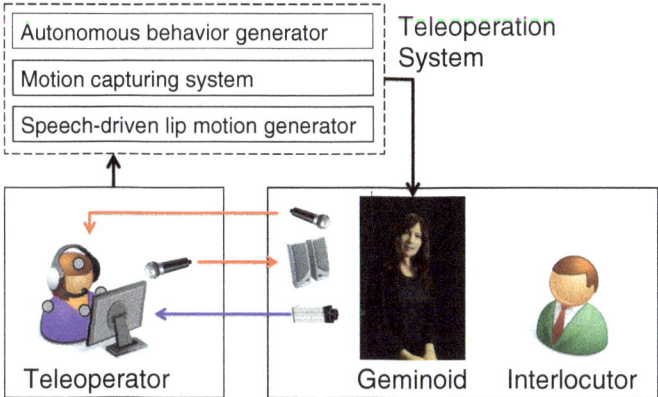

**Fig. 20.2**  Geminoid teleoperation system

The teleoperation system provides three means of controlling the robot: (1) the operator's head motions are captured and regenerated as the robots motions, (2) the operator's voice is processed and regenerated to provide the robot's lip/jaw motions [14], and (3) operators can press buttons to play pre-defined robot motions, such as smiling or bowing. As the pneumatic actuators of the geminoid take approximately 300 ms to move, the audio transmitted from the operator side is delayed to achieve synchronicity with the motion, in particular for lip synchronization.

In this experiment, however, we used only (1) head motion synchronization and (2) lip motion generation, and did not use (3) pre-defined motion, in order to have more control of the participants' teleoperation actions.

### 20.2.2   Conditions

As mentioned previously, we designed the experiment to include three factors: operation, partner presence, and response reaction, as described below. Hereafter, we represent a condition as $^{Op^{s_1}}C^{Pre^{s_2}}_{Res^{s_3}}$, where $s_1, s_2$, and $s_3$ ($s1, s2, s3 = \{+(ON)| - (OFF)\}$) indicate the existence of the operation ($Op$), partner presence ($Pre$), and response reaction ($Res$) factors, respectively. For example, $^{Op^+}C^{Pre^-}_{Res^+}$ denotes a condition where the operation is performed, the partner is not present (not visible), and conversation responses were provided. In addition, we use "$*$" (asterisk) as a *do not care* symbol to show that we do not consider a certain factor.

(*a*) *Operation* (*Op*): We investigated the influence of teleoperating the geminoid on the operators sense of body ownership using two levels of the partner presence factor: Whether the teleoperator controls the geminoid's movements or not. In the control condition ($^{Op^+}C^*_*$), the participants' movements were reflected in the geminoid. In the without-control condition ($^{Op^-}C^*_*$), the geminoid was motionless.

with reaction (*Res+*)      without reaction (*Res-*)

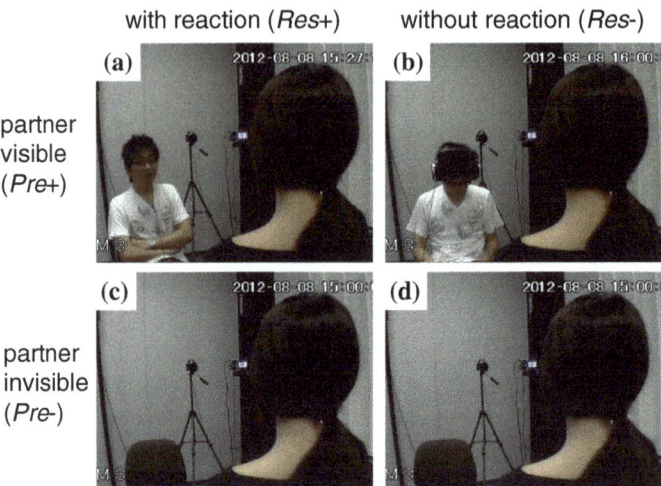

Fig. 20.3 Experimental conditions. The rows indicate the *partner presence* factor and the columns show the *response reaction* factor. These two factors were within-participant factors. In addition, another between-participant factor was used, the *operation* factor

(b) *Partner presence (Pre)*: We prepared two levels of the partner presence factor to examine whether a sense of body ownership is enhanced when teleoperators can see their interaction partners. One level indicates that the teleoperators could see their interaction partner, $^*C_*^{Pre^+}$, and the second that they could not, $^*C_*^{Pre^-}$.

(c) *Response reaction (Res)*: We examined the influence of the partner's reaction by creating two conditions: conversing with a partner ($^*C_{Res^+}^*$), and speaking without any interaction ($^*C_{Res^-}^*$). In the former condition, a teleoperator and a partner discussed given topics. If the teleoperator could see the partner ($^*C_{Res^+}^{Pre^+}$), he/she could also observe the partner's nonverbal information, such as body gestures, head nodding, and facial expressions (Fig. 20.3a). When the partner was not visible ($^*C_{Res^+}^{Pre^-}$), the participants could only hear the voice of the partner (Fig. 20.3c).

In the latter condition, a teleoperator talked without receiving any response reactions from his/her partner. When the partner was visible ($^*C_{Res^-}^{Pre^+}$), we asked the partner to look down, read a book, and wear headphones so that he/she could not hear the participant's speech, and so that the partner would not show any reactions to the speech (Fig. 20.3b). When the partner was not visible ($^*C_{Res^-}^{Pre^-}$), the participants could neither see the partner nor hear any voice (Fig. 20.3d).

## 20.2.3 Procedure

Participants completed a consent form. They were given written instructions that explained that their task was to talk with/without another participant as a partner who

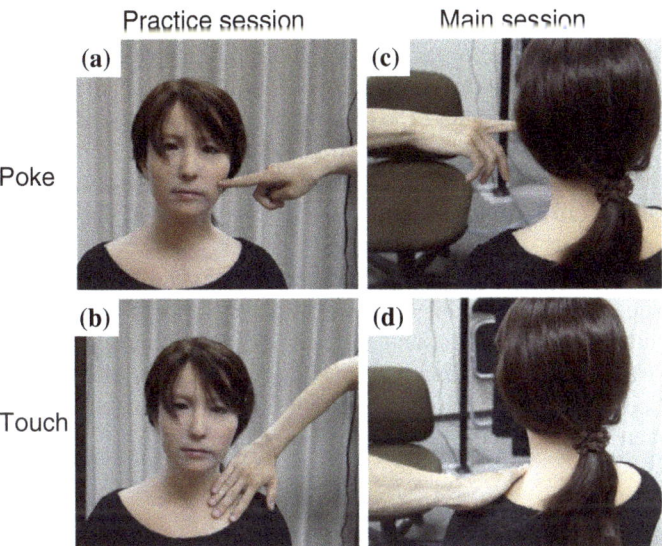

**Fig. 20.4** Images and stimuli displayed during practice and test phases. Participants watched Geminoid F's frontal face to ensure that their movements were or were not reflected by those of Geminoid F during the practice phase (a) and (b). They also watched an experimenter **a** poking and **b** stroking the geminoid. In the test phase, participants watched the geminoid and their partner from behind Geminoid F (c) and (d). They also watched an experimenter **c** poking and **d** stroking the geminoid in the same manner as in the practice phase

was in fact involved in the experiment. The participants were also given explanations about Geminoid F and its control.

From the teleoperation room, the participants assigned to the operation condition ($^{Op^+}C_*^*$) ensured that their movements were reflected by those of the geminoid by slowly shaking their heads and counting to thirty. The participants in the no-operation condition ($^{Op^-}C_*^*$) ensured that the geminoid was motionless by following the same procedure. In this phase, they watched the geminoid's face in their display monitor to ensure that their movements were or were not reflected by those of the geminoid. After they finished counting, an experimenter, who was out of their sight in the experiment room, raised his hand for 5 s to display it in the monitor, poked the geminoid's cheek three times, and stroked its neck three times (Fig. 20.4). These stimuli were presented for about 20 s. The participants then responded to the following questions on a 7-point Likert scale:

Question 1 (Q1)   Did you feel that you were controlling the robot?
Question 2 (Q2)   Did you feel as if the robot was yourself while talking?
Question 3 (Q3)   Did you feel as if you were touched when the robot was touched?
Question 4 (Q4)   Did you feel as if you were in the remote room?

After the practice phase, the participants were asked to talk about the given topics under four different conditions ($^*C_{Res^+}^{Pre^+}$, $^*C_{Res^-}^{Pre^+}$ $^*C_{Res^+}^{Pre^-}$, and $^*C_{Res^-}^{Pre^-}$) the order of which was counter balanced. Topics were chosen from two different types: common and controversial. Common topics included those that people frequently introduce in conversation, e.g., "How do you spend your holidays?"; others were related to favorite TV programs or animals. The controversial topics included capital punishment and the one thing the participant would take to a desert island. The interaction with a partner depended on the assigned condition. For example, in condition $^{Op^+}C_{Res+}^{Pre+}$, the participants conversed with the partner, while in condition $^{Op^+}C_{Res-}^{Pre-}$, the participants gave a speech about a topic.

After they talked for 4 min and 50 s, an experimenter out of the participant's sight in the experiment room raised his hand and displayed it for 5 s in the monitor and started touching the geminoid for about 20 s in the same manner as in the practice phase. The participants then filled the same questionnaire for the the condition as in the practice phase, and then they continued to the next condition. After all the conditions were completed, the participants were debriefed.

### 20.2.4 Participants

We recruited 36 Japanese university students (20 males and 16 females) and divided them into two groups: one for the operation condition $(^{Op^+}C_*^*)$ and the second for the no-operation condition $(^{Op^-}C_*^*)$. The average age of the participants was 20.2 years (S.D. = 1.9).

We used a between-participants design for the operation factor $(Op)$, and within-participants design for the partner presence $(Pre)$ and the response reaction $(Res)$ factors. We ensured that this design avoided the operation condition $(^{Op^+}C_*^*)$ influencing the no-operation condition $(^{Op^-}C_*^*)$. Moreover, performing every condition for the three factors took a considerable amount of time and degraded the concentration of the participants.

## 20.3 Results

Since the experiment was designed with three factors (operation as between-participant factor and partner presence and response reaction as within-participant factors), we used a three-way ANOVA to investigate the differences in the responses to the questions between the conditions.

### 20.3.1 Effect of the Operation Factor

The results showed that the main effect of the operation factor was significant in the responses to Q1, Q2, and Q3 (Fig. 20.5a). The participants felt more strongly

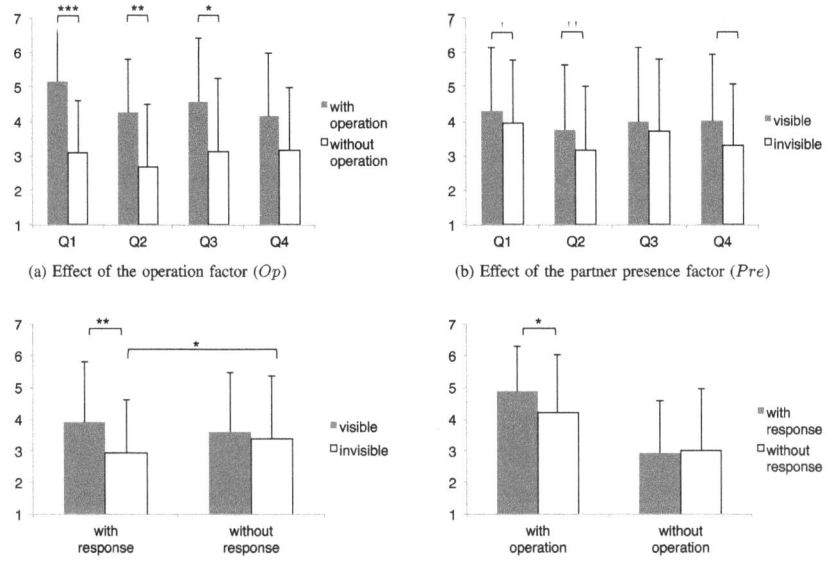

(a) Effect of the operation factor ($Op$)

(b) Effect of the partner presence factor ($Pre$)

(c) Interaction effect between the partner presence factor and the response reaction factor as shown in Q2 responses as shown in Q2 responses

(d) Interaction effect between the operation factor and the response reaction factor when the partner was visible ($Pre^+$)

**Fig. 20.5** Results of responses to questions. The graphs show averages and standard deviations. Differences having a statistical significance are marked as follows: *: $p < 0.05$, **: $p < 0.01$, ***: $p < 0.001$

as if they were managing the robot's movements in the operation condition than in the no-operation condition ($F(1, 34) = 20.68, p < 0.001$ for Q1). They strongly felt as if they were the geminoid while talking about the given topics ($F(1, 34) = 10.60, p < 0.01$ for Q2). They also strongly felt as if they were being touched when the geminoid was touched ($F(1, 34) = 5.61, p < 0.05$ for Q3). These results indicate that the operation of the geminoid caused the participants to feel a strong sense of body ownership of it, as found in previous studies.

### 20.3.2   Effect of the Partner Presence Factor

The results for Q1, Q2, and Q4 (Fig. 20.5b) showed that the main effect of the partner presence factor was significant. The participants felt more strongly as if they were managing the robot's movements when they could see their partner than when they could not ($F(1, 34) = 6.95, p < .05$ for Q1). They strongly felt as if they were the geminoid while talking when they could not see their partner ($F(1, 34) = 9.21, p < .01$ for Q2). They also strongly felt as if they were in the same room as the geminoid while talking ($F(1, 34) = 9.67, p < .01$ for Q4). These results, in particular those for Q2, show that people have a strong sense of body ownership of the geminoid when they see their partner with it. The results for Q4 also showed that participants experienced a strong feeling of telepresence when they could see their interaction partners.

### 20.3.3 Effect of the Response Reaction Factor

The responses to the questions showed no significance in the main effect of the response reaction factor. However, they showed interaction effects between this factor and the other factors.

### 20.3.4 Interaction Effects

The responses to Q2 showed a significant interaction effect between the partner presence and response reaction factors ($F(1, 34) = 8.127$, $p < 0.01$). Further analysis of the simple main effects showed that the participants strongly felt as if they were the robot when they conversed with a partner whom they could see ($^*C_{Res+}^{Pre+} > {}^*C_{Res+}^{Pre-}$, $F(1, 68) = 16.87$, $p < 0.01$) or when they spoke without any interaction with the partner ($^*C_{Res-}^{Pre-} > {}^*C_{Res+}^{Pre-}$, $F(1, 68) = 4.12$, $p < 0.05$) (Fig. 20.5c). These results indicate that the response reaction factor alone is not sufficient to enhance the sense of body ownership.

The results for Q2 also showed a weak effect between the three factors ($F(1, 34) = 3.22$, $p < 0.10$). The analysis of the simple main effect revealed that the participants felt as if they were the robot when they performed the operation with a visible participant more strongly when the partner reacted ($^{Op+}C_{Res+}^{Pre+} > {}^{Op+}C_{Res-}^{Pre+}$, $F(1, 34) = 5.19$, $p < 0.05$).

In addition, the responses to Q4 showed significant interaction effects between the operation and partner presence factors ($F(1, 34) = 6.25$, $p < 0.05$). Analysis of the simple main effect revealed that the participants felt more strongly as if they were in the remote room when the partner was not visible, if they were operating the geminoid ($^{Op+}C_*^{Pre-} > {}^{Op-}C_*^{Pre-}$, $F(1, 34) = 8.25$, $p < 0.05$). Further, even when they were not operating the geminoid, they had a stronger sense of being in the remote location because of the existence of the partner ($^{Op-}C_*^{Pre+} > {}^{Op-}C_*^{Pre-}$, $F(1, 68) = 8.16$, $p < 0.01$).

## 20.4 Discussion

The results showed that the partner's presence, as well as the act of teleoperation, enhances the sense of body ownership transfer of the android robot. This tendency was stronger when the partner reacted (interactivity of the conversation), in particular when the teleoperation was performed. In addition, the partner's presence factor also showed a tendency to enhance the feeling of remote presence. Although we could not observe a significant effect of the single factor of the partner's reaction, the result that the mere presence of the partner enhances the body ownership transfer illusion, regardless of whether the participant operated the robot, suggests that social factors have a strong influence on BOT. The finding that the partner's reaction showed a stronger effect with the presence of the partner or with the act of teleoperation seems to show that the naturalness of the situation also affects the feeling of BOT.

These findings imply that the current model of eccentric projection and body schema, including the BOT and RHI, should be reconsidered. That is, top-down factors, such as social interaction, which have been considered a rather low-level, built-in functionality in the cortex, may have a strong influence on the body schema construction. Moreover, by changing the conditions of the teleoperation task, it is possible that we can improve the efficiency of the teleoperation. However, we should note that this study had some limitations. Only the performances of the conversation tasks were tested using evaluation by subjective questionnaires. The factors for social interaction were limited. Only a single robot was used for teleoperation. By extending our study to obviate these limitations, we should be able to obtain a considerably clearer general view of the manner in which people recognize their own body and practical methods for enhancing teleoperation performance.

The result that the partner's presence enhanced BOT may also be explained from another viewpoint. In previous studies, we attempted to realize various teleoperated conversations in various situations. In some cases, we attempted to realize a conversation between two or more teleoperated androids; that is, a situation where two or more operators each operate their own android, and the androids are collected to converse in a single location. The results show that they had no difficulty talking with each other. However, some operators recalled that the situation "lacked reality." One operator said, "sometimes I felt like I was watching a computer graphics scene and that the robots were talking by themselves, not being teleoperated by someone." When we added another person to the group of androids, this feeling seemed to disappear. This past observation may also be explained by the results of this experiment that show that remote presence was enhanced by the partner's presence.

## 20.5 Conclusion

This study examined the influence of social interaction on the sense of BOT of a teleoperated android. We found that the presence of an interaction partner enhances the sense of BOT, in particular in combination with the act of teleoperation and the partner's appropriate reaction. The results suggest that human higher-level behavior may operate as a factor in establishing and transfiguring how we perceive our bodies. In addition, our findings may lead to a new methodology for performing the teleoperation of robots in remote locations more effectively.

In future work, we will verify whether a strong sense of body ownership increases a feeling of telepresence at a remote location. We will also investigate how a strong sense of body ownership of a telerobot facilitates human–human telecommunication and how this will result in better quality communication.

# References

1. Nishio, Shuichi, Koichi Taura, Hidenobu Sumioka, and Hiroshi Ishiguro. 2013. Effect of social interaction on body ownership transfer to teleoperated android. In *22nd IEEE international symposium on robot and human interactive communication (RO-MAN 2013)*, 565–570.
2. Maravita, Angelo, and Atsushi Iriki. 2004. Tools for the body (schema). *Trends in Cognitive Sciences* 8 (2): 79–86.
3. Sonoda, Kohei, Akira Asakura, Mai Minoura, Robert W. Elwood, and Yukio P. Gunji. 2012. Hermit crabs perceive the extent of their virtual bodies. *Biological Letters* 8 (4): 495–497.
4. Ladd, George Trumbull. 1887. *Elements of physiological psychology: A treatise of the activities and nature of the mind from the physical and experimental point of view*, ed. C. Scribner's sons.
5. Colman, Andrew M. 2008. *A Dictionary of psychology*. Oxford University Press.
6. Ramachandran, V.S., and W. Hirstein. 1998. The perception of phantom limbs. *Brain* 121 (9): 1603–1630.
7. Botvinick, Matthew. 1998. Rubber hands 'feel' touch that eyes see. *Nature* 391 (6669): 756.
8. Tsakiris, M., L. Carpenter, D. James, and A. Fotopoulou. 2010. Hands only illusion: Multi-sensory integration elicits sense of ownership for body parts but not for non-corporeal objects. *Experimental Brain Research* 204 (3): 343–352.
9. Nishio, Shuichi, Hiroshi Ishiguro, and Norihiro Hagita. 2007. Geminoid: Teleoperated android of an existing person. In *Humanoid robots: New developments*, ed. Armando Carlos de Pina Filho, 343–352. Vienna, Austria: I-Tech Education and Publishing.
10. Watanabe, Tetsuya, Shuichi Nishio, Kohei Ogawa, and Hiroshi Ishiguro. 2011. Body ownership transfer to android robot induced by teleoperation (in Japanese). *IEICE Transaction* J94-D (1): 86–93.
11. Nishio, Shuichi, Tetsuya Watanabe, Kohei Ogawa, and Hiroshi Ishiguro. 2012. Body ownership transfer to teleoperated android robot. In *International conference on social robotics (ICSR 2012)*, 398–407, Oct 2012, Chengdu, China.
12. Ogawa, Kohei, Koichi Taura, Shuichi Nishio, and Hiroshi Ishiguro. 2012. Effect of perspective change in body ownership transfer to teleoperated android robot. In *IEEE international symposium on robot and human interactive communication (RO-MAN)*, 1072–1077, Sept 2012, Paris, France.
13. Alimardani, Maryam, Shuichi Nishio, and Hiroshi Ishiguro. BMI-teleoperation of androids can transfer the sense of body ownership. In *Cognitive neuroscience society's annual meeting (CNS2012)*, Apr 2012, Chicago, Illinois, USA.
14. Ishi, C.T., L. Chaoran, H. Ishiguro, and N. Hagita. 2011. Speech-driven lip motion generation for tele-operated humanoid robots. *International conference on auditory-visual speech processing*, 131–135.

# Chapter 21
# Exploring Minimal Requirement for Body Ownership Transfer by Brain–Computer Interface

**Maryam Alimardani, Shuichi Nishio and Hiroshi Ishiguro**

**Abstract** Operators of a pair of robotic hands report ownership of those hands when they hold an image in their mind of a grasp motion and watch the robot perform it. We present a novel body ownership illusion that is induced by merely watching and controlling a robot's motions through a brain–machine interface. In past studies, body ownership illusions were induced by the correlation of sensory inputs, such as vision, touch, and proprioception. However, in the presented illusion none of these sensations was integrated, except vision. Our results show that the interaction between the motor commands and visual feedback of an intended motion is sufficient to evoke the illusion that non-body limbs are incorporated into a person's own body. In particular, this work discusses the role of proprioceptive information in the mechanism of agency-driven illusions. We believe that our findings can contribute to the improvement of tele-presence systems in which operators perceive tele-operated robots as themselves.

**Keywords** Proprioception · Android robot · Tele-operation
Body ownership

This chapter is a modified version of a previously published paper [1], edited to be comprehensive and fit with the context of this book.

M. Alimardani · S. Nishio (✉) · H. Ishiguro
Advanced Telecommunications Research Institute International,
Keihanna Science City, Kyoto, Japan
e-mail: nishio@botransfer.org

M. Alimardani · H. Ishiguro
Department of Systems Innovation, Graduate School of Engineering Science,
Osaka University, Osaka, Japan

## 21.1 Introduction

The mind–body relationship has always been an appealing subject for human beings. How we identify our body and how we use it to perceive our "self" are questions that have fascinated many philosophers and psychologists throughout history. However, only in the last few decades has it become possible to investigate the mechanism of self-perception by using empirical approaches.

Our recently developed human-like androids have become new tools for investigating the question of how humans perceive their own "body" and correlate it to their "self." Operators of these tele-operated androids report unusual feelings of transforming into the robot's body [2]. This sensation of owning a non-body object, in general termed the "illusion of body ownership transfer," was first scientifically reported as the "rubber hand illusion" (RHI) [3]. Following the introduction of the RHI, many researchers studied the conditions under which the illusion of body ownership transfer can be induced [4, 5]. In these studies, the feeling of owning a non-body object was examined mainly through the manipulation of sensory inputs (vision, touch, and proprioception) supplied to the subject by his own body and the fake body in a congruent manner. In previously reported studies on illusions, the correlation of at least two channels of sensory information was found to be indispensable. The illusion was either passively evoked by synchronized visuo-tactile [3] or tactile-proprioceptive [6] stimulation or by synchronized visuo-proprioceptive [7] stimulation in voluntarily performed actions.

However, the question that remained was whether it is possible to induce body ownership illusions without the correlation of sensory modalities. We were specifically interested in the role of sensory inputs in evoking motion-involved illusions. In these illusions that are aroused by triggering a sense of agency toward the actions of a fake body, afferent signals, such as vision and proprioception, need to be integrated with efferent signals in order to generate a coherent self-presentation. A recent study addressed the contribution of proprioceptive feedback in the inducement of the ownership illusion for an anesthetized moving finger [7]. In contrast to that study, in the current study we hypothesized that, even in the absence of proprioceptive feedback, the match between efferent signals and visual feedback is alone sufficient to trigger a sense of agency toward the robot's motion and therefore induce the illusion of ownership of the robot's body.

The challenging part of this work was the removal of proprioceptive feedback from the operators' sensations. Since proprioception is an internal sense of body posture that is constantly updated with movement, we employed a brain–computer interface (BCI) system for translating the operator's thoughts into the robot's motions and removed real motions during operation. The subjects conducted a set of motor imagery tasks of grasping with the right or left hand, and their online encephalography (EEG) signals were classified into two classes corresponding to the right- and left-hand motions performed by the robot. During tele-operation, the subjects wore a head-mounted display (HMD) and watched real-time first-perspective images of the robot's hands (Fig. 21.1).

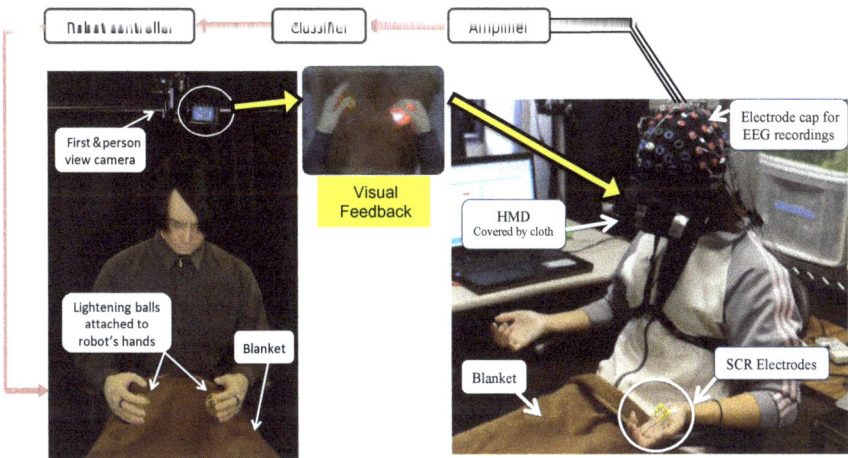

**Fig. 21.1** Experiment setup. EEG electrodes installed on the subject's sensorimotor cortex recorded brain activities during a motor imagery task. The subjects wore a head-mounted display (HMD) through which they had a first-person view of the robot's hands. They received a cue from a lighted ball in front of the robot's hands and held images in their mind of a grasp for their own corresponding hand. The classifier detected two classes of results (right or left) and sent a motion command to the robot's hand. Skin conductance response (SCR) electrodes, attached to the subject's left hand, measured physiological arousal during the session. Identical blankets were laid on both the robot and the subject's legs to match the background view

The operators' aroused sense of body ownership was evaluated in terms of their subjective assessment and physiological reaction to an injection applied to robot's body at the end of each test session.

## 21.2  Results

Forty subjects participated in the BCI-operation experiment. They operated the robot's hands while watching them through an HMD. Each participant performed two randomly conditioned sessions as follows.

1. Still condition: The robot's hands were motionless, although the subject attempted to perform the imagery task and operate the hands. (Control)
2. Match condition: The robot's hands performed the grasp motion according to the classification results and only when the results were correct. If the subject failed the trial, the hands did not move.

To facilitate the participants' imagination and give a visual cue for the motor imagery task, two balls were installed in front of the robot's hands that lit up randomly and indicated the hand that the subjects were required to move (Fig. 21.2a). At the end of each test session, an injection syringe was inserted into

(a)                                                          (b)

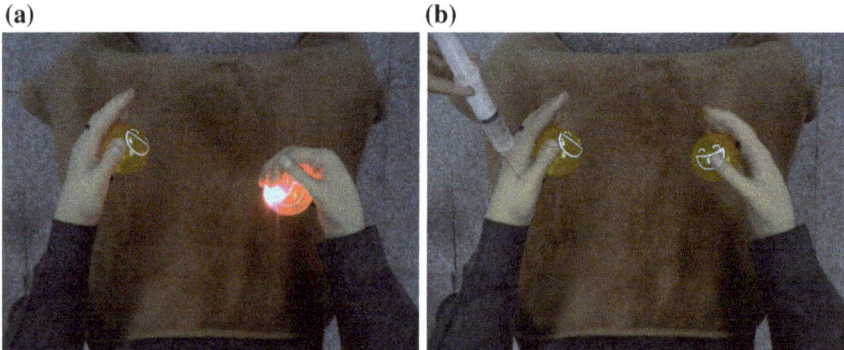

**Fig. 21.2** Participant's view in HMD. **a** The robot's right hand grasps the lighted ball according to the classification results of the subject's EEG patterns. **b** The robot's left hand receives an injection at the end of each test session. The subject's reaction to this painful stimulus was subjectively and physiologically evaluated

the thenar muscle of the robot's left hand (Fig. 21.2b). Immediately after the injection, the session was terminated and the participants were orally asked the following questions. Question 1 (Q1): When the robot's hand was injected, did you feel as if you own hand was being injected? Question 2 (Q2): Throughout the session, while you were operating the robot's hands, did you feel as if they were your own hands? Participants scored their responses to Q1 and Q2 based on a seven-point Likert scale, where 1 meant "Did not feel at all" and 7 meant "Felt very strongly."

The scores acquired for each condition were averaged and compared within subjects by a paired t-test (Fig. 21.3a). The results for Q1 showed a significant difference between the Match (M = 4.38, SD = 1.51) and Still (M = 2.83, SD = 1.43) conditions; [Match > Still, $p < 0.001$, t(39) = −4.33]. Similarly, there was a significant difference in Q2 scores between the Match (M = 5.15, SD = 1.10) and Still (M = 2.93, SD = 1.25) conditions; [Match > Still, $p < 0.001$, t(39) = −9.97].

In addition to the self-assessment measurement, the body ownership illusion was physiologically measured by recording skin conductance responses (SCR). The SCR recordings of the responses of only 34 participants were evaluated, as six participants showed unchanged responses during the experiment, and therefore, their data were excluded from the analysis. The peak response value within a 6-sec interval post-injection [8] was selected as the SCR reaction value. The difference in the SCR results was significant between the Match (M = 1.68, SD = 1.98) and Still (M = 0.90, SD = 1.42) conditions ([Match > Still, $p < 0.01$, t(33) = −3.29]), although the subjects' responses were spread over a large range of values (Fig. 21.3b).

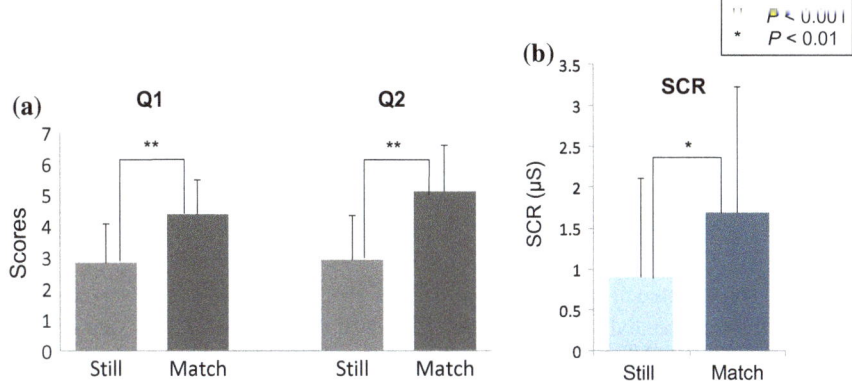

**Fig. 21.3** Evaluation results. **a** Participants responded to Q1 and Q2 immediately after watching the injection. Q1: When the robot's hand was injected, did you feel as if your own hand was being injected? Q2: Throughout the session while you were performing the task, did you feel as if the robot's hands were your own hands? The mean score value and standard deviation for each condition is plotted. A significant difference between conditions ($p < 0.001$; paired t-test) was confirmed. **b** The skin conductance response (SCR) peak value after injection was assigned as the reaction value. The mean reaction value and standard deviation are plotted; the results show a significant difference between conditions ($p < 0.01$; paired t-test)

## 21.3 Discussion

From both the questionnaire and SCR results, we can conclude that the operators' reaction to a painful stimulus (injection) was significantly stronger in the Match condition in which the robot's hands followed the operators' intentions. This reaction constitutes evidence that the body ownership illusion was evoked, and therefore, our hypothesis in this study was supported. We showed that body ownership transfer to a robot's moving hand can be induced when proprioceptive feedback from the operator's actual movements is excluded. This is the first report of body ownership transfer to a non-body object that is induced in the absence of integration between multiple afferent inputs. In the presented illusion, the correlation between efferent information (operator's plan of motion) and a single channel of sensory input (visual feedback of intended motion) was sufficient to trigger the illusion of body ownership. Since this illusion occurs in the context of action, we estimate that the sense of ownership of the robot's hand here is modulated by the sense of agency generated for the robot's hand motions. Although all participants were completely aware that the congruently placed hands they watched through the HMD were non-body human-like objects, the explicit sense that it was they who were causing these hands' motions, together with a lifelong experience of performing their own body's motions, modulated the sense of body ownership toward the robot's hands and evoked the illusion.

The original ownership illusion for tele-operated androids [2] (mentioned at the beginning of this paper) was a visuo-proprioceptive illusion evoked by the motion synchronization between the operator and the robot's body [8]. The mechanism behind this illusion can be explained based on the cognitive model according to which a person's body is integrated in him/her self in a self-generated action [9]. When operators move their body and watch a robot copying them, the match between motion commands (efferent signals carrying both raw and processed predictive information) and sensory feedback from the motion (visual afference from the robot's body and proprioceptive afference from the operator's body) modulated a sense of agency over the robot's actions and ultimately resulted in the integration of the robot's body into the operator's sense of self-body (Fig. 21.4). In this paper, we targeted in particular the role of proprioception in this model and showed that the presented mechanism remains valid even when the updating of the proprioceptive feedback channel is blocked.

Finally, the authors conclude from the results of the present study that in inducing illusions of body transfer, congruency between only two channels of information, either efferent or afferent, is sufficient to integrate a non-body part into one's own body, regardless of the context in which the experience of body transfer occurs. In passive or

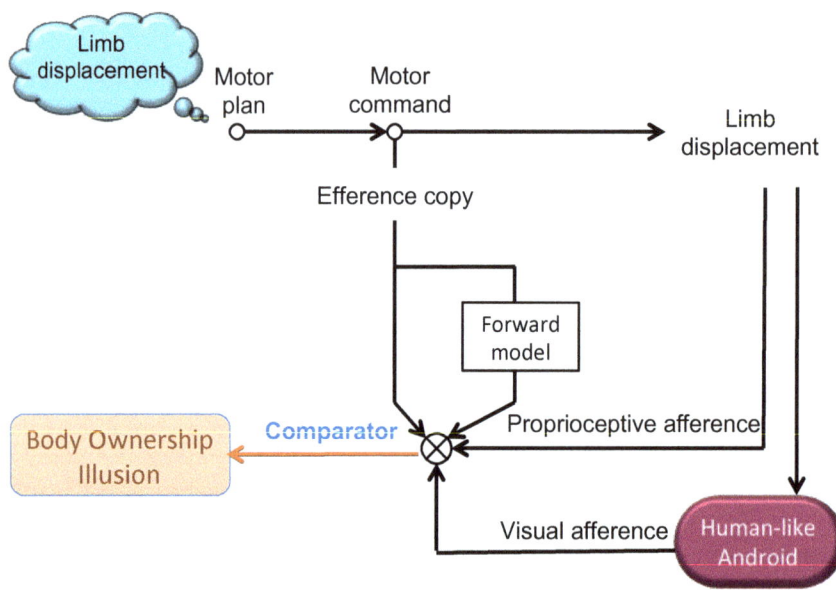

**Fig. 21.4** Body recognition mechanism. The mechanism of the body ownership illusion for an operated robotic body is explained based on Tsakiris' cognitive model of self-recognition. During tele-operation of a very human-like android, the match between the efferent signals of a motor intention and afferent feedback of the performed motion (proprioceptive feedback from the operator's body and visual feedback from the robot's body) evokes the illusion that the robot's body belongs to the operator. However, the role of proprioceptive feedback in modulation of this feeling has remained unclear. This work confirmed that the body ownership illusion is elicited in the absence of proprioceptive feedback and by modulation of only motor commands and visual inputs

externally generated experiences, the integration of two sensory modalities from both non-body and body parts is indispensable. However in voluntary actions, efferent signals play a critical role in the recognition of one's own motions and therefore their congruency with only a single channel of visual feedback from non-body part motions is adequate to override the internal mechanism of body ownership.

## 21.4 Methods

**Subjects**: Forty healthy participants (26 male and 14 female) in the age range of 18–28 years (M = 21.13, SD = 1.92), most of whom were university students, were selected for the experiment. Thirty-eight participants were right-handed and two left-handed. All the participants were naïve to the research topic and received an explanation prior to the experiment. Subsequently, they signed a consent form in accordance with the guidelines of the ATR ethical committee. At the end of the experiment, all participants received payment for their participation.

**EEG recording**: Cerebral activities were recorded by g.USBamp biosignal amplifiers developed by Guger Technologies, Graz, Austria. The subjects wore an electrode cap, and 27 EEG electrodes were installed over their primary sensorimotor cortex. Electrode placement was according to the 10–20 system. A reference electrode was placed on the right ear and a ground electrode on the forehead.

**Classification**: The data that were obtained were processed online using Simulink/MATLAB (Mathworks) for real-time parameter extraction. This process included band-pass filtering between 0.5 and 30 Hz, sampling at 128 Hz, removing artifacts by using a notch filter at 60 Hz, and adopting a common spatial pattern (CSP) algorithm for discriminating event-related desynchronization (ERD) and event-related synchronization (ERS) patterns associated with motor imagery tasks [10]. The results were classified using weight vectors that weighted each electrode according to its importance for the discrimination task and suppressed noise in individual channels by using the correlations between neighboring electrodes. During each right or left imagery movement, the decomposition of the associated EEG led to a new time series, which was optimal for the discrimination of the two populations. The patterns were designed such that the signal that resulted from the EEG filtering with the CSP had maximum variance for left trials and minimum variance for right trials and vice versa. Thus, the difference between the left and right populations was maximized and the only information contained in these patterns was where the variance in the EEG was greatest in the course of the comparison of the two conditions. Finally, the left and right imagery were discriminated, and the classification block outputs a linear array signal in the range of [−1, 1], where −1 means the extreme left and 1 means the extreme right.

**Motor imagery task**: The participants were asked to imagine their own hand performing a grasping or squeezing motion. In both the training and experiment sessions, a visual cue specified the timing and the hand that they should imagine moving.

**Training**: Participants practiced the motor imagery task of controlling a feedback bar on a computer screen to the left or right side. They sat in a comfortable chair in front of a 15-in. laptop computer and were asked to remain motionless. The first run consisted of 40 trials conducted without feedback. The subjects were required to watch a cue in the form of an arrow pointing to the left or right side of the bar randomly, and imagine gripping or squeezing the corresponding hand. The duration of each trial was 7.5 s and started with the presentation of a fixation cross on the display. After 2 s, an acoustic warning was given in the form of a "beep." From second 3 to second 4.25, an arrow pointing to the left or right side was shown. Depending on the arrow's direction, the participants were instructed to perform the motor imagery. They continued the imagery task until the screen content was erased (at second 7.5). After a short pause, the next trial started. The recorded brain activities in this non-feedback run were used to set up a subject-specific classifier for the classification in the following feedback runs. In the feedback runs, participants performed similar trials; however, this time, after the appearance of the arrow and execution of the motor imagery task, the classification results were shown in the form of a horizontal feedback bar on the screen. The subject's task was to hold the imagery immediately after the arrow and extend the feedback bar as far as possible in the same direction. The feedback runs, as well as the non-feedback runs, consisted of 40 trials with 20 trials per class (left/right) presented in a randomized order.

Participants performed two sessions of training with feedback until they became familiar with the motor imagery task. The subjects' performance during each session was recorded to evaluate their improvement. At the end of the training sessions, most participants could reach a performance level of 60–100%.

**Experimental setup**: The subjects wore an HMD (Vuzix iWear VR920) through which they had a first-person view of the robot's hands. Since this HMD design is not protected from environmental light, we wrapped a piece of cloth around the HMD frame to block the surrounding light. Two balls with lights were installed in front of the robot's hands to facilitate the imagery task during the experiment sessions. The participants received visual cues when the balls lit up randomly and they then imagined grasping with their own corresponding hand. The classifier detected two classes of results (right or left) from the EEG patterns and sent a motion command to the robot's hand. Identical blankets were laid on both the robot and the subject's legs to match the background view of the robot's body with that of the subject's own body.

SCR electrodes were attached to each subject's hand to measure the physiological arousal during each session. A bio-amplifier recording device (Polymate II AP216, TEAC, Japan) with a sampling rate of 1000 Hz was used for the SCR measurements. The participants were asked to rest their hands with their palms upward on the chair arms, and SCR electrodes were attached to the thenar and hypothenar eminences of their left palm. As the robot's hands were configured in an inward position for the grasping motion, there was a tendency among participants to alter their hands' position to an inward configuration similar to that of the robot's hands. However, this could have caused difficulties in reading the SCR electrodes and moreover could have given the participants the space and comfort to perform unconscious grasping motions during the task. Therefore, we asked participants to keep their hands and elbows motionless on the chair arms with their palms upward.

**Testing**: The participants practiced operating the robot's hand by motor imagery in one session and then performed two test sessions. All the sessions consisted of 20 imagery trials. Test sessions were randomly given the condition "Still" or "Match." In the Still condition, the robot's hands remained motionless, although the subjects performed the imagery task according to the cue stimulus. In the Match condition, the robot's hands performed the grasping motion but only in those trials in which the classification result was correct and the same as the cue. If the subject failed the trial, the robot's hands did not move.

At the end of each test session, an injection syringe was inserted into the thenar muscle of the robot's left hand, the same hand on which the subject wore the SCR electrodes. Immediately after the injection, the session was terminated and the participants were orally asked: (Q1) When the robot's hand was injected, did you feel as if your own hand was being injected? (Q2) Throughout the session, while you were performing the task, did you feel as if robot's hands were your own hands? They scored their response to each question based on a seven-point Likert scale, where 1 meant "did not feel at all" and 7 meant "felt very strongly."

**SCR measurements**: The peak value of the SCR responses was selected as the reaction value. In general, the SCR value starts to rise 1–2 s following a stimulus and returns to normal 5 s after that [11]. The moment at which the injection syringe appeared in the participant's view was selected as the start point for the evaluation, because some participants showed a reaction to the syringe itself, even before it was inserted into the robot's hand, as a result of the body ownership illusion [11]. Therefore, SCR peak values were sought in the range from 1 s after the syringe appeared in the participant's view to 5 s after the injection was in fact applied.

**Data analysis**: The scores that were obtained in the questionnaire and the SCR peak values for each condition were averaged and compared within subjects. A statistical analysis was performed using a paired t-test. A significant difference between the two conditions was revealed in the responses to Q1 and Q2 (Match > Still, $p < 0.001$) and the SCR values (Match > Still, $p < 0.01$).

# References

1. Watanabe, T., S. Nishio, K. Ogawa, and H. Ishiguro. 2011. Body ownership transfer to android robot induced by teleoperation. *Transaction on IEICE* J94 (1): 86–93 (in Japanese).
2. Tsakiris, M., P. Haggard, N. Franck, N. Mainy, and A. Sirigu. 2007. A specific role for efferent information in self-recognition. *Cognition* 96 (3): 215–231.
3. Alimardani, M., S. Nishio, and H. Ishiguro. 2013. Humanlike robot hands controlled by brain activity arouse illusion of ownership in operators. *Scientific Reports* 3: 2396.

4. Nishio, S., and H. Ishiguro. 2008. Android science research for bridging humans and robots. *Journal of IEICE* 91 (5), 411–416 (in Japanese).
5. Botvinick, M. 1998. Rubber hands 'feel' touch that eyes see. *Nature* 391 (6669): 756.
6. Ehrsson, H.H. 2007. The experimental induction of out-of-body experiences. *Science* 317: 1048.
7. Petkova, V.I., and H.H. Ehrsson. 2008. If I were you: Perceptual illusion of body swapping. *PLoS ONE* 3 (12): e3832.
8. Ehrsson, H.H., N.P. Holmes, and R.E. Passingham. 2005. Touching a rubber hand: Feeling of body ownership is associated with activity in multisensory brain areas. *Journal of Neuroscience* 25 (45): 10564–10573.
9. Walsh, L.D., G.L. Moseley, J.L. Taylor, and S.C. Gandevia. 2011. Proprioceptive signals contribute to the sense of body ownership. *The Journal of Physiology* 589 (12): 3009–30021.
10. Neuper, C., G.R. Muller-Putz, R. Scherer, and G. Pfurtscheller. 2006. Motor imagery and EEG-based control of spelling devices and neuroprostheses. *Progress in Brain Research* 159: 393–409.
11. Armel, K.C., and V.S. Ramachandran. 2003. Projecting sensations to external objects: Evidence from skin conductance response. *Proceedings: Biological Sciences* 270 (1523): 1499–1506.

# Chapter 22
# Regulating Emotion with Body Ownership Transfer

**Shuichi Nishio, Koichi Taura, Hidenobu Sumioka and Hiroshi Ishiguro**

**Abstract** In this study, we experimentally examined whether changes in the facial expressions of teleoperated androids can affect and regulate their operators' emotion, based on the facial feedback theory of emotion and the phenomenon of body ownership transfer to the robot. Twenty-six Japanese participants conversed with an experimenter through a robot in a situation where the participants were induced to feel anger, and during the conversation, the android's facial expression was changed according to a pre-programmed scheme. The results showed that facial feedback from the android did occur. Moreover, a comparison of the results of two groups of participants, one of which operated the robot and the second did not, showed that this facial feedback from the android robot occurred only when the participants operated the robot, and that when an operator could effectively operate the robot, his/her emotional states were more affected by the facial expression change of the robot.

**Keywords** Emotion · Teleoperation · Android robot · Facial feedback hypothesis

## 22.1 Introduction

Emotion regulation, which is the process through which people control, as well experience and express, their emotions, is a critical skill in our social lives [2]. The importance of this skill is growing with the rapid development of telecommunications technology that increases the opportunities for non-direct conversation

---

This chapter is a modified version of a previously published paper [1], edited to be comprehensive and fit with the context of this book.

---

S. Nishio (✉) · K. Taura · H. Sumioka · H. Ishiguro
Advanced Telecommunications Research Institute International,
Keihanna Science City, Kyoto, Japan
e-mail: nishio@botransfer.org

K. Taura · H. Ishiguro
Department of Systems Innovation, Graduate School of Engineering Science,
Osaka University, Osaka, Japan

© Springer Nature Singapore Pte Ltd. 2018
H. Ishiguro and F. Dalla Libera (eds.), *Geminoid Studies*,
https://doi.org/10.1007/978-981-10-8702-8_22

(e.g., [3]). When people do not meet in a face-to-face setting, most of the nonverbal information that is normally obtained when directly meeting other persons are lost, which complicates their emotion self-regulation according to the cues of others that reflect upon their behavior. One typical and extreme example is the uninhibited behaviors frequently seen in computer-mediated communication, such as Internet defamation or harassment. This occurs in the case not only of asynchronous media, such as e-mails, but also of synchronous media, such as Internet chat applications or TV conference systems [4].

The definition of "emotion" has been long discussed in various fields, including psychology and cognitive psychology [5–7], and recently also in the field of affective computing [8]. Most theories of emotion seem to accept emotion as comprising flexible response sequences that occur under challenging, difficult, or critical situations. Although emotion has positive aspects, such as allowing a person to prepare for rapid motor responses [5], sometimes emotional responses may be misleading and do more harm than good. When peoples emotions are not suitable for the situation, they attempt to regulate their emotional responses [2]. This act of emotion regulation is defined as a process by which individuals influence which emotions they have, when they have them, and how they experience and express these emotions [9]. Gross proposed a process model of emotion regulation in which the timeline of an emotional response sequence is divided into stages [2]. Thus, this process model consists of five stages: (1) situation selection, (2) situation modification, (3) attentional deployment, (4) cognitive change, and (5) response modulation. While the first four stages occur before an emotional response is prepared, the last stage, response modulation, refers to what people do when an emotional response is running. The emotion regulation process is known to operate consciously or unconsciously in one or more of these stages. It is known that in the last stage the suppression of negative emotional expression affects physiological responses, such as blood pressure, but not subjective reports, whereas the suppression of positive emotional expression affects both [2, 10].

Although many studies have been conducted in which proposed advisory agents were implemented in remote communication systems (e.g., [11]), few have addressed emotion regulation using information technologies. One possible reason is that it is difficult to measure emotional states; however, their measurement is becoming more practical with the progress in pattern recognition technologies [12]. Another important issue is the methods for providing feedback to the speaker and regulating his/her emotions. In advisory systems, a virtual avatar typically appears on the communication screen to provide advice. When an avatar provides advice, such as "calm down," to regulate the users' emotion, the user may ignore the advice if he/she has already become too emotional. Thus, a more effective method that operates directly and subconsciously is required. If a device that implements such a method were available, people could choose to use it to effectively regulate their emotional state and regain control.

In the past, psychologists studied a phenomenon known as the "facial feedback hypothesis," which is based on the well-known idea of William James, who claimed that the awareness of bodily changes activates emotion. Many studies have addressed the question whether facial movement influences emotional experience [13–16]. These studies clarified that facial movements do affect emotional states. Although researchers have proposed models for utilizing the facial feedback hypothesis for emotion regulation (e.g., [17]), it has not been possible to actively control emotion by applying this theory without physical equipment that provides physical stimulation.

Recently, we developed a series of teleoperated androids [18]. The appearance of these robots very closely resembles that of real people and, until now, several types of android have been created using Japanese and Danish persons as model sources. These androids are teleoperated using intuitive interfaces. Using these teleoperated androids, we found an interesting phenomenon that we call "body ownership transfer" [19]. This phenomenon is that, as the operator controls the android robot, he/she gradually feels as if the robot is his/her own body and starts to respond, without any haptic feedback, to the physical stimuli given to the robot's body. A similar phenomenon called the "rubber hand illusion" (RHI) has also been actively studied [20]. While the RHI requires tactile sensation, the body ownership transfer we found requires only the coordination of visual stimuli and the act of teleoperation. An interesting conjecture arises here: when the operator feels that the android's face is his/her own face, will the facial feedback phenomenon occur? When the android's facial expression changes autonomously, without being controlled by the operator, will the operator's emotion be affected? If the facial feedback from the teleoperated robot could positively affect the operator's emotion, we may be able to implement a new communication medium that can support its users' emotion regulation so that they can avoid uninhibited behaviors. In addition, by integrating the facial feedback effect and the body ownership transfer phenomenon, we can implement a system that does not require physical stimuli to be applied to the user, which results in a more comfortable and simpler interface. When utilizing such a simple interface, users may not be aware of the facial feedback effect. However, as the cost of this system is higher than that of normal TV conference systems and it shows the explicit appearance of the robot, we assume that users should have a strong motivation and intention to control themselves.

In this study, we experimentally tested the assumption that changes in the facial expressions of a teleoperated android can affect and regulate an operator's emotion. For this purpose, we created a conversational situation where the participants felt anger. In this situation, the experimenter, beyond the operators' control, was responsible for the facial changes of the android. If the facial feedback phenomenon occurs from the teleoperated android to the operator, the operator's anger should decrease when the android shows a positive expression, such as a smile. In addition, we examined the relation between body ownership transfer and the effect of facial feedback from the android. We divided the participants into an Operation group and a No-operation group in order to examine this relationship. The participants who teleoperated the android were expected to be more affected by the android's facial expression than those who did not.

In the rest of the paper, we first describe the experimental settings that we used to clarify the extent to which the facial expression of the android affected the participants' emotions. Next, we present results showing that these effects depended on the experimental conditions. Finally, we discuss a technique for regulating the operators' emotions and means of improving this technique. Further, issues such as ethical concerns related to this technique are also discussed.

## 22.2 Methods

Twenty-six university undergraduate students participated in our experiment (14 males, 12 females; average age = 20.7 years, S.D. = 1.2). All the participants were Japanese students in disciplines that did not include engineering and had no experience of using any type of robot. We explained the experimental procedure to them, and they signed an informed consent form.[1] Communicating through a teleoperated android, participants conversed with a female member of the experiment's staff. The robot was placed alongside the staff member, and the participants teleoperated the robot using their own motions, which were motion captured. While the participants were conversing through the teleoperated android, we changed the android's facial expression gradually. While controlling the android's facial changes, we measured the extent of the influence of this change on the emotional states of the participants.

### 22.2.1   Conversation Scenario

To observe the changes in the emotional states of the participants, we created a situation where the participants became emotional and emotion regulation was required. Although we daily feel many types of emotion, one of the most common emotions is anger. Anger is considered to have remained during the evolution of the species from our biological past, when it was an adaptive function, and therefore it frequently appears in modern human beings [21]. When people encounter unreasonable or unfair situations, they tend to feel negative emotions such as anger or sadness. Among the negative emotions, anger is most likely to lead to hostile behaviors and thus requires regulation [4]. In this experiment, we recreated a situation where people frequently became angry: calling a customer service center with a problem.

The participants played the role of customers who had experienced inferior products or unsatisfactory services and are thus complaining to the customer service center. A female staff member of the experiment group played the role of the customer service representative. The conversation flow proceeded as follows. First, the participant described his/her situation and problem. Then, the participant negotiated with the customer service representative about possible solutions. When they

---

[1]This experiment was approved by the ethical committee of Advanced Telecommunications Research International Institute (No. 12-506-1).

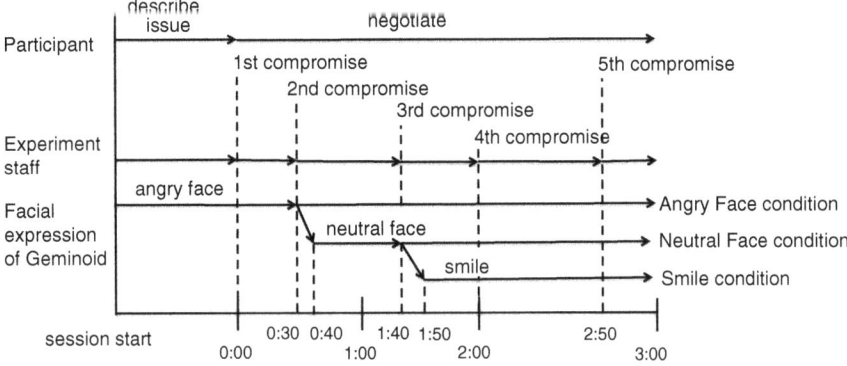

**Fig. 22.1** Flow of experimental sessions: while participants negotiate with the member of the experiment staff, the facial expression of the teleoperated android changes according to the time schedule and the condition

agreed on a solution, or when the three-min time limit was reached, the session ended (Fig. 22.1). An example of such a conversation is as follows:

```
Service rep.: Thank you for calling the customer service desk.
    How may I help you?
Customer: I bought a used computer at your shop, and in fact, your
    salesperson recommended it. I was planning to buy a new one, but
    since he recommended a used one, saying it gave a better performance,
    was cheaper, and even had a warranty, I bought it for 30,000 yen.
    When it was delivered, I found that the monitor was broken. I asked
    the store to replace the monitor, but they said I needed to pay
    23,000 yen for a replacement because that part was not covered by
    the warranty. Since I needed the computer for work, I bought a new
    monitor, but the delivery actually took two weeks...

(omitted)

Customer: I've been having lots of problems. I want a new computer.
Service rep.: I'm really sorry for all your trouble. But since you
    bought a used computer, there may be problems. How about replacing
    your computer with another one from our store? This one is also used,
    but it's been completely checked.

Customer: Resolving this problem has already dragged on long enough,
    and I don't want any more trouble. I definitely want a new computer...

(rest omitted)
```

Prior to each session, the participants were given a description sheet of the issue and five potential solutions. We pre-defined three issues and provided one each to the participants for each session in counterbalanced order. We selected issues that show problems typically encountered by university students. The solutions were numbered from 1 to 5 to indicate the degree of customer benefit (5 = maximum benefit). Participants were instructed to obtain as much benefit as possible. For example, in the above scenario, the following solutions were provided:

(1) Repair the computer at a cost of 30,000 yen
(2) Repair the computer at a cost of 15,000 yen (half price)
(3) Replace with another used computer
(4) Refund the monitor cost and replace the computer
(5) Refund the entire purchase amount and provide a new computer at a discount.

The customer service representative gradually compromised based on the solutions proposed by the customer. The staff member was given the same list of possible solutions, and she could see a laptop monitor that showed the acceptable solution at each period and negotiated with the participant accordingly. The acceptable solution changed according to the timetable shown in Fig. 22.1. After 2 min and 50 s, measured from the start time of the negotiation, she could agree to the best solution for the participants, but at the 3 min mark, the experimenter terminated the session.

At each stage, the staff member did not actively propose the best possible solution. We pre-defined acceptable solutions for each stage of the experiment. She agreed when the participant proposed such a solution. Only when the participants failed to propose acceptable solutions, did she propose one that could be accepted. Some participants were content with the offered solution, which was not the best one, while others did not accept this solution before the session was terminated. Therefore, not all the participants reached the best solution.

Our original intention was to provoke the strongest possible anger level. Thus, in the pretest, which we conducted with different participants prior to the experiment described in this paper, the customer service representative never changed her mind and agreed only with the worst solution. However, some participants became so upset that they completely forgot the experimental procedures, whereas some participants found that the customer service representative never compromised, and they became reluctant to continue negotiating. To maintain the motivation of the participants, we changed the customer service representative's strategy such that she compromised more and accepted better solutions.

Here, the time schedule of providing acceptable solutions is an important factor for maintaining the participants' motivation. In an additional pretest, we found that participants' motivation was decreased when the customer service representative compromised at fixed intervals. This was because the participants could easily realize that she had rules for negotiating and that they did not need to negotiate hard to obtain better solutions. To solve this problem, we used varying time intervals between solutions. We assumed that the participants would feel their negotiation skill had affected the service representatives compromising to reach a better solution when they experienced success in reaching better solutions both easily and with difficulty. Although the staff member negotiated based on the same rules in each session, most participants did not notice these rules.

## 22.2.2 Hypotheses

Our assumption is based on the body ownership transfer illusion. The results of previous studies showed that this phenomenon occurs when the android's motion is highly synchronized with its operator's motion [19], and therefore, the operator feels that the android's face is his/her own face under this condition. This implies that the operator considered the change in the android's facial expression as his/her own change, although his/her facial expression did not actually change. Facial feedback hypothesis suggests that changes in a person's facial expression activate his/her emotion. However, an actual change in the operator's facial expression may not be needed to activate emotion; instead, the android's change in expression activates the operator's emotion when he/she experiences body ownership transfer. Accordingly, our first hypothesis was as follows:

H1 People who conduct a conversation through an android are affected by the android's facial feedback.

Since the facial feedback from the android affects the participants' emotions, their emotions will change according to the changes in the android's facial expression.

One might argue that the influence of the facial changes of others could be a result of empathy with them. Although people may empathize with the android according to its facial expressions even when they do not operate it, we assumed that the operator would be affected more strongly as a result of body ownership transfer. That is, if the participants who operated the android were more strongly affected than the participants who did not operate the android, we could infer that the emotional change in the participants was induced by the body ownership transfer effect and not only by empathy. Therefore, our second hypothesis was as follows:

H2 People who operate the android are affected more than people who do not operate it.

In order to verify H2, we divided the participants into two groups, the Operation and the No-operation group. We considered that if the facial feedback from the android was not a result of empathy, there would be a difference in the emotional responses under the Operation and the No-operation conditions.

## 22.2.3 Equipment

As the teleoperated robot, we used Geminoid F, which is one of our Geminoid series (Fig. 22.2). Geminoids are teleoperated androids that resemble real existing persons [18]. Geminoid F was modeled on an existing Japanese female. It has nine pneumatic actuators, most of which are in its face, and therefore it can produce various facial expressions.

People can control the android and talk with a conversation partner by using it as a communication medium. In this study, participants performed teleoperation with a

**Fig. 22.2** Teleoperated
android robot Geminoid F

simplified version of the Geminoid teleoperation system [18]. The body motions of
the participants were captured, converted, and sent to the android robot. At the same
time, the voices from both conversation sides were captured by microphones and
transmitted to each side. In addition, the participants viewed a teleoperation console
that showed two transmitted video screens, one showing the service representative
and second showing the android robot's face (Fig. 22.3a). The participants wore six
motion capture markers for tracking their head motion (Fig. 22.3b). The android's
lip motion was generated automatically from the voice recognition system [22], and
the data were sent to the robot so that its motions would be synchronized with the
participant's motions. Thus, the android's motions comprised the participant's head
motions and the generated lip motions, and the facial expression change was controlled by the experimenter's staff.

The staff member who played the service representative was only able to hear
the participant's voice, and the android was hidden from her view (Fig. 22.3c). This
prevented her from being affected by the android's facial expressions. The staff member saw only the partition that hid the android, a stopwatch on the computer monitor

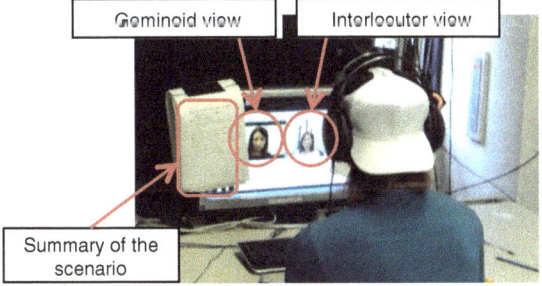

(a) Teleoperation console

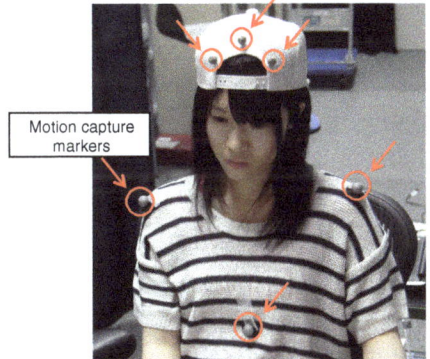

(b) Motion capture marker alignment

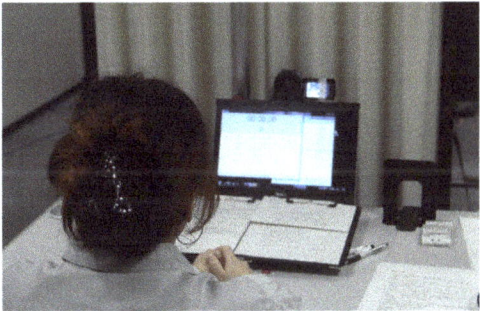

(c) Scene of the interlocutor (service representative) side

**Fig. 22.3  a** Teleoperation console: on the display screen, views of the Geminoid robot and inter-locutor (service representative) were shown. In addition, we provided a summary sheet of the scenario. **b** Participants wore a cap with markers to capture their motions for the robot teleoperation. Additional markers were also placed on their clothes to allow stable positioning. **c** A female experiment staff member performed the role of interlocutor (service representative). The staff member viewed a summary sheet of the solutions while conducting conversations. In addition, a stopwatch was shown on the computer screen so that she would adhere to the time schedule. So that the staff member would not be affected by the android's facial expression change, we placed a partition between the android robot and the staff member. The fact that the android was not visible to the staff member was concealed from the participants

to show the timings of the compromises, and a summary of the script that listed possible solutions. Although the staff member could not see the android, the participants were told that the android was in front of her.

## 22.2.4  Experimental Design

To test our hypotheses, we designed this experiment with two factors: the Operation factor and the facial expression factor. The design of this experiment was mixed: Operation was a between subjects factor while facial expression was a within subject factor.

### 22.2.4.1  Operation Factor

For the Operation factor, we set up an Operation condition and a No-operation condition. The participants controlled the android in the Operation condition but did not control it in the No-operation condition. We divided the participants into the Operation group (13 participants) and the No-operation group (13 participants). In the No-operation condition, the android did not receive the data of the participants' motions and did not move its head or lips. The android performed only expression changes during the conversation in this condition.

### 22.2.4.2  Facial Expression Factor

For the facial expression factor, we used three expressions: smiling, neutral, and angry (Fig. 22.4). Accordingly, we set up three conditions: the Smile condition,

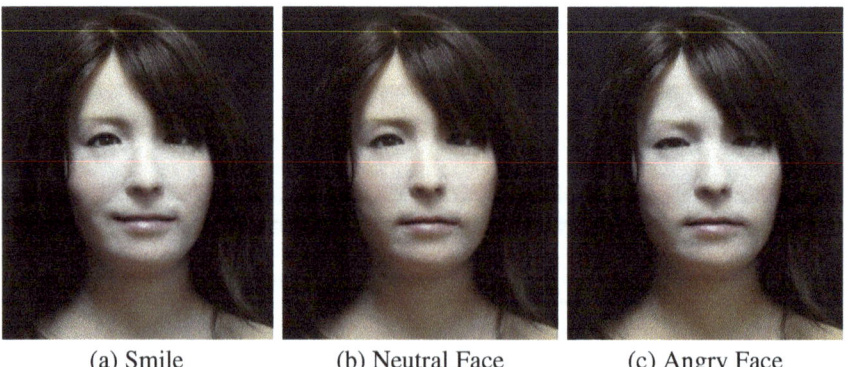

(a) Smile                    (b) Neutral Face                    (c) Angry Face

**Fig. 22.4**  Facial expressions of Geminoid F in each condition

Neutral Face condition, and Angry Face condition. At the beginning of the session, the android showed the angry expression in all conditions. In the Angry Face condition, the android's face did not change during the session. In the Neutral Face condition, we changed the android's face from angry to neutral during the conversation. In the Smile condition, however, we divided the change in expression into two steps, because the gap between the angry and the smiling faces seemed to be too large. Accordingly, we changed the android's face from angry face to neutral face in the first step and from neutral face to smiling face in the second step. These facial expressions were used to change the android's emotional appearance while the participants conversed with the staff member.

If we had changed the facial expressions rapidly, the participants would have recognized this and felt less body ownership. Therefore, we programmed the duration of a change to be 10 s. During these changes in the facial expression, the actuators were controlled by linear interpolation between the initial facial expression and the target expression. In addition, we designed each facial expression to be rather weak so that the participants would not notice the changes.

### 22.2.5   Procedure

This experiment consisted of one practice trial and three main trials. The participants answered questions on a pre-experiment questionnaire to measure their initial emotional states.[2] First, the participants performed the practice trial to practice and understand the procedural flow. Then, they performed the three main trials involving different situations. They experienced all three facial expression conditions, the order of which was counter balanced.

Each trial consisted of preparation, a session, filling a post-session questionnaire, and a two-min break. Each trial was conducted as follows. Prior to the session, participants were given the description sheet and they were given time to read and understand it. Since it was sometimes difficult to remember all the details, we placed a summary sheet next to the teleoperation console (Fig. 22.3a). Each session was started when the participants completely understood the situation. The session began after the teleoperation console was activated and the participants started to describe their problem to the customer service representative. When the participants started to negotiate, we changed the facial expression of the teleoperated android robot (Fig. 22.1). At the beginning of the session, the android's facial expression was angry (Fig. 22.4c). In the Smile and the Neutral Face conditions, the android's facial expression started to change to the neutral one (Fig. 22.4b) at the time when the acceptable compromise became the second one (at 0:30). In the Smile condition, the android's facial expression then started to change to smiling (Fig. 22.4a) at the time when the acceptable compromise became the third solution (at 1:40). After the

---

[2]We used the same effect scale for the pre-experiment questionnaire. However, the results of the pre-experiment questionnaire were not used in this study.

facial expression reached its maximum state, the android continued to show the same facial expression until the end of the session. After the session, the participants filled a post-session questionnaire and were given a two-min break.

### 22.2.6 Measures

Before the experiment and after each session, the participants answered a set of questions to measure their emotional state.[3] We used the General Affect Scales of Ogawa et al. [23], which measure mood states according to 3 factors (positive affect, negative affect, and calmness) using 8 subscales each, that is, 24 subscales in total. Each subscale indicates an adjective (such as "excited," "guilty," and "hostile"), and the participants were asked to rank how accurately the adjective fit their emotional state on a seven-point Likert scale (1 = Not at all, 7 = Extremely). The values of the three factors were derived from the total of eight subscale responses that corresponded to each factor. Therefore, the value ranges of the three factors were from 8 to 56. This instrument was validated with more than 200 participants with a Cronbach's alpha reliability score of 0.86–0.91. In addition, it has a high correlation with the Japanese version of the Profile of Mood States (POMS). In this experiment, we used only the positive effect factor for analysis.

We also asked the Operation group the following question after each session.

Question 1 (Q1)    How well could you operate the robot?

We asked the participants to rank their response to Q1 on a seven-point Likert scale (1 = operated poorly to 7 = operated well).

We also asked all participants the following questions after all the sessions were complete:

Question 2 (Q2)    Did you notice that the robot's facial expression changed?
Question 3 (Q3)    If you noticed a change, how did you feel?

## 22.3   Results

The Shapiro–Wilk normality test for a positive score of the General Affect Scales showed rejection in some conditions (Operation—Smile: $p = 0.10$, Operation—Angry Face: $p < 0.05$). Therefore, we analyzed the results using nonparametric methods. We conducted a Friedman test for the facial expression factor in the Operation group and the No-operation group; the results showed significant differences in both the Operation group ($\chi^2 = 31.02, d.f. = 12, p < 0.01$) and the No-operation group ($\chi^2 = 22.63, d.f. = 12, p < 0.05$). For the negative and calmness scores, no

---

[3] In this study, we used only the post-session results for analysis.

significant difference was shown. However, the average value of the negative score was 26.15 (S.D. = 9.91) and was higher than the average negative score of 18.08 (S.D. = 12.04) resulting from the testing in the original study [23]. The higher negative score indicates that the participants did feel a negative emotion induced by the experimental scenario.

We compared conditions by performing multiple comparisons using Ryan's method that uses the Wilcoxon signed-ranked test (Fig. 22.5). The positive score for the Operation—Smile condition was higher than that for the Operation—Neutral Face condition or the Operation—Angry Face condition (both $p < 0.05$). The positive score for the No-operation—Neutral Face condition and the No-operation—Smile condition was higher than that for the No-operation—Angry Face condition (both $p < 0.05$). The positive score for the No-operation—Neutral Face condition was marginally higher than that for the No-operation—Smile condition ($p < 0.10$).

To examine the effect of body ownership transfer, we compared the Operation and No-operation conditions for each of the facial expression conditions using the Mann–Whitney U test (Fig. 22.5). The results showed marginal significant differences between the Operation group and the No-operation group under the Smile and the Angry Face condition. Under the Smile condition, the score of the Operation group was higher than that of the No-operation group (U = 51.0, $p < 0.10$). Under the Angry Face condition, the score of the Operation group was higher than that of the No-operation group (U = 47.5, $p < 0.10$).

We now describe the results of the responses to the questions of the post-session and post-experiment questionnaires. For Q1 (How well could you operate the robot?), the average score for each evaluated condition was over 4, which is the median score of a seven-point Likert scale. There was no significant difference in the facial expression factor.

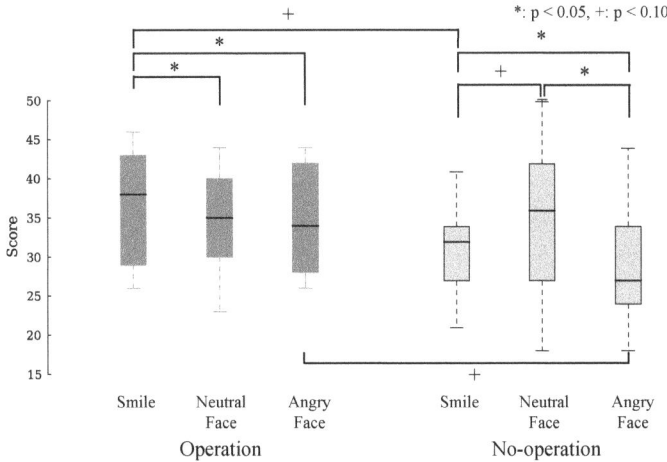

**Fig. 22.5** Box plot of positive scores from the results of the general affect scales for the operation group (left) and No-operation group (right)

According to their responses to Q2 (Did you notice that the robot's facial expression changed?) and Q3 (If you noticed a change, how did you feel?), 18 participants were aware of the smile expression and 2 were aware of the neutral expression (Operation: 8, No-operation: 10). Among the participants who were aware of the facial expression changes, seven (Operation: 3, No-operation: 4) felt a sense of discomfort because of mismatch between the androids expression and their own. They commented, for example, "I felt a sense of discomfort because of a mismatch with my own expression" or "I felt a strange sensation because I thought the smiling face didn't match the complaint." In addition, eight participants (Operation: 4, No-operation: 4) said that they felt changes in their own emotional state, such as "I felt the facial expression of the robot changed because of my feeling." The remaining participants who were aware of the android's facial expression change (Operation: 1, No-operation: 2) gave neutral answers to Q3, such as "I felt nothing when I recognized that the android's facial expression changed."

In addition, we conducted a correlation test of the scores for Q1 ("How well did you operate the android?") and the positive score for each condition. For the three facial expression conditions in the Operation group, we found two significant correlations. Under the Smile condition, there was a high correlation between the scores for Q1 and a positive score for the General Affect Scales ($r = 0.74, p < 0.01$) (Fig. 22.6a). In the Angry Face condition, there was a moderate correlation between the scores for Q1 and a positive score ($r = 0.60, p < 0.05$) (Fig. 22.6b). In the Neutral Face condition, there was no correlation between the scores for Q1 and a positive score ($r = 0.46, p = 0.11$) (Fig. 22.6c). However, in the No-operation group, no significant correlation was found (Fig. 22.6d–f).

## 22.4 Discussion

The results showed that participants' positive scores were affected by the facial expression factor both in the Operation and in the No-operation group. This supports our first hypothesis, H1: "People who conduct a conversation through an android are affected by the android's facial feedback." The results of the comparison of the Operation group and No-operation group in the Smile and the Angry Face conditions showed marginal significant differences between the Operation group and the No-operation group (Fig. 22.5). Furthermore, the scores of the Operation group were higher than those of the No-operation group. These results showed that the observed facial feedback effect depends on the act of operation, and not merely on observing the facial expression changes. That is, this effect was the result of a phenomenon that was different from empathy. In particular, we can state that the higher positive scores in the Operation group than in the No-operation group in the Smile condition support our second hypothesis H2: "People who operate the android are affected more than people who do not operate it." The results of the correlation analysis that showed a significant positive correlation between the participants better impressions of their operation and positive scores also supports the hypothesis H2 (Fig. 22.6a).

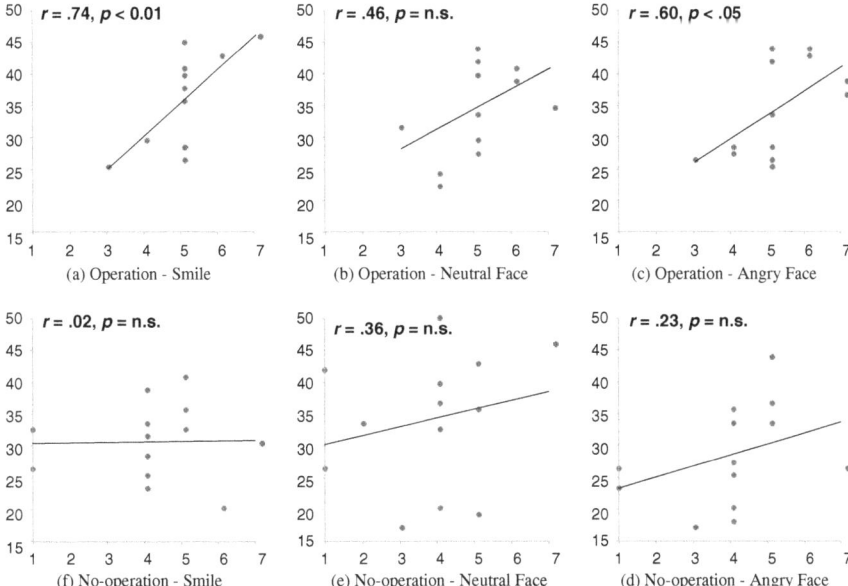

**Fig. 22.6** Scatter plots and linear regression lines for the relationship between the scores for the post-sessions question Q1 ("How well did you operate the android?") (horizontal axes) and the positive score of general affect scales (vertical axes). **a–c** Plots for the operation condition, and **d–f** plots for the No-operation condition

However, we cannot state that H2 was completely supported by the results. Participants in the No-operation group seemed to be affected differently from those in the Operation group. There seems to be an additional effect that we had not expected. In the No-operation group, the positive scores were lower under the Smile and Angry Face conditions than under the Neutral Face condition (right side of Fig. 22.5), although the score for the Neutral Face condition was not different from that in the Operation group. In the No-operation group, the positive scores (right side of Fig. 22.5) were higher under the Smile and Angry Face conditions than under the Neutral Face condition, although the scores under the Neutral Face condition were not different from those in the Operation group.

In our opinion, these results show negative effects of the android's facial expression change, as follows. In the No-operation group, because they did not control the robot, the participants felt little body ownership transfer to the android. As a result, they may have felt that the android was a bystander in the conversation; that is, the participants may have felt that the android showed mocking smiles in the Smile condition and hostile anger in the Angry Face condition toward them.

In the Operation group, we found an effect that was due to the facial feedback from the android. However, not all of the participants were affected in the same manner. The positive scores for the Smile and Angry Face conditions in the Operation group seem to have two peaks (Fig. 22.7a, c). In these two conditions, the lower peaks

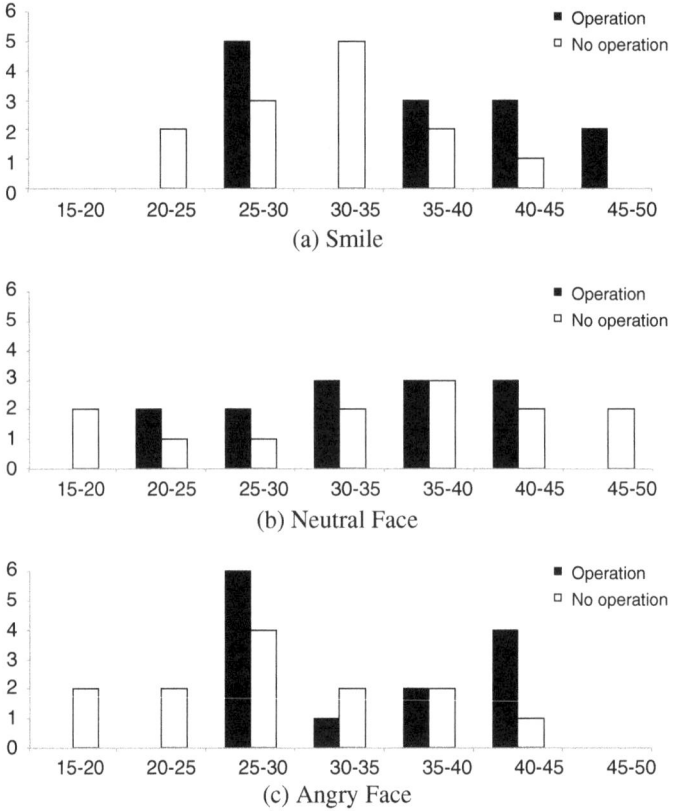

**Fig. 22.7** Histograms of positive scores for each condition

were around 25–30, which was lower than the median of the scores in the Neutral Face condition (35–40), and the higher peaks were around 40–45, which was higher than the median in the Neutral Face condition. In our prediction, we assumed that the emotions of the participants who operated the android would become positive when the android's facial expression showed a smile and negative when the android's facial expression showed anger. However, the lower peak in the Smile condition and higher peak in the Angry Face condition showed an effect opposite to that which we had predicted. According to the results of the post-experiment questions, participants felt comfortable when the mood suggested by android's facial expression was close to their own emotion. In contrast, they felt uncomfortable when the android's facial expression did not coincide with their emotion. We drew the conclusion that the participants felt the android's facial expressions as both positive and negative, according to the condition or timing of the facial expression change. In line with one participant's comment, when the android's facial expression did not match the context of the conversation, the participant may have experienced an uncanny feeling with regard to the android robot. Perhaps the negative effect occurred as a result of a large gap between the android's facial expression and the participant's emotional state. This gap needs to be reduced, which would require determining the participant's emotional state by using, for example, physiological measures such as brain activity or heart rate.

The results of the correlation analysis (Fig. 22.6) seem to support our idea. We can state that these results show that when participants could operate the android well, they were more affected by the facial expression changes in the android. As previously mentioned, the participants in the No-operation group may have identified the android as a bystander in the conversation. In contrast, the participants in the Operation group may have felt the android as themselves; that is, they may have experienced strong feelings of body ownership transfer to the android when they could operate it well. Therefore, in our opinion participants who could operate the android well had a stronger sense of body ownership transfer, which led to stronger facial feedback effect induced by the facial expression change of the android.

## 22.5  Conclusion

In this study, we experimentally examined whether participants were affected by facial feedback received from an android. We changed the android's facial expression to examine whether such changes affected the operators' emotions during conversation. We divided the participants into an Operation group and a No-operation group to examine the relationship between the facial feedback from the android and the phenomenon of body ownership transfer. The results show that the participants who operated the android were affected by the android's facial expression.

This phenomenon seems to be different from that of only feeling empathy when observing a facial expression change, as the participants who rated their operating skill higher seemed to be more strongly affected by the android's facial feedback. We assume that these participants had a stronger sense of body ownership transfer, and that they were more affected by the facial expression of the android.

In future studies, we should consider the negative effect caused by the android's facial expression. The negative effect perhaps occurred because of a gap between the operator's emotion and the android's facial expression. To reduce this gap, we need to measure each participant's emotional state with finer resolution in both its degree and real time. We may utilize advanced sensing of physiological features, such as brain activity and heart rate, and use pattern recognition technology. Moreover, we should also improve our operating interface. Although some operators felt body ownership transfer and were affected by the facial feedback, others could not sufficiently attain this feeling. When we have enhanced the interface, the operator will be more strongly affected. Such an enhanced system will enable us to telecommunicate through an android comfortably.

In addition, we should also consider how the social and cultural context affected the results shown here. In this study, we used Japanese participants and an android robot, the appearance of which was based on a Japanese female. Further investigation is necessary in different social and cultural contexts.

Moreover, for practical applications for emotion regulation, we should consider ethical issues. Although more than half of the participants were aware of the android's facial expression change in this experiment, the original design in this study was to change the facial expression of the communication avatar silently and automatically so that the users would be affected unconsciously. Such a manipulation may lead to a subconscious effect of which the users would not be aware. Although the proposed system for emotion regulation may be beneficial in many cases, and users of the system will be aware of its subconscious effect, the possibility exists that the same system may be abused for negative effects. We shall have a careful discussion and examine such ethical issues in depth.

# References

1. Nishio, Shuichi, Koichi Taura, Hidenobu Sumioka, and Hiroshi Ishiguro. 2013. Teleoperated android robot as emotion regulation media. *International Journal of Social Robotics* 5 (4): 563–573.
2. Gross, James J. 2002. Emotion regulation: Affective, cognitive, and social consequences. *Psychophysiology* 39: 281–291.
3. Botherel, V., and V. Maffiolo. 2006. Regulation of emotional attitudes for a better interaction: Field study in call centres. In *Proceedings of the 20th international symposium of human factors in telecommunication*, Mar 2006.
4. Derks, Daantje, Agneta H. Fischer, and E.R. Arjan. 2008. Bos. The role of emotion in computer-mediated communication: A review. *Computers in Human Behavior* 24 (3): 766–785.
5. Nico, H. 1986. *Frijda*. New York, USA: The emotions. Cambridge University Press.
6. Lewis, Jeannette M. Haviland-Jones, and Lisa Feldman Barrett (eds.). 2008. *Handbook of emotions*. New York, USA: Guilford Press.
7. Scherer, Klaus R., and Paul Ekman (eds.). 1984. *Approaches to emotion*. New Jersey, USA: Lawrence Erlbaum Associates.
8. Calvo, Rafael A., and Sidney D'Mello. 2010. Affect detection: An interdisciplinary review of models, methods, and their applications. *IEEE transactions on affective computing* 1 (1): 18–37.
9. Gross, James J. 1998. The emerging field of emotion regulation: An integrative review. *Review of General Psychology* 2 (3): 271–299.
10. Harris, Christine R. 2001. Cardiovascular responses of embarrassment and effects of emotional suppression in a social setting. *Journal of Personality and Social Psychology* 81 (5): 886–897.
11. Nakanishi, Hideyuki. 2004. Freewalk: A social interaction platform for group behaviour in a virtual space. *International Journal of Human-Computer Studies* 60 (4): 421–454.
12. Zeng, Zhihong, Maja Pantic, Glenn I. Roisman, and Thomas S. Huang. 2007. A survey of affect recognition methods: Audio, visual and spontaneous expressions. In *Proceedings of the 9th international conference on multimodal interfaces*, 126–133.
13. Kleinke, C., T. Peterson, and T. Rutledge. 1998. Effects of self-generated facial expressions on mood. *Journal of Personality and Social Psychology* 74: 272–279.
14. MacIntosh, D. 1996. Facial feedbck hypotheses: Evidence, implications, and directions. *Motivation and Emotion* 20: 121–147.
15. Soussignan, R. 2002. Duchenne smile, emotional experience, and autonomic reactivity : A test of the facial feedback hypothesis. *Emotion* 2: 52–74.

16. Strack, F., L. Martin, and S. Stepper. 1988. Inhibiting and facilitating conditions of the human smile: a nonobtrusive test of the facial feedback hypothesis. *Journal of Personality and Social Psychology* 54: 768–77.

17. Izard, Carrol E. 1990. Facial expressions and the regulation of emotions. *Journal of Personality and Social Psychology* 58 (3): 487–498.

18. Nishio, S., Hiroshi Ishiguro, and Norihiro Hagita. 2007. Geminoid: Teleoperated android of an existing person. In *Humanoid robots: New developments*, ed. Armando Carlos de Pina Filho, 343–352. Vienna, Austria: I-Tech Education and Publishing.

19. Nishio, Shuichi, Tetsuya Watanabe, Kohei Ogawa, and Hiroshi Ishiguro. 2012. Body ownership transfer to teleoperated android robot. In *International conference on social robotics (ICSR 2012)*, 398–407, Oct 2012, Chengdu, China.

20. Botvinick, Matthew. 1998. Rubber hands 'feel' touch that eyes see. *Nature* 391 (6669): 756.

21. Averill, James R. 1983. Studies on anger and aggression: Implications for theories of emotion. *American Psychologist* 38 (11): 1145–1160.

22. Ishi, C.T., C. Liu, H. Ishiguro, and N. Hagita. 2012. Evaluation of a formant-based speech-driven lip motion generation. In *13th Annual conference of the international speech communication association (Interspeech 2012)*, P1a.04, Portland, Oregon, Sept 2012.

23. Ogawa, Tokihiro, Rie Monchi, Mami Kikuya, and Naoto Suzuki. 2000. Development of the general affect scales (in Japanese). *The Japanese Journal of Psychology* 71 (3): 241–246.

# Chapter 23
# Adjusting Brain Activity with Body Ownership Transfer

Maryam Alimardani, Shuichi Nishio and Hiroshi Ishiguro

**Abstract** Feedback design is an important issue in motor imagery brain–computer interface (BCI) systems. However, extant research has not reported on the manner in which feedback presentation optimizes coadaptation between a human brain and motor imagery BCI systems. This study assesses the effect of realistic visual feedback on user BCI-performance and motor imagery skills. A previous study developed a teleoperation system for a pair of humanlike robotic hands and showed that the BCI control of the hands in conjunction with first-person perspective visual feedback of movements arouses a sense of embodiment in the operators. In the first stage of this study, the results indicated that the intensity of the ownership illusion was associated with feedback presentation and subject performance during BCI motion control. The second stage investigated the effect of positive and negative feedback bias on BCI-performance of subjects and motor imagery skills. The subject-specific classifier that was set up at the beginning of the experiment did not detect any significant changes in the online performance of subjects, and the evaluation of brain activity patterns revealed that the subject's self-regulation of motor imagery features improved due to a positive feedback bias and the potential occurrence of ownership illusion. The findings suggest that the manipulation of feedback can generally play an important role with respect to training protocols for BCIs in the optimization of the subject's motor imagery skills.

**Keywords** Body ownership illusion · BCI-teleoperation · Motor imagery learning · Feedback effect · Training

This chapter is a modified version of a previously published paper [2], edited to be comprehensive and fit with the context of this book.

M. Alimardani · S. Nishio (✉) · H. Ishiguro
Advanced Telecommunications Research Institute International,
Keihanna Science City, Kyoto, Japan
e-mail: nishio@botransfer.org

M. Alimardani · H. Ishiguro
Department of Systems Innovation, Graduate School of Engineering Science,
Osaka University, Osaka, Japan

© Springer Nature Singapore Pte Ltd. 2018
H. Ishiguro and F. Dalla Libera (eds.), *Geminoid Studies*,
https://doi.org/10.1007/978-981-10-8702-8_23

## 23.1  Introduction

Brain–computer interfaces (BCIs) are widely used in several fields as a new communication and control channel between the human brain and an external device. However, the application of this technology is not as simple and intuitive as suggested by its concept. The operation of a BCI requires subjects to perform certain tasks and to learn how to intentionally modulate certain characteristics of their brain activities to express their intentions. For example, motor imagery is one of the most commonly employed methods for BCI control of intended motions [5]. Subjects imagine the movement of a certain limb of their own body to induce changes in mu and beta rhythms over the corresponding subregion of the sensory-motor cortex. These changes are detected by the BCI and translated into control commands. The motor imagery task requires relatively longer training when compared to that of other BCI paradigms, such as P300 or SSVEP, because the mental rehearsal of a movement without actual execution is not a normal and daily practice for subjects, and thus, the task of motor imagery is an unfamiliar experience to most subjects.

The importance of an individual's motor imagery skills in BCIs is well recognized. However, most extant studies focused on the computer side and improving classification algorithms. There is a paucity of research examining the human side and training paradigms that can facilitate the skill acquisition process for subjects [8]. In a manner similar to other forms of interfaces, BCI users learn to coadapt with the system based on the feedback received regarding their performance. Therefore, feedback design is especially influential in the process of motor imagery learning and performance improvement. Standard BCI protocols typically provide online visual feedback in the form of a moving cursor or target on a computer screen. Neuper et al. compared the realistic presentation of feedback (in the form of a grasping hand) relative to abstract feedback (in the form of an extending bar) on a computer screen [11]. Nevertheless, they did not find any evidence of a significant difference between the performances of two feedback groups. Another study investigated the influence of motivation on BCI-performance by biasing feedback accuracy [4]. The results indicated that subjects exhibiting a poor performance benefitted from positive biasing, while subjects with a relatively better performance were impeded by inaccurate feedback. A similar study [6] involved providing subjects with fake negative and positive feedback of their performance and reported that negative feedback resulted in a greater learning effect on motor imagery BCIs.

Although the above works examined the effect of feedback presentation and accuracy, they did not specifically discuss the direct interaction between subjects and the BCI system. Motor imagery task requires the subjects to imagine their own body movements, while the output is provided as feedback in the form of movements of objects that are separate from the subject bodies. This mismatch and dissociation between the life experience of a subject and the BCI task could even interfere with imagination and impair the performance of motor imagery (especially for novice users).

The goal of the present study includes exploring the influence of feedback design on enhancing user performance and interaction with a BCI system. A previous study demonstrated that the BCI-operation of a pair of humanlike robotic hands by motor imagery while watching first-person perspective images of robot movement could induce an illusion of body ownership transfer (BOT) for the operators [1, 12]. In the post-experiment interview, a few subjects revealed that when the robot moved as per subject intentions, the feeling experienced was similar to that of the movement of the subjects' own hands and motor imagery became easier. In the present study, it is hypothesized that the inducement of the fore-mentioned feelings of ownership and the sense of agency driven toward the observed motions may have a positive loop effect on the execution of motor imagery during the BCI-operation. Hence, it is speculated that the thought of "I am the one moving the hands" leads to the feeling of "These hands are mine," and the illusion of owning hands enhances the imagery ability in subjects and boosts the inverse thought corresponding to "These are my hands so I can move them."

To this end, in the present study, the same BCI-teleoperation paradigm is used while exposing naïve subjects to different feedback conditions to probe the relationship between subject BOT experience and BCI-performance. Two experiments were performed. The first experiment involved manipulating the presentation of mis-performance to survey the manner in which subjects' perception of their own performances could affect the BOT intensity. The second experiment involved examining the manner in which this effect influenced the real performance of subjects and the trend of motor imagery learning.

Both experiments in this study were approved by the Ethics Committee of Advanced Telecommunications Research Institute International (12-506-3). All subjects read and signed a written consent form prior to the experiment and received payment for their participation.

## 23.2  Experiment 1

This experiment was designed to investigate the inducement of body ownership illusion for a pair of BCI-operated humanlike robotic hands under different feedback presentations.

### 23.2.1  Participants

Forty healthy participants (26 male, 14 female, age $M = 21.13$, $SD = 1.92$) were selected for the experiment. Thirty-eight participants were right-handed, and two participants were left-handed. All participants had no prior knowledge of the research topic and received an explanation prior to the experiment.

## 23.2.2 Method

Each participant was seated in a comfortable chair and asked to remain motionless. The participants wore an EEG electrode cap, and twenty-seven EEG electrodes were installed over their primary sensory-motor cortex based on a 10–20 system. The reference electrode was placed on a participant's right ear and a ground electrode on the forehead. Participants were asked to imagine a grasp or squeezing motion with respect to their own hands, while their cerebral activities were recorded by g.USBamp biosignal amplifiers (Guger Technologies). An initial training session involved the participants practicing in a motor imagery task by extending a feedback bar to the left or right side on a 15-in. laptop computer screen. A visual cue, in form of a horizontal pointing arrow, specified the timing and the hand with the participants were supposed to hold the imagery. Each trial lasted 7.5 s and commenced with the presentation of a fixation cross on the display. An acoustic warning was given in the form of a "beep" after 2 s. An arrow randomly pointing to the left or right side was shown in the duration from 3 to 4.25 s. The participants were instructed to perform motor imagery based on the arrow's direction. The participants watched the feedback bar and continued the imagery task until the fixation cross was erased. The next trial started after a short pause. The first run consisted of 40 trials (20 trials per class left/right presented in a randomized order) that were conducted without feedback. The recorded brain activities in the initial non-feedback run were used to set up a subject-specific classifier for the classification in the following feedback runs. In the feedback runs, participants performed similar trials, although they received the online classification results of their performance in the form of a horizontal feedback bar on the screen. The subject's task involved extending the feedback bar in the correct direction.

Recorded brain signals were processed under Simulink/MATLAB (MathWorks) for offline and online parameter extraction. This process included band-pass filtering between 0.5 and 30 Hz, sampling at 128 Hz, cutting off artifacts with a notch filter at 60 Hz, and adopting a common spatial pattern (CSP) algorithm for the discrimination of Event-Related Desynchronization (ERD) and Event-Related Synchronization (ERS) patterns associated with a motor imagery task [7]. The CSP found weight vectors that weighed each electrode based on its importance with respect to the discrimination task. The spatial filters were designed such that the resulting signal displayed the maximum variance for left trials and minimum variance for right trials. Therefore, the difference between the left and right populations was maximized to show the location at which the EEG variance displayed maximum fluctuation. Finally, the classifier outputted a linear array signal in the range of $[-1, 1]$ when the discrimination between left and right imaginations was made in which $-1$ denotes the extreme left and 1 denotes the extreme right. Negative values and positive values were then translated as the left-hand grasp motions and right-hand grasp motions, respectively, of the robot.

After the training sessions, the experiment was continued to the main test sessions in which subjects wore a head-mounted display (Vuzix iWear VR920) and teleoperated the robot's hands using the same BCI system. The subjects performed a motor imagery task for their right hand or left hand, while they watched first-person images of the robot's hands performing the corresponding motions (Fig. 23.1a). During the experiment, the subjects were asked to look down imagining that they were watching their own hands and identical blankets were laid on the robot's legs as well as the participants' legs to provide a similar view of a subject's own body. Participants placed their arms in positions and orientations similar to those of a robot's arms. Skin conductance response (SCR) electrodes were installed on the left palms of subjects to measure subjects' physiological reaction to a threatening stimulus. A bio-amplifier recording device (Polymate II AP216, TEAC, Japan) with a sampling rate of 1000 Hz was used to record SCR measurements. Prior to the testing sessions, the participants watched an act in which the robot's hand was injected (painful stimulus) through a head-mounted display, which was explained to them as a part of the robot adjustment procedure. The injection continued until the subjects' SCR disappeared [3]. This was followed by conducting testing sessions in a random order under the following three conditions (Fig. 23.1b):

(1) Still (no feedback): The robot's hands did not move throughout the whole session, although the subjects performed motor imagery based on a cue.

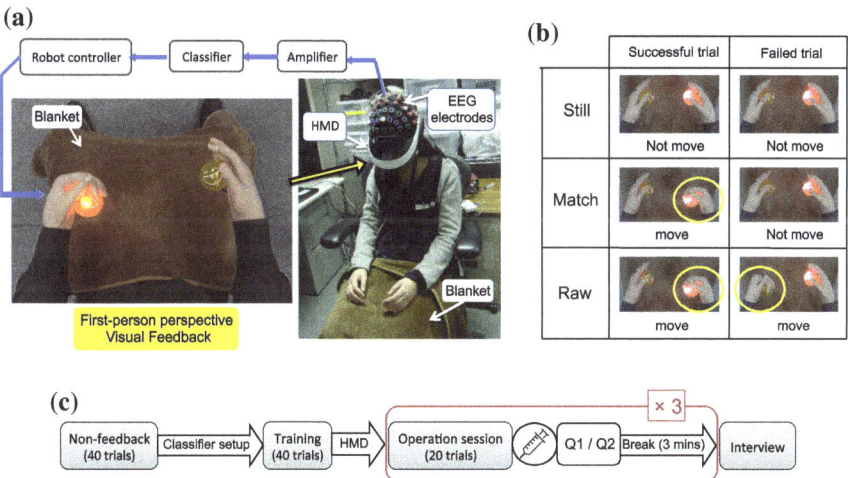

**Fig. 23.1** Experiment setup. **a** EEG electrodes installed on a subject's sensory-motor cortex recorded brain activities during a motor imagery task. Subjects watch first-person images of robot through a head-mounted display. A lighting ball in front of the robot's hands provides motor imagery cue, and the subject holds images of a grasp with respect to his/her own corresponding hand. The classifier detects two classes of results (right or left) and sends a motion command to the robot's hand. **b** The subjects repeat the experiment under three different conditions; Still condition (in which the robot's hand did not move), Match condition (in which the robot's hand only moved in successful trials), and Raw condition (in which the robot's hand performed the failed trial by using the wrong hand)

(2) Match (no negative feedback): The robot's hands moved only if the classification result was correct and identical to the cue.
(3) Raw: The robot's hands moved based on the classification results even if the result was wrong and opposite to the cue.

The last session is termed as Raw since the unprocessed values obtained from the classifier were input as the robot's motion parameter. In all the above conditions, participants performed trials that were designed in a manner similar to training sessions with respect to duration and stimulus timing. After completing 20 trials that lasted for a total of 2 min and 40 s, the robot's left hand was injected to examine if the illusion of ownership could cause a response to a pain-causing stimulus. The session was terminated immediately following the injection, and the participants were orally asked the following questions: (Q1) When the robot's hand was injected, did it feel as if your own hand was receiving the injection? (Q2) Throughout the entire session while you were operating the robot's hands, did it feel as if they were your own hands? Participants scored Q1 and Q2 based on the seven-point Likert scale with 1 corresponding to "Didn't feel such thing at all" and 7 corresponding to "Felt it very strongly." In addition to the self-assessment, the body ownership illusion was physiologically measured by recording the skin conductance responses (SCRs).

### 23.2.3 Result

The response variables for 40 participants were obtained from the questionnaires and SCR recordings. Participant responses in three conditions (Still, Match, and Raw) were averaged and compared using the Tukey-HSD multiple comparison method. The mean value, standard deviation, and $p$-value are depicted on each graph.

For both Q1 and Q2, the average value of the Match condition exceeded those of the other two conditions (Fig. 23.2a, b). The results for Q1 were significant between Match (M = 4.38, SD = 1.51) and Still (M = 2.83, SD = 1.43); [Match > Still, $p < 0.0001$] and between Raw (M = 3.15, SD = 1.57) and Still; [Raw > Still, $p < 0.01$]. Similarly, there was a significant difference in the Q2 scores for Match (M = 5.15, SD = 1.10) and Still (M = 2.93, SD = 1.25); [Match > Still, $p < 0.0001$] and between Raw (M = 4.18, SD = 1.38) and Still; [Raw > Still, $p < 0.05$].

The SCR peak value within a 6 s interval (1 s after the appearance of the syringe in the participant's view to 5 s after the injection) was selected as the reaction value [1]. In this experiment, the response values of only 35 participants were evaluated since five participants showed unchanged responses during the experiment and were excluded from the analysis. The results of the SCR measurements (Fig. 23.2c) confirmed significant differences only between Match (M = 1.68, SD = 1.98) and

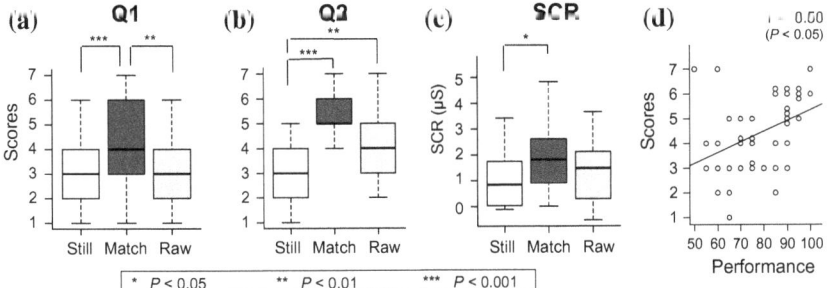

**Fig. 23.2** Results for experiment 1. **a** Mean value and standard deviation for Q1 in each session, and the values in Match condition are significantly higher than those in the Still and Raw conditions; **b** mean value and standard deviation for Q2 in each session, and the values in the Match and Raw conditions are significantly higher than those in the Still condition; **c** mean value and standard deviation of SCR values for 35 subjects, the value in the Match showed a significantly higher response to injection than that in Still condition; and **d** subject performances versus illusion scores in the Match condition. For subjects with the same performance and score, the score is slightly modified to a non-integer neighbor value to avoid the overlap of the markers. A positive correlation is indicated between BCI-performance and the illusion intensity

Still (M = 0.90, SD = 1.42); [Match > Still, $p < 0.05$]. Additionally, a positive correlation was found between subjects' performance and the Q1 score only in the Match condition (correlation coefficient r = 0.56, $p < 0.001$, Fig. 23.2d). The term performance refers to the rate of trials in which the subjects could successfully grasp the lightened ball out of the total twenty trials conducted in each run scaled in terms of percentages.

### 23.2.4  Discussion

In study 1, the inducement of body ownership illusion was investigated for a pair of BCI-operated humanlike robotic hands under different feedback conditions. Results from both measurement methods (Q1 and SCR) indicated significantly high responses to the injection in the Match condition in which a robot's hands moved only if the classification result was correct and identical to the cue. This indicated that the feeling of receiving an injection was significantly stronger when the robot's hands moved exclusively in agreement with the operator's intentions than when the robot did not perform any motion (Still) or when the robot performed the wrong motion in the case of errors (Raw). This corresponds to a feeling aroused due to the illusion of ownership over the robot's body, and thus, it could be stated that the transfer of body ownership could be more reliably evoked by precise mind control of a robot's hands.

Conversely, in the case of Q2, participants directly scored their sensation of ownership with respect to the robot's body during the entire operation time.

The assessment of the participants indicated that the feeling of ownership was significantly stronger in both the Match and Raw conditions (in which the robot dynamically moved and reacted to the participant's intentions) when compared to that in the Still condition (in which the robot did not display any motion). Although the Match condition showed a higher average response when compared to that of the Raw condition, a significant difference between these two conditions was not confirmed in Q2. This could imply that the robot's successive motions in both the Raw and Match conditions followed the participant's act of motor imagery and raised a sense of agency during the session that led to a perception of hand ownership in the participants.

Meanwhile, the results of this experiment showed a wide dispersion over the response values of illusion in each condition. It was presumed that this was due to the difference in the performance levels among the subjects. A positive correlation was confirmed between the participants' performance and the score given by the participants with respect to Q1 (Fig. 23.2d). This indicated that the participants with a better operational performance experienced a stronger illusion of BOT. Therefore, this suggested that a subject's skill in the motor imagery task and BCI-performance are associated with the intensity of ownership illusion in this type of a teleoperational system.

The obtained results in the experiment suggested that the feedback presentation could affect the eliciting of ownership illusion over the controlled hands in a BCI-teleportation system for humanlike hands. The illusion was augmented when negative feedback of a subject's mis-performance was eliminated. Additionally, a positive correlation was revealed between the BOT intensity and subject performance, which suggested that subjects with better BCI-performance experienced stronger illusion. Therefore, BOT was affected by a subject's BCI-performance as well as feedback design, which regulated the subjects' perception of their own performance. Conversely, although an intuitive conclusion of this experiment involved the idea that better BCI-performance was a cause for stronger illusion in subjects, the reverse could also be claimed wherein higher BOT was the motive for the better performance of subjects with respect to a motor imagery task. Therefore, it is necessary to clarify the manner in which the mutual interaction between performance and BOT is formed and the manner in which feedback design contributes to improving each element and their interaction (Fig. 23.3). Thus, experiment 2 focused on the effect of feedback design on a subject's BCI-performance and examined the manner in which the manipulation of subjects' perception of

**Fig. 23.3** Model diagram for effect of feedback design. Feedback bias can affect the interaction between BCI-performance and body ownership transfer illusion

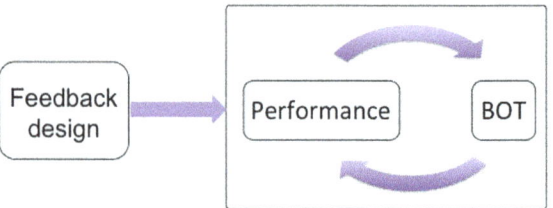

self-performance can affect the trend of their motor imagery learning and BCI-performance.

## 23.3  Experiment 2

Experiment 1 indicated that subjects' perception of their own performance was important in the inducement of ownership illusion. Moreover, a close relationship between the intensity of illusion and a subject's performance was also indicated. It is also important to understand the manner in which subjects' self-evaluation and subsequent inducement of BOT can directly affect subject skills in a motor imagery operational system. Therefore in this experiment, the presentation and accuracy of subjects' performance were manipulated to design four different feedback conditions for the performance of each subject that was positively and negatively biased in the first half of the session. This was followed by examining the manner in which conditioning feedback affected the learning trend by using two methods, namely (1) measuring the online performance of subjects in the second half of the sessions and (2) comparing the time-variant distribution of EEG features with respect to right-hand imagery and left-hand imagery in each half of a session.

### 23.3.1  Participants

Sixteen healthy subjects (6 male and 10 female, age M = 21.1, SD = 1.4) participated in the experiment. None of the subjects had participated in the previous experiments, and all the subjects were unfamiliar with the BCI research topic. The participants received an explanation prior to the experiment.

### 23.3.2  Method

The BCI devices, preparation procedure, and session paradigms of experiment 2 were identical to those in experiment 1 with the exception that experiment 2 involved the usage of a head-mounted display (Sony HMZ-T1) for first-person visual feedback (Fig. 23.1a).

Participants performed four experimental sessions that consisted of 40 imagery trials. The first half of each session (20 trials) was randomly conditioned as follows:

(1) Raw: The performance of the participants was not biased. The robot's hands grasped the ball according to the classification result.

(2) Match: The performance of the participants was not biased. However, the robot's hands only grasped the lighted ball when the classification results matched the cue.
(3) Positive Feedback (Fake-P): The performance of the participants was positively biased. The robot's hands grasped the lighted ball correctly in 90% of trials irrespective of a subject's real performance.
(4) Negative Feedback (Fake-N): The performance of the participants was negatively biased. The robot's hands grasped the lighted ball correctly only in 20% of trials irrespective of a subject's real performance.

In the first two conditions, Raw and Match (Fig. 23.1b), the performance of a subject was not biased. However, the presentation of mistaken trials was different in which a trial included the execution of incorrect hand motion and a trial did not include robot motion. It was previously revealed that the Raw and Match conditions corresponded to different levels of illusion due to negative feedback presentation. Thus, two more sessions (Fake-P and Fake-N) were designed in which performance feedback was biased irrespective of subjects' real performance accuracy to deliberately enhance or decrease their self-evaluation.

In the second half of all the sessions, the subjects received feedback of their real performance. The goal involved seeking changes in BCI-performance and motor imagery skills in the second half of each session based on the positive or negative feedback biases. The performances of the subjects were registered in the second half of all sessions.

### 23.3.3   Result

#### 23.3.3.1   Online Performance

The performance of 16 subjects in the second half of each session was averaged and compared by using the Tukey-HSD multiple comparison method (Fig. 23.4a). The term performance refers to the percentage of successful trials among the post 20 trials. Fake-P (M = 60.78, SD = 10.24) showed the highest performance when the Raw (M = 49.22, SD = 9.07), Match (M = 54.37, SD = 10.89), and Fake-P (M = 50.47, SD = 10.58) conditions were compared. However, the results did not indicate any significant differences between the sessions.

Conversely, the average score of participants for body ownership illusion was significantly higher in Fake-P condition (M = 4.44, SD = 1.01) when compared to the remaining three conditions (Raw (M = 3.25, SD = 1.03), Match (M = 3.38, SD = 1.11), and Fake-N (M = 2.75, SD = 0.90); [Fake-P > Raw, Match, Fake-N, $p < 0.000$]). Significant differences were also found between Match and Fake-N; [Match > Fake-N, $p < 0.05$].

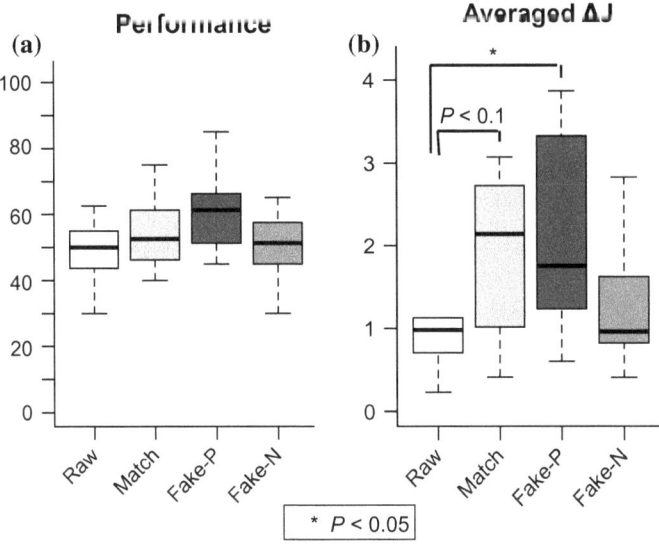

**Fig. 23.4** Results for experiment 2. **a** Mean value of subject performance in the second half of each session is demonstrated. Significant differences are not found. **b** Mean value of the ratio $J_2/J_1$, which is an identifier of motor imagery quality, showed significantly higher values in Fake-P and Match conditions when compared to that in the Raw condition

### 23.3.3.2 Offline Classification

As the results did not reveal any significant differences with respect to online performance in the second half of the sessions, it was estimated that this could be due to the mis-classification of the initially set classifier. The classifier did not use a learning algorithm. Thus, when the classification boundary for two classes right/left was defined within the feature space in the initial training session, the same classifier and parameters were used till the end of the experiment. However, it was speculated that subjects consciously or unconsciously modified the generation of their brain activity patterns during motor imagery while receiving biased feedback or experiencing illusion, and the classifier did not accurately detect the same. Therefore, the original brain signals were used, and the feature distribution of right and left motor imagery was extracted through offline processing.

The features used for classification were obtained by the CSP method following artifact removal and temporal filtering [7]. The CSP built an N × N projection matrix **W** with N channels of EEG for each left and right trial corresponding to **X**. With the projection matrix **W**, the mapping of a trial is given as follows:

$$Z = WX$$

The columns of $\mathbf{W}^{-1}$ denote the common spatial patterns and can be observed as time-invariant EEG source distribution vectors. The construction was conducted such that the variance for left movement imagination was maximum in the first row of $\mathbf{Z}$ and decreased with the increases in number of subsequent rows. It is not necessary to calculate the variances of all N time series to obtain reliable features. The optimal number of common spatial patterns used to build the feature vector corresponds to 4 [10]. Therefore, only the first and last two rows ($p = 4$) of $\mathbf{W}$ were used to filter data $\mathbf{X}$ and build new signals $\mathbf{Z_p}$ ($p = 1 \ldots 4$). The variance of the resulting four time series is obtained for a time window $T = (t_0, t_1)$ as follows:

$$var_p = \sum_{t=t_0}^{t_1} \left( \mathbf{Z}_{p(t)} \right)^2$$

where window length was set to 1 s, beginning 1500 ms after the presentation of the cue [13]. Feature vectors are obtained after normalizing and log transforming as follows:

$$f_p = \log \left( \frac{var(Z_p)}{\sum_{i=1}^{p} var(Z_p)} \right)$$

The online classifier uses the feature vector $f_\mathbf{p}$ of each trial to categorize it into two classes of right and left. Further, Fisher's solution is used in a linear discriminant analysis to observe the distribution of two classes feature vectors in a four-dimensional space. Fisher's parameter $J$ is defined as follows:

$$J = \frac{|\tilde{\mu}_R - \tilde{\mu}_L|^2}{\tilde{s}_R^2 + \tilde{s}_L^2}$$

where $\tilde{\mu}_R$ and $\tilde{\mu}_L$ denote the means of feature vectors for two right and left classes, respectively, and the quantity $|\tilde{\mu}_1 - \tilde{\mu}_2|^2$ corresponds to the distance between the means of the two classes. With respect to each class, $\tilde{s}_R^2$ and $\tilde{s}_L^2$ were defined as the scatter, which is an equivalent of the variance that is given by the following expression:

$$\tilde{s}_i^2 = \sum_{x \epsilon f_i} (x - \tilde{\mu}_i)^2$$

The quantity $\tilde{s}_R^2 + \tilde{s}_L^2$ denotes the within-class scatter. In the performance of motor imagery, a larger $J$ corresponds to a closer dispersion of feature vectors per each class and increased distance between two class means, which represents a better feature distribution for classification and therefore better execution of a motor imagery task.

The $J$ parameter was calculated for the first 20 conditioned trials ($J_1$) and for the second 20 test trials ($J_2$) in each session. The initial skills of the subjects were diverse, and thus for every subject the order of sessions was considerable with respect to the amount of motor imagery skills. The ratio $\Delta J = J_2/J_1$ was selected as a measurement of subject motor imagery learning in a specific session. Additionally, $\Delta J$ was calculated for 16 subjects. The interquartile range (IQR) for statistical dispersion in each condition [9] was used to detect two outliers in the Fake-N condition (S2 and S4) and an outlier in the Raw condition (S15). The data of the fore-mentioned three subjects were discarded from further analysis. The $\Delta J$ for the remaining 13 subjects was averaged and compared using the Tukey-HSD multiple comparison method (Fig. 23.4b). The mean value was highest in the Fake-P condition (M = 2.12, SD = 1.78) when compared to those in the other three conditions Raw (M = 1.07, SD = 0.75), Match (M = 1.87, SD = 0.95), and Fake-N (M = 1.35, SD = 0.84). A significant difference was obtained between Fake-P and Raw; [Fake-P > Raw, $p < 0.05$] and between Match and Raw; [Fake-P > Raw, $p < 0.1$].

## 23.3.4  Discussion

In experiment 2, the visual feedback of performance was biased in a BCI-teleoperation system of a humanlike robot to probe the effect of positive and negative feedback on the BCI-performance of subjects and motor imagery skills.

The online results did not reveal any significant changes in the real-time performance of the subjects, and the mean value of performance remained at the chance level for all conditions. It was assumed that the classifier defect in detecting the correct class for each feature vector as the classification parameters was set at the beginning of the experiment, and an offline process of the original brain activities was adopted to determine changes in the distribution of right and left motor imagery features. The results revealed that the ratio $J_2/J_1$, which is an identifier of class separation between two halves of sessions, was significantly higher in the Fake-P condition when compared to that in the Raw condition. This indicated that subjects could generate motor patterns that could be classified better by a CSP algorithm by receiving positive feedback with respect to their performance in the Fake-P condition. A statistical significance level of 10% was used to confirm a similar relation between Match and Raw conditions, thereby indicating that motor imagery improved in the Match condition in which subjects did not receive negative feedback of their failed performance, and the subjects could produce more separable activity patterns for the two classes of right- and left-hand movements. Both results implied that positive feedback bias corresponded to an enhancing effect on motor imagery learning, and this is consistent with the findings of a few previous studies [8]. A potential cause for the fore-mentioned finding could relate to the inducement of a stronger BOT due to biased feedback, which facilitates imagination related to movement in the motor imagery task and eventually enhances the self-regulation of brain patterns in subjects (Fig. 23.3).

In contrast to previous reports on biased BCI feedback, significant improvements [6] or impediments [4] were not revealed in the Fake-N condition when compared to other conditions. However, S2 and S4 were discarded from analysis as outliers and showed a drastic $\Delta J$ increase in the Fake-N condition. A significant enhancement in learning was noted in the Fake-P condition, and thus, it was assumed that the effect of biasing is closely relevant to the subject's personality and the influence of motivation on different individuals. Although there are learners that benefit from encouragement and positive feedback of their performance, a few learners benefit more from negative feedback and try harder when the feedback informs the subjects regarding their poor performance. Future experiments will involve personality tests to categorize subjects in two groups and to independently conduct the survey.

Finally, in the experiment in the present study, it was hypothetically assumed that enhancement of motor imagery learning due to positive bias of feedback was associated with ownership illusion over controlled robot's hands (Fig. 23.3). However, a further survey is required to accurately measure the intensity of illusion at the end of each conditioned section. In the experiment in the present study, it was assumed that it was unlikely that pausing the session and asking an assessment question would shatter the illusion. A future study is necessary to compare humanlike and non-humanlike visual feedback under biased feedback to precisely verify as to whether the illusion of body ownership influences the motor imagery learning trend.

## 23.4 Conclusion

In the present study, two experiments were designed to answer the following questions: (1) How does the presentation of visual feedback affect the inducement of body ownership illusion in the BCI-operators of humanlike hands and (2) How does positively and negatively biased feedback in this type of a system influence operators' interaction with the system and improve their BCI-performances. Results of the first experiment revealed that negative feedback of subjects' errors impeded the intensity of ownership illusion. Additionally, BOT was correlated with subjects' performance in BCI and the extent to which subjects felt that they were in control of the robot hands. In the second experiment, it was revealed that biasing feedback could not immediately boost subjects' performance in the same session. However, the analysis of brain patterns showed that it could in fact change the motor imagery learning trend.

With respect to feedback design for future BCI systems, it is conceivable that a more realistic feedback presentation can assist novice users to train and adapt faster and more efficiently to a system. Furthermore, BCI users could also benefit from feedback positive bias in training sessions, although it is necessary to consider their personalities. Meanwhile, since subjects' motor imagery skills dynamically change during a session based on the subjects' state of mind, it is necessary to further develop sophisticated classifiers that customize classification parameters in an online session.

**Acknowledgements** This work was supported by a Grant-in-Aid for Scientific Research (S) KAKENHI (24650114).

# References

1. Alimardani, M., S. Nishio, and H. Ishiguro. 2013. Humanlike robot hands controlled by brain activity arouse illusion of ownership in operators. *Science Reports* 3. https://doi.org/10.1038/srep02396.
2. Alimardani, M., S. Nishio, and H. Ishiguro. 2014. Effect of biased feedback on motor imagery learning in BCI-teleoperation system. *Frontiers in Systems Neuroscience* 8: 52.
3. Armel, K.C., and V.S. Ramachandran. 2003. Projecting sensations to external objects: Evidence from skin conductance response. *Proceedings of the Royal Society of London: Biological* 270: 1499–1506.
4. Barbero, Á., and M. Grosse-Wentrup. 2010. Biased feedback in brain-computer interfaces. *Journal of Neuroengineering and Rehabilitation* 7 (34): 1–4.
5. Curran, E.A., and M.J. Stokes. 2003. Learning to control brain activity: A review of the production and control of EEG components for driving brain–computer interface (BCI) systems. *Brain and Cognition* 51 (3): 326–336.
6. Gonzalez-Franco, M., Y. Peng, Z. Dan, H. Bo, and G. Shangkai. 2011. Motor imagery based brain-computer interface: A study of the effect of positive and negative feedback. In *Engineering in medicine and biology society, EMBC, Annual international conference of the IEEE.*
7. Guger, C., H. Ramoser, and G. Pfurtscheller. 2000. Real-time EEG analysis with subject-specific spatial patterns for a brain-computer interface (BCI). *IEEE Transactions on Rehabilitation Engineering* 8 (4): 447–456.
8. Lotte, F., F. Larrue, and C. Muhl. 2013. Flaws in current human training protocols for spontaneous brain-computer interfaces: Lessons learned from instructional design. *Frontiers in Human Neuroscience* 7 (568). https://doi.org/10.3389/fnhum.2013.00568.
9. Moore, D.S., and G.P. McCabe. 1998. *Introduction to the practice of statistics*, 3rd ed. New York: W. H. Freeman.
10. Müller-Gerking, J., G. Pfurtscheller, and H. Flyvbjerg. 1999. Designing optimal spatial filters for single-trial EEG classification in a movement task. *Clinical Neurophysiology* 110: 787–798.
11. Neuper, C., R. Scherer, S. Wriessnegger, and G. Pfurtscheller. 2009. Motor imagery and action observation: Modulation of sensorimotor brain rhythms during mental control of a brain–computer interface. *Clinical Neurophysiology* 120 (2): 239–247.
12. Nishio, S., T. Watanabe, K. Ogawa, and H. Ishiguro. 2012. Body ownership transfer to teleoperated android robot. In *International conference on social robotics, ICSR 2012*, 398–407.
13. Pfurtscheller, G., and C. Neuper. 2001. Motor imagery and direct brain-computer communication. *Proceedings of the IEEE* 89 (7): 1123–1134.

# Chapter 24
# At the Café—Exploration and Analysis of People's Nonverbal Behavior Toward an Android

**Astrid M. von der Pütten, Nicole C. Krämer, Christian Becker-Asano, Kohei Ogawa, Shuichi Nishio and Hiroshi Ishiguro**

**Abstract** Current studies investigating natural human–robot interaction (HRI) in the field concentrate on the analysis of automatically assessed data (e.g., interaction times). What is missing to date is a more qualitative approach to investigate the natural and individual behavior of people in HRI in detail. In a quasi-experimental observational field study, we investigated how people react to an android robot in a natural environment according to the behavior it exhibits. We present data on unscripted interactions between humans and the android robot "Geminoid HI-1" in an Austrian public café and subsequent interviews. Data related to the participants' nonverbal behavior (e.g., attention paid to the robot and proximity) were analyzed. The results show that participants' behavior toward the android robot, as well as their interview answers, was influenced by the behavior the robot exhibited (e.g., eye contact). In addition, huge inter-individual differences existed in the participants' behavior. Implications for HRI research are discussed.

**Keywords** Human–robot interaction · Field study · Observation
Multimodal evaluation of human interaction with robots · Uncanny Valley

This chapter is a modified version of a previously published paper [1], edited to be comprehensive and fit with the context of this book.

A. M. von der Pütten · N. C. Krämer
University of Duisburg Essen, Forsthausweg 2, 47048 Duisburg, Germany

C. Becker-Asano
Albert-Ludwigs-Universität Freiburg, Georges-Köhler-Allee 52, 79110 Freiburg, Germany

K. Ogawa (✉) · S. Nishio · H. Ishiguro
Osaka University, 1-3 Machikaneyama, Toyonaka, Osaka 560-8531, Japan
e-mail: ogawa@irl.sys.es.osaka-u.ac.jp

© Springer Nature Singapore Pte Ltd. 2018
H. Ishiguro and F. Dalla Libera (eds.), *Geminoid Studies*,
https://doi.org/10.1007/978-981-10-8702-8_24

## 24.1 Introduction

Robotics is a vastly expanding field of research, which is becoming increasingly relevant for various fields as it is considered to have the potential to solve major societal problems (e.g., to compensate for decreasing numbers of healthcare employees by providing support in low-priority tasks [2] and the rehabilitation of post-stroke participants [3, 4]). Thus, the field of robotics is also becoming more salient in the public discourse. In contrast, the majority of the existing research on human–robot interaction (HRI) has been limited to laboratory experiments and did not investigate how humans interact with robots in natural unscripted situations. Therefore, the results of HRI experiments may be subject to demand characteristics; that is, participating in an experiment can be an irritating experience, and participants attempt to discern the experimenter's hypotheses and behave or evaluate accordingly [5]. In addition, some experiments in the area of social robotics were conducted with participants who had an interest in the topic of robotics and considerable knowledge about robotics, or at least about technology in general, thus impeding the generalizability of the experimental findings.

Although laboratory experiments are able to show statistically significant differences between groups according to the manipulation of the study, they assess only specifically chosen dependent variables. Thus, the investigator is able to draw general conclusions regarding these specific dependent variables, but is certainly unable to make any statements about individuals' natural behavior toward robots, which can indeed vary greatly. In order to gain knowledge of the topic of natural HRI, more field research is required. However, this is difficult to realize as most humanoid robots (robots with uniquely humanlike characteristics, such as a humanlike appearance with arms, legs, and head, upright stance, bipedalism) and android robots (a robot that closely resembles a human, foremost in appearance but also in behavior) are still in the development stage and are not easily displayed in real-life scenarios. Thus, only a small number of studies have investigated users' behavior toward humanoids or androids in a natural environment.

There is a growing body of literature that includes observational research studies featuring commercial robots that can more easily be displayed (e.g., toy-like robots). These studies have revealed that participants show huge inter-individual differences in their behavior toward robots. For instance, von der Pütten et al. [6] showed that participants differed greatly in terms of their verbal and nonverbal behavior toward a robot rabbit placed in their homes. Although the robot did not understand natural speech or perceive nonverbal behavior, some participants frequently talked to the robot, smiled at it, and also showed other signs of bonding (e.g., name-giving), while others adhered to the interaction possibilities offered by the system. It seems clear that these inter-individual differences will also be found in interactions with humanoid and android robots.

What is lacking thus far is a more qualitative approach for investigating the natural and individual behavior of people in HRI in detail. The example of von der Pütten et al. [6], but also research in the field of virtual agents (addressing, e.g., proximity [7, 8] eye gaze [7] and mimicry [9, 10]), underlines the importance of considering new data analysis approaches, such as the analysis of nonverbal behavior, and of going beyond the assessment of the number and length of inter-actions. We therefore conducted a quasi-experimental observational field study to investigate how people react toward an android robot in a natural environment according to the behavior displayed by the robot. We present data of unscripted interactions between humans and the android robot "Geminoid HI-1," which we analyzed in terms of the participants' nonverbal behavior.

When investigating people's interactions with humanoid or android robots, the Uncanny Valley hypothesis [11] and its assumptions should be taken into account. According to the Uncanny Valley theory, people will react increasingly positively toward increasingly humanlike robots until a certain peak is reached, after which the effect is reversed, resulting in negative reactions. Although the present study was focused on the qualitative examination of people's natural behavior toward robots, we also analyzed the collected data in terms of the Uncanny Valley effect. Therefore, in the following, we introduce related work on previous observational studies using robots, as well as an introduction to, and previous research on, the Uncanny Valley theory.

## 24.2  Related Work

### 24.2.1  Challenges for Observational Field Studies and the Resulting Shortcomings of Previous Research

The most common method of investigating HRI consists of laboratory experiments, which can be affected by the artificial situation in which the participant is placed or by a sample bias (e.g., the subjects are very interested in robots). Thus, studies that analyze behavior toward robots in a natural environment are lacking. Real-life scenarios, however, are not easy to observe, and the resulting data are difficult to analyze, because they lack the quality that can be reached in laboratory settings (e.g., the audio recording is affected by noise or the video quality is degraded). Therefore, it is common practice to focus on data that can be automatically col-lected during a field trial and which minimize the effort required for the subsequent analysis. Some studies, for instance, utilize automatic technology (e.g., motion sensors) to record and analyze behavior, resulting in an impressive number of logged interactions (see [12, 13]). The data thus gathered do not go beyond the duration of interactions or the identification of people who repeatedly visit the robot (gathered by using additional logging equipment, such as ID cards). The purpose of these studies was not to analyze in more detail the manner in which people interact with the robot, e.g., whether or not people talk to the robot and their nonverbal behavior. For example, in their report of an 18-day field trial held at a Japanese elementary school, Kanda et al. [13] included the interaction duration in an experiment in which they introduced two English-speaking "Robovie" robots [14] to first- and sixth-grade pupils. They collected data on the interaction duration and the interaction situation, i.e., whether pupils interacted alone with the robot or in groups with other children, and also conducted an English test as a performance measure. Although they collected a large amount of video data, the researchers did not analyze more profoundly the material related to verbal and nonverbal behavior toward the robots. Thus, many questions remain unanswered, including, for instance, how the children treated the robots during the interactions.

An improvement in terms of more detailed information of users' behavior can be found in a report on a quasi-experimental field study conducted by Hayashi et al. [15], in which either a single robot or two robots communicating with each other were present at a train station where the passengers could see them. The video analysis showed that 5,900 passengers passed by the robot(s) during the 8 days of data collection, with 900 in fact stopping and interacting with the robot(s). In a more fine-grained analysis of these 900 videos, the experimenters observed varying behaviors, which were coded into inductively developed categories. They found that some people changed their course to stop and watch the robot(s), some touched the robot(s) or took pictures, others talked to strangers about the robots, etc. These behaviors occurred more frequently when two robots, as compared to only one robot, were present and when the robots were more interactive. The study showed that people react very differently toward robots in a natural environment and that their behavior does not depend only on the experimental manipulation. For instance, it was also important at what time of day or on which day of the week people encountered the robot.

Similarly, the aforementioned longitudinal field study by von der Pütten et al. [6] using a rabbit-shaped companion robot revealed that participants showed huge inter-individual differences in their verbal and nonverbal behavior. Differences were also found in the behavior change over time, as well as in the post-trial evaluation of the companion and self-reported relationship building. For instance, some of the total of six participants frequently talked to the robot companion throughout all three field trials, while others did not address it at all. Similar findings were observable for the participants' smiling behavior. Moreover, some behaviors, e.g., speech, seemed to be stable over time, whereas others changed, e.g., smiling behavior. This type of analysis is very time-consuming, which is reflected in the small number of participants, but it is able to reveal huge inter-individual differences for specific behaviors in a natural environment that would not be possible in a laboratory setting. The robot used in this study was a modified off-the-shelf product and thus inexpensive and more convenient to use than an android robot.

In summary, a general shortcoming of research appears to be that in many studies robots were applied in a field trial, but they were either limited by automatically collected interaction times, and sometimes performance measures, or restricted by small sample sizes. In the present work, we therefore focus on the assessment of the nonverbal behavior toward an android robot in the field on the basis of a larger sample. Because of the setting of the present study, we were unable to assess a great number of personality characteristics to shed light on which of these factors influence the behavior toward the robot. However, we do include the participants' gender as a possible influencing factor.

### 24.2.2  The Uncanny Valley

Recent advances in robotics in the field of humanoid robots render the investigation of the Uncanny Valley increasingly relevant and possible. Ishiguro [16] stated that "humanoids and androids have a possibility to become ideal human interfaces accepted by all generations" (p. 2), because, with their humanlike appearance, they offer the human interaction partner all the interactive possibilities to which he or she is accustomed in human–human interaction. The Uncanny Valley theory, however, restricts these possibilities, as described in the following. In particular, in the present study, Mori's Uncanny Valley theory [11] is relevant, because, according to this theory, the more humanlike these robots become, the more people feel familiar with them and the more they are willing to accept them. However, shortly before the ideal of perfect humanness, the curve breaks inward; familiarity is reversed into uncanniness (see Fig. 24.1) and people react negatively toward the robot (e.g., they are disgusted or distressed).

Mori describes this effect for *still* as well as for *moving* objects, although the latter are considered as eliciting stronger effects. Although—or because—Mori's theory lacks precise definitions for realism or human-likeness and familiarity, it is suitable for various applications, and despite the early introduction of the Uncanny

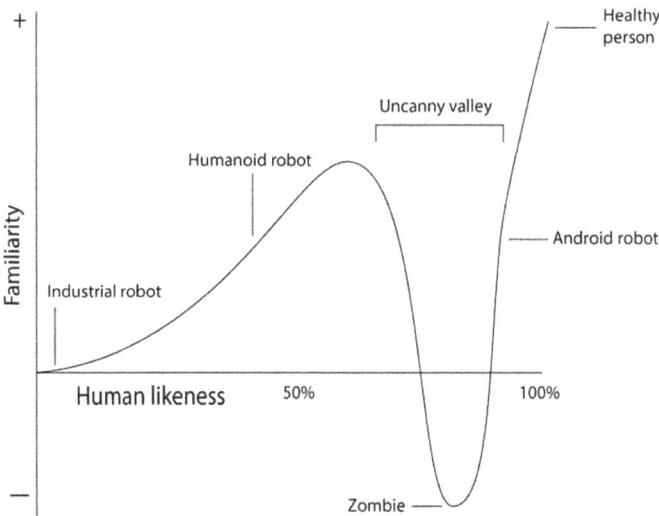

**Fig. 24.1**  The Uncanny Valley

Valley theory into scientific discourse, researchers mostly refer to it anecdotally as an explanation for their unexpected results [e.g., 17, 18]. Thus, since the 1970s, the Uncanny Valley has been a frequently cited explanation in science as well as in the public sphere, which has barely been tested empirically. However, in the last five years, scientists in the field of robots have addressed the task of investigating the Uncanny Valley more systematically. The first attempts were aimed to provide empirical evidence for its existence [19–21] or to explain the Uncanny Valley effect using classical work from communication theories [19]. According to the Uncanny Valley hypothesis, the android robot used in this observational field study may suffer from the Uncanny Valley effect, since the android looks strikingly humanlike. The displayed behavior, however, is very limited and does not match the quality of the android's outer appearance. It was therefore considered worthwhile to analyze the gathered data also in light of the Uncanny Valley effect. The following section describes previous studies within a similar setting using the same android.

### 24.2.3 Does Geminoid HI-1 Suffer the Uncanny Valley Effect? Previous Observations on Geminoid HI-1 in the Field

In fall 2009, the annual "ARS Electronica" festival in Linz, Austria, featured the android Geminoid HI-1 and its creator, Hiroshi Ishiguro, as a special attraction in the overall context of the arts festival. Before the official opening of the festival, Geminoid HI-1 was placed in Café CUBUS in the ARS Electronica building. The android robot sat behind a table, with a laptop computer in front of it and an information desk about Kyoto and its attractions beside it (see Sect. 3 for a more detailed description of the setting). During the festival itself, it was installed as an exhibit in the basement of the ARS Electronica building. Within both settings, different studies (including the study reported in the current paper) were conducted that investigated diverse research questions. Two of these studies are presented in the following.

Thirty visitor dialogues with Geminoid HI-1 (teleoperated by a fellow experimenter) within the Café CUBUS setting were analyzed in terms of the identity perception of the interlocutor facing Geminoid HI-1 [see 22]. The results show a tendency to ascribe an identity to the android robot Geminoid HI-1 that is independent of the identity of the person controlling the robot. The authors concluded that the humanoid features of Geminoid HI-1 elicit assumptions on the part of the user about the robot's (humanlike) character, expected reactions, and conversational skills, and it is therefore treated as a social actor. However, the analysis of the speech sequences also revealed that Geminoid is categorized as an "entity 'in-between'" [22], and references to it include anthropomorphizing words and human characteristics, as well as robotic characteristics. This finding is in line with the argument in [23] that the categorization of objects and experiences is imperative for humans. Robots, however, cannot be categorized easily and reliably as either "alive" or "not alive," because they are on the boundary between these categories.

Becker-Asano et al. [24] reported interviews conducted with 24 visitors to the festival who had previously interacted with Geminoid HI-1 within the exhibition. When asked to describe the android robot, the visitors gave more positive descriptions of Geminoid HI-1 (e.g., very humanlike, striking verbal skills, terrific, very likable) than negative descriptions (e.g., surreal, quite thick fingers, too obviously robot-like, a bit scary). When asked directly about their emotional reactions toward the robot, only 37.5% of the interviewed visitors reported an uncanny (or strange or weird) feeling, and 29% stated that they enjoyed their conversation with the robot. In five cases, the interviewees' feelings changed during the interaction. For example, one reported that the interaction was "amusing" at first, but that he experienced a "weird" feeling when he discovered that his inter- action partner was actually a robot. Most descriptions of Geminoid HI-1 were related to its outward appearance, with negative attributes being voiced here in particular. Negative descriptions also referred to the imperfections of its move- ments. With respect to emotional reactions, fear was found to be the predominant emotion. According to the Uncanny Valley[1] hypothesis, this is because fear is regarded as indicating a person's submissive behavioral tendency to withdraw from a threatening or unfamiliar situation (cf. [24]), an assumption that is in line with the studies reported in [19, 25]. The authors concluded that "Geminoid HI-1's inade- quate facial expressivity as well as its insufficient means of producing situation-appropriate social signals (such as a smile or laughter, […]) seems to impede a human's ability to predict the conversation flow," [24] thus causing a feeling of not being in control of the situation.

## 24.2.4  Research Questions

The research mentioned above provides interesting insights into the manner in which people interact with robots in natural environments, although all of these studies had certain limitations. Some studies did not go beyond the mere logging of interactions, and their authors failed to analyze how people react toward or interact with robots. In other studies, as in the Geminoid HI-1 studies presented previously, the visitors were prompted to talk about the robotic nature of Geminoid HI-1 or about the experimental situation, because it was explicitly located at the festival as an exhibit or subjects were invited to converse with a telepresence robot, respec- tively. The location of the present study was also in the previously described setting of the ARS Electronica Festival. In contrast to the previous studies, however, in this study participants were not informed about the robot. Thus, it was considered that an analysis of these unprompted, unscripted interactions might reveal different reactions toward the robot. Furthermore, interviews allow only for the assessment of reactions that are under conscious control or upon which the participants are able to reflect. What has not yet been analyzed in depth is the human interlocutors' nonverbal behavior, which is more direct and spontaneous. It was therefore considered that an analysis of participants' nonverbal behavior might offer valuable clues as to the nature of human–robot interaction. Finally, we used a quasi-experimental setting and varied the behavior displayed by the android robot in order to investigate whether and how the robot's behavior influences the participants' nonverbal behavior.

Against this background, the research questions were as follows:

RQ1: How do people behave toward robots in a natural environment in unscripted and unprompted situations? Which different nonverbal behaviors can be observed? Is the participants' nonverbal behavior influenced by the robot's behavior? Can gender differences be found?
RQ2: Do people recognize the android as a robot? Is the recognition influenced by the robot's behavior?
RQ3: Do people report unprompted feelings related to the Uncanny Valley effect?

## 24.3 Method

### 24.3.1 General Setup

Prior to the Ars Electronica Festival 2009, Geminoid HI-1 was placed in the Café CUBUS, which is part of the Ars Electronica Center (AEC) in Linz, Austria, from August 10–30, 2009. Unlike other studies using Geminoid HI-1 during the festival, in this setting, the android sat on a chair behind a small table with a laptop computer in front of it in order to give the impression of a working visitor; see Fig. 24.2. Next to the robot, people could find information material about traveling to Japan. The scene was video-recorded from five camera perspectives: One camera was placed behind the android to record participants immediately in front of it (see Fig. 24.3) and one facing and recording the android (see Fig. 24.4); three additional cameras covered the rest of the room. Visitors were not given any hints that they might encounter a robot within the café.

Most of the visitors entered the café via an elevator. They then passed the robot on their way to the bar, where the interviewer asked them to participate in a short interview. Some visitors came from the stairway ("behind" the yellow wall in the direction of the bar) or from the outside patio.

### 24.3.2 The Android Geminoid HI-1 in Different Conditions

Geminoid HI-1 is an android robot and a duplicate of its scientific originator Hiroshi Ishiguro (HI). Geminoid HI-1 was designed to function as an embodied

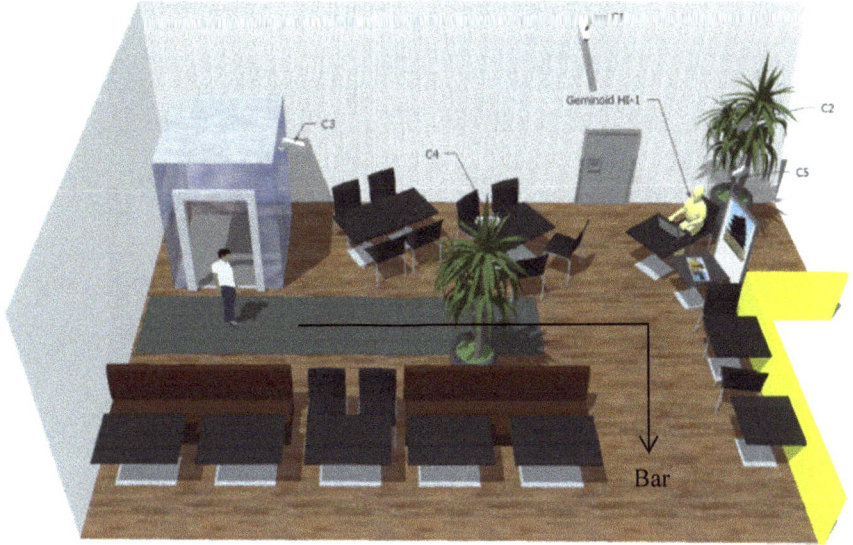

**Fig. 24.2** Setup with Geminoid H1 in the Café CUBUS

**Fig. 24.3** View from camera behind Geminoid HI-1

interaction tool in human shape that can be controlled from a remote location by teleoperation. The robot is covered by silicone skin with a natural skin tone, wrinkles, etc. Geminoid HI-1 is able to show facial expressions and move its head, torso, arms, and legs driven by 50 pneumatic actuators [26]. For teleoperation, the robot is connected via Internet to an interface through which the operator can observe the robot and its environment and control the robot's speech and movements. Via microphone and speakers, the speech is transmitted and synchronized with lip and head movements. The robot features idle behavior, which includes

**Fig. 24.4** Camera view recording Geminoid HI-1

breathing, eye blinking, and in some cases posture shifts, and was designed to give the impression of being alive. Further expressive movements can either be separately programmed or executed in real time via teleoperation. During the 11 days of data collection for this analysis, Geminoid HI-1 was presented in different conditions, which are listed in the following.

**No eye contact condition**. Geminoid HI-1 was presented in the *no eye contact mode*, in which the robot was only looking down at the laptop computer in front of it, ignoring passersby. It showed the idle behavior mentioned above.

**Eye contact condition**. In the *eye contact mode*, Geminoid HI-1 looked up when the participant looked straight in its direction, or more precisely in the direction of the camera behind Geminoid HI-1, which used face-tracker software. The face-tracker software module was an extension of the OpenCV-based "Face Detection" source code [27]. It tracked one or more frontal faces over a sequence of images taken from a video stream, allowing Geminoid HI-1 to follow one particular visitor with its gaze, even if he or she was moving slightly. Eight to ten times per second the algorithm checked with the previously stored information about recognized faces at the identical position. When a previously recognized face remained in the same position and continued looking frontally into the camera, the algorithm gave this face a one-step-higher priority. If a new person joined the situation and faced the camera frontally, the algorithm started the same procedure of counting up (beginning with priority zero) for the new face. A priority shift occurred when the "old" face broke eye contact with Geminoid HI-1, because, for instance, the subject turned to face the newcomer. In this case, the "old" face was no longer observed. The new face then had a higher priority, and Geminoid HI-1 faced the newcomer.

**Table 24.1** Distribution of subjects across conditions

| Geminoid | | | Hiroshi Ishiguro |
|---|---|---|---|
| No eye contact | Eye contact | Remote-controlled with both eye contact and speech | N = 7 |
| N = 30 | N = 45 | N = 16 | |
| N = 91 | | | |
| Total N = 98 | | | |

When nobody interacted with Geminoid HI-1 and the algorithm did not recognize any faces, Geminoid HI-1 returned to looking down at the laptop computer.

**Remote control condition**. In the third condition, Geminoid HI-1 was *remote-controlled* by a human experimenter who was able to control the eye contact. It also addressed passersby proactively by saying "Hello, would you like to have a brochure?" and was able to engage in conversations. In this condition, the experimenter's head and lip movements were captured using facial feature-tracking software and transmitted alongside the experimenter's speech utterances to the android.

**Real Hiroshi Ishiguro condition**. Furthermore, participants interacted with the human counterpart of Geminoid HI-1, Hiroshi Ishiguro, who sat at the same table as Geminoid HI-1 working quietly until he was addressed by the subjects.

Table 24.1 shows the distribution of subjects across conditions. As people visited the café to enjoy their free time and not with the intention of participating in an experiment, we were dependent on their willingness to participate. Thus, an equal distribution was not possible. The experimental design, however, was not the most important aspect of this study, because the investigation of the participants' general reactions toward Geminoid HI-1 was more important. We thus decided to include all the data sets.

### 24.3.3  Interviews

On 10 of the 20 days, passersby were asked to participate in an interview. The interviewer addressed the passersby with the following standardized sentences: "Hello we would like to conduct a brief interview with you if you have a moment. It will not take longer than ten minutes and you would be volunteering to contribute to scientific research." If they asked what the interview was about, the interviewer told them the following: "The interview is about your personal experience within the last five minutes since you entered the café. We won't ask personal questions and you may withdraw from the interview at any time without giving reasons." In total, 107 visitors agreed to participate, of whom 98 gave consent to be audio- and videotaped. The interviewer stood at the entrance to the bar of the Café CUBUS, recruited participants, and guided them into a relatively quiet corner of the bar.

From the moment of recruitment, participants could not look back into the room where Geminoid HI-1 and the information desk were located. First, the participants were asked whether they noticed the information table next to Geminoid HI-1. If they gave a positive answer, they were requested to describe the table. If they had not noticed the table, they were asked whether they had perceived something unusual or special when they entered the café. From these follow-up questions, we derived whether the interviewees recognized the robot as such or mistook it for a human being. If people recognized Geminoid HI-1 as a robot, they were asked whether they thought it might be good for advertisements (for instance, as in the setting in the café). The participants were then debriefed and thanked for their participation.

## 24.3.4   Analysis of Videos

We identified the 98 subjects who agreed to be interviewed and videotaped within the video material and extracted the interaction sequences starting from the moment of the subjects' first appearance until they were recruited and guided into the next room by the interviewer. The video material was annotated in ELAN (Max Planck Institute for Psycholinguistics, Nijmegen, The Netherlands; http://www.lat-mpi.eu/tools/elan/, [28]). We assessed the different behaviors of the participants. First, we assessed the total time for which people appeared in the videotaped area (coded as *appearance time*) and then annotated the participants' attention directed to Geminoid HI-1. In the latter category, we subsumed two different behaviors: When subjects were in the direct vicinity of Geminoid HI-1, we were able to observe gaze behavior and coded the eye contact established with Geminoid HI-1; when participants were further away, we assessed the amount of time for which they faced Geminoid HI-1 frontally. Both times were summed as *attention directed toward Geminoid HI-1*. In addition, we coded the *participants' verbal behavior*. We assessed whether, and if so, for how long they talked to Geminoid HI-1, to third persons, or said something unrelated to the situation. Furthermore, we looked for *behaviors that attempt to test Geminoid HI-1's vividness* and reactions, e.g., waving in front of the robot's face, pulling a face, raising eyebrows, or taking a picture. Finally, we coded the participant's *proximity* to Geminoid HI-1. Proximity was always coded using the same camera perspective (camera 1) in order to guarantee reliable coding and was based on previously defined proximity areas, namely the "outer area," the "vicinity area," the "adjacent or table area," and "touch," which are illustrated in Fig. 24.5. We assessed the amount of "eye contact" established by Geminoid HI-1 (looking up and facing a participant) for use as an independent and control variable. Furthermore, it was coded whether there was a group situation when the participants encountered the robot (Geminoid HI-1 was alone vs. Geminoid HI-1 was surrounded by visitors when the subject arrived) and whether the participants were in company (subject was alone vs. in company) when they entered the café.

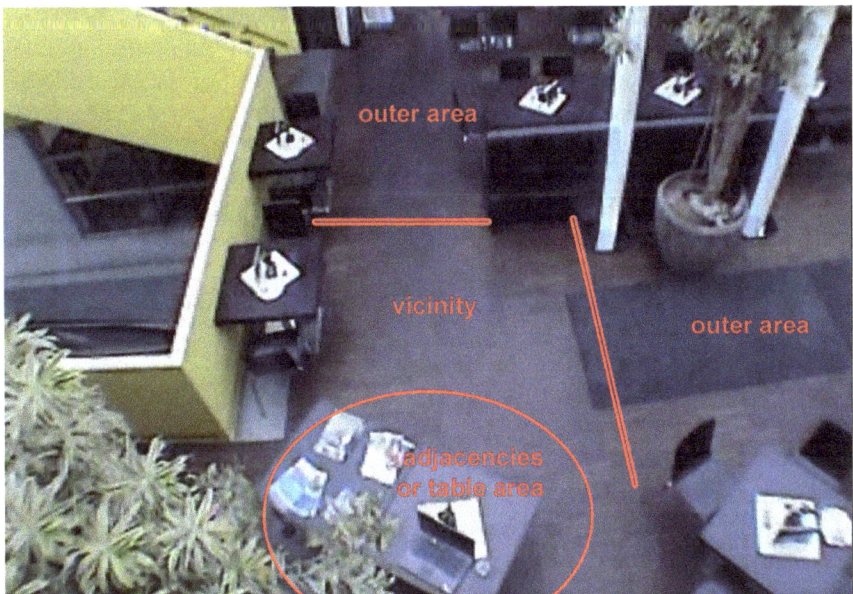

**Fig. 24.5** Proximity areas

Ten percent of the coded video material (10 participants) was coded by a second rater. The ratings were checked for agreement between the two raters using the in-built function in ELAN for comparing annotators. Within this function, the start time and end time of each annotation was given and the amount of overlap and the total extent (time from lowest start time until highest end time); a value indicating the agreement was calculated (overlap/extent) for each category in each video, which was then averaged over all videos. The inter-rater reliability values were as follows: 96% for the category "appearance time of participants," 86% for the category "attention to Geminoid HI-1," 86% for category "Geminoid HI-1's eye contact with participant," and 94% for category "proximity." The agreement of the annotators was 100% on the occurrences of specific behaviors (such as talking to Geminoid HI-1, waving hands in front of the robot's face, taking pictures, and grimaces).

## 24.3.5 Sample

In total, 107 guests were interviewed, of whom 98 (38 males, 60 females) agreed to be audio- and videotaped. Their age ranged from 8 to 71 years (M = 38.43, SD = 14.98). The majority were Austrians (81), followed by visitors from Germany (12), Italy (2), Spain (1), Belgium (1), and the Netherlands (1). Nine participants were retired, 20 were school pupils or university students, and 61 were employed.

Eight participants did not indicate their profession. Although the Café CUBUS is part of the AEC, it does not attract only visitors to the center. As it offers upscale cuisine, unique architecture, and an outstanding view over the Danube and the local recreation area next to the river, it has many frequent visitors and tourists also visit. Fourteen participants stated that they had visited the café before, while some stated that their travel guide recommended the café. We thus assumed that our interviewees were not predominantly interested in the center or the festival, which began two weeks after the data collection.

## 24.4   Results

### 24.4.1   Interviews

To answer research question 1, we derived from several open questions (see above) whether participants recognized the robot or mistook it for a human being. Of the seven participants who met Prof. Ishiguro, five reported that they had noticed a (human) man behind the table and two did not notice anything special. Of the remaining 91 participants, 23 made no comments in response to this question related to the robot, because they had noticed neither the table nor the robot. They interpreted the questions as addressing the architecture of the café, which indicates that Geminoid HI-1 was not recognized as uncommon and was not sufficiently salient for the participants. Eighteen people mistook Geminoid HI-1 for a human being. Most of these participants did not believe that it was indeed a robot, even after the experimenter had told them, and some announced that they would return for a second interaction. Fifty participants clearly stated that they had seen a robot, although 18 of these mentioned that they initially mistook the robot for a human.

To answer research question 3, that is, whether participants make Uncanny Valley-related statements, we analyzed the reasons that people gave for recognizing Geminoid HI-1 as a robot and whether they stated that they experienced negative emotions. In order to avoid artificially prompting or suggesting experiences of uncanniness, no explicit questions were asked about possible (negative) emotional experiences related to the Uncanny Valley effect. However, we found that only three participants mentioned without being prompted that Geminoid HI-1 gave them an uneasy feeling (N = 4; "it looks so real, a little uncanny"; "I think that might be unpleasant" (to use Geminoid HI-1 for advertising)). When asked why they recognized that Geminoid HI-1 was not human, most participants referred to the stiff posture and abrupt movements (N = 21, e.g., "he sits there in a weird way"; "his movements are too jerky"; "I recognized no movement, the hands…"; "his restricted motor activity") or the lack of movements (N = 2, e.g., "We waved, but he didn't wave back."). Others mentioned that his face seemed like a mask (N = 5) and his hands looked unnatural (N = 14). Two mentioned that they recognized that the "man" sitting there was jacked up in some way (e.g., "he was jacked up"; "I saw cables"). One participant initially concluded that Geminoid HI-1 might be a disabled person, and three thought it was a wax figure. Some participants had difficulty in formulating their first impression of Geminoid HI-1 and eventually described it as "kind of artificial being," "extraterrestrial," or just a "weird person." In summary, different aspects seemed to have influenced the participants' perception of Geminoid HI-1. As suggested by the Uncanny Valley theory, the crucial factors were the robot's movement, as well as its unnatural hands and inexpressive face.

With respect to the participants' interest in the robot, we found that of the 50 participants who recognized Geminoid HI-1 as being a robot, 12 stated that they did not engage in longer interactions because purchasing a coffee or something to eat was a higher priority than dealing with a robot. However, they suggested that they planned to go back and examine the robot more closely later on.

## 24.4.2  Videos of Interactions

Only the videos of those participants (N = 91) who were confronted with the android robot Geminoid HI-1 were included in the analysis, because the analysis focused on HRI and the participants' behavior toward the android. To answer research question 1, we analyzed different nonverbal behaviors, such as the time at which people appeared at the café, whether they performed actions to test the robot's capabilities, whether they spoke to Geminoid HI-1, and their nonverbal behavior (proximity to the robot, attention directed toward the robot).

### 24.4.2.1  Appearance Time and Testing Actions

The amount of time for which people interacted with (or passed by) Geminoid HI-1 lays between 9.25 and 277.44 s ($M = 58.11$; $SD = 52.07$). The high standard deviation shows the scale of the individual differences in terms of the amount of time people spent in the café before they were asked to be interviewed. Some people quickly passed by the robot, whereas others spent several minutes exploring the robot's capabilities.

People attempted different actions to test whether Geminoid HI-1 would react to them. Only one person touched the robot, although we observed that some accompanying persons touched Geminoid HI-1 while the interviewees watched or took a picture of the interaction. Seven subjects waved (partially in front of Geminoid HI-1's face), one stuck out her tongue, one pulled a face, and two persons raised their eyebrows in an exaggerated manner. Five subjects took a picture or videotaped the interaction (for the distribution across conditions, see Table 24.2). Interestingly, these actions performed by the participants to test Geminoid HI-1 occurred not in the *no eye contact* condition, but rather in the *eye contact* and *remote control* conditions. A chi-square test for all testing reactions over the conditions showed that the occurrence of "talk to Geminoid HI-1" was

**Table 24.2** Distribution of testing actions across conditions

| Condition | Touch | Wave | Grimace | Tongue | Picture | Raise eyebrows | Talk to Geminoid |
|---|---|---|---|---|---|---|---|
| No eye contact | 0 | 0 | 0 | 0 | 0 | 0 | 0 |
| Eye contact | 1 | 4 | 1 | 1 | 2 | 1 | 5 |
| Remote control | 0 | 2 | 0 | 0 | 3 | 1 | 7 |

higher than predicted in the *remote control condition* ($c^2(2, N = 91) = 17.78$, $p \leq 0.000$, partial $\eta^2 = 0.414$; refer to Table 24.3). Neither appearance times nor the occurrence of testing actions were significantly influenced by gender.

### 24.4.2.2 Proximity

We wished to ascertain how closely people approached Geminoid HI-1 and the duration of their stay in the different proximity areas. Here, too, we found great inter-individual differences, indicated by the high standard deviations over all the proximity categories. The time spent within the "outer area" varied between 7.48 and 147.25 s ($M = 31.49$; $SD = 25.34$). Most participants at least briefly went through the "vicinity area," and only three of the visitors (3.3%) chose a different path to cross the dining area to reach the bar. The remaining 88 participants were in the vicinity area for between 1.36 and 243.31 s ($M = 17.45$; $SD = 31.57$). Twenty-eight subjects (30.8%) entered the "table area" and stood close to the table in front of the robot or surrounded the table to examine the robot more closely.

**Table 24.3** Chi-square test for "Talking to Geminoid HI-1"

| Condition | | "Talk to Geminoid HI-1" | | Total |
|---|---|---|---|---|
| | | No | Yes | |
| No eye contact | Count | 30 | 0 | 30 |
| | Predicted | 26.0 | 4.0 | 30.0 |
| | Standardized residuals | 0.8 | −2.0 | |
| Eye contact | Count | 40 | 5 | 45 |
| | Predicted | 39.1 | 5.9 | 45.0 |
| | Standardized residuals | 0.1 | −0.4 | |
| Remote control | Count | 9 | 7 | 16 |
| | Predicted | 13.9 | 2.1 | 16.0 |
| | Standardized residuals | −1.3 | 3.4 | |
| Total | Count | 79 | 12 | 91 |
| | Predicted | 79.0 | 12.0 | 91.0 |

| Table 24.1 Time spent in the vicinity area | Mean | SD | N |
|---|---|---|---|
| No eye contact | 5.13 | 7.61 | 30 |
| Eye contact | 16.99 | 36.27 | 45 |
| Remote control | 38.58 | 33.01 | 16 |
| Total | 16.88 | 31.19 | 91 |

Subjects remained in the "table area" from 1.14 to 214.36 s ($M = 41.24$; $SD = 45.86$). Only one person touched Geminoid HI-1.

One-factorial ANOVAs with condition as the independent variable and the proximity categories as dependent variables showed that participants encountering Geminoid HI-1 in the *no eye contact* condition spent less time in the vicinity area than participants in the *eye contact* or *remote control* conditions (F(1, 91) = 6.768, $p = 0.002$, $\eta^2 = 0.133$; see Table 24.4 for means and SDs). Post hoc comparisons using the Scheffé test indicated that the mean score for the *no eye contact* condition was significantly different from that for the *remote control* condition ($p = 0.002$; $SE = 9.09$). In addition, there was a significant difference between the *eye contact* condition and the *remote control* condition ($p = 0.046$; $SE = 8.45$). A two-factorial ANOVA with condition and gender as independent variables revealed no gender effects.

### 24.4.2.3   Participants' Attention to Geminoid HI-1

We coded the attention that participants paid to the robot as described above. The results of a one-factorial ANOVA showed that participants encountering Geminoid HI-1 in the *no eye contact* condition paid significantly less attention to the android than participants in the *eye contact* or *remote control* conditions (F(1, 91) = 10.246, $p = 0.000$, $\eta^2 = 0.189$; see Table 24.5 for means and SDs). The results of post hoc comparisons using the Scheffé test indicated that the mean score for the *no eye contact* condition differed significantly from that for the *eye contact* condition ($p = 0.052$; $SE = 1.56$) as well as the *remote control* condition ($p = 0.000$; $SE = 2.04$). In addition, there was a significant difference between the *eye contact* condition and the *remote control* ($p = 0.024$; $SE = 1.92$) condition.

| Table 24.5 Participants' attention directed toward Geminoid HI-1 in seconds | Mean | SD | N |
|---|---|---|---|
| No eye contact | 4.09 | 7.90 | 30 |
| Eye contact | 22.72 | 35.65 | 45 |
| Remote control | 48.58 | 46.15 | 16 |
| Total | 21.12 | 35.05 | 98 |

#### 24.4.2.4  Speech

We analyzed whether or not people addressed Geminoid HI-1 verbally and how this behavior was distributed across conditions. None of the people in the *no eye contact condition* and 11% in the *eye contact condition* spoke to Geminoid HI-1. In the *remote control condition*, in which the robot itself addressed the participants, 44% of the people talked to Geminoid HI-1 (see also Table 24.2). Whereas in the *eye contact* condition the participants' utterances were quite short (three said only "hello," one said "I may take this [brochure], right?" and one said "I am going to take a picture, now. No! Please look up again"), people engaged in longer conversations with Geminoid HI-1 in the *remote control* condition. People introduced themselves and had short chats related mainly to the brochures, or took a picture.

#### 24.4.2.5  Detection of the Robot

Furthermore, we wished to know whether Geminoid HI-1's behavior (see RQ 2) or other factors influenced whether people were able to recognize Geminoid HI-1 as a robot. We calculated a stepwise regression analysis with the following predictors: Geminoid HI-1's eye contact with the participant, group situation (Geminoid HI-1 is alone vs. Geminoid HI-1 is surrounded by visitors when subject arrives), company (subject is alone vs. in company), participant's gender, and the duration of the subject's attention paid to Geminoid HI-1. This resulted in a valid regression model for Geminoid HI-1's eye contact ($r = 0.267$, $r^2 = 0.072$, $p = 0.027$). The more eye contact the robot showed, the more easily people detected it as a robot. We assumed this effect to be mediated by the time people spent in the vicinity area. Indeed, the relationship between Geminoid HI-1's eye contact and the participants' ability to detect that Geminoid HI-1 was a robot was fully mediated by the duration of their stay in the vicinity area. As Fig. 24.6 illustrates, the standardized regression coefficient between Geminoid HI-1's eye contact and detection decreased substantially when we controlled for the duration of the participants' stay in the vicinity area. The other conditions of mediation were also met: Geminoid HI-1's eye contact

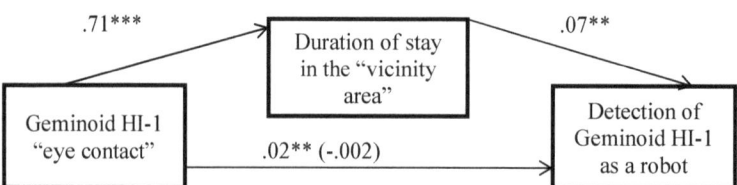

**Fig. 24.6** Standardized regression coefficients for the relationship between Geminoid HI-1's eye contact and the participants' ability to detect that Geminoid HI-1 was a robot as mediated by the duration of their stay in the vicinity area (proximity). The standardized regression coefficient between Geminoid's eye contact and the detection of Geminoid HI-1 as a robot controlling for time spent in the vicinity area is in parentheses. *$p < 0.05$, **$p < 0.01$, ***$p < 0.000$

**Table 24.6** Logistic regression for *Detection of Robot* with the predictor *Geminoid HI-1's eye contact*

|  | B (SE) | 95% CI for odds ratio | | |
|---|---|---|---|---|
|  |  | Lower | Odds ratio | Upper |
| Included |  |  |  |  |
| Constant | −0.787 (0.28) |  |  |  |
| Geminoid HI-1's eye contact | 0.02** (0.01) | 1.00 | 0.455 | 1.04 |

*Note*  $R^2 = 0.12$  (Hosmer & Lemeshow); 0.13 (Cox & Snell), 0.18 (Nagelkerke). Model $x^2(1) = 12.66$, $p < 0.000$. **$p < 0.01$

**Table 24.7** Logistic regression for *Detection of Robot* with the predictor *Proximity (vicinity area)*

|  | B (SE) | 95% CI for odds ratio | | |
|---|---|---|---|---|
|  |  | Lower | Odds ratio | Upper |
| Included |  |  |  |  |
| Constant | −1.086 (0.32) |  |  |  |
| Proximity | 0.07** (0.02) | 1.03 | 1.07 | 1.12 |

*Note* $R^2 = 0.21$ (Hosmer & Lemeshow); 0.21 (Cox & Snell), 0.29 (Nagelkerke). Model $x^2(1) = 21.09$, $p < 0.000$. **$p < 0.01$

was a significant predictor of the participants' ability to detect that Geminoid HI-1 was a robot (see Table 24.6) and of the time spent in the vicinity area ($\beta = 0.71$, $t(88) = 9.53$, $p < 0.000$, and also explained a significant proportion of variance, $R^2 = 0.51$, F(1.90) = 90.81, $p < 0.000$). The duration of the participants' stay in the vicinity area was a significant predictor of their ability to detect that Geminoid HI-1 was a robot while controlling for Geminoid HI-1's eye contact (see Table 24.7).

## 24.5  Discussion

This report of an observational field study presented data of unprompted and unscripted interactions between humans and the android robot Geminoid HI-1 in an Austrian café. The study was aimed to open up a new perspective on the investigation of HRI in the field, thus going beyond the assessment of mere interaction times and qualitatively analyzing how people behave toward robots in a natural environment. Ninety-eight people agreed to be interviewed, and their interactions with Geminoid HI-1 were analyzed with regard to the following dimensions: the appearance time, proximity to the robot, attention paid to the robot, actions to test the robot's capabilities, and verbal addressing of the robot.

Unlike in previous research investigating HRI in laboratory settings or in field trials, in which the investigative nature of the study was obvious to the participants [22, 24], the participants in this study did not expect to encounter a robot. There were no hints that they would interact with an android robot and the interactions did not follow any script. Given this very free situational context, we were interested in answering the question of whether people would recognize the android as a robot and whether this recognition would be mediated by different degrees of displayed behavior. We found that 43 participants either mistook Geminoid HI-1 for a human or even did not notice it at all, because it did not seem to appear conspicuously non-human. This effect was mediated by the displayed behavior of the android. People in the conditions with eye contact were most able to reliably recognize that Geminoid HI-1 was a robot, which may have been caused by the robot's rather jerky movements when looking up from the table to the participant or redirecting its attention to another participant. The effect was fully mediated by the time people spent in the immediate area around Geminoid HI-1. This means that Geminoid HI-1's eye contact caused people to spend more time in its close vicinity and thus they clearly also had more time to explore the robot's capabilities.

When subjects became aware of the robot, they examined it more closely and some explored Geminoid HI-1's capabilities. In particular in the conditions with eye contact, participants tested Geminoid HI-1's capabilities by waving their hands in front of its face, saying hello to it, pulling a face, or sticking out their tongue in anticipation of an appropriate reaction. It is interesting that these actions were not performed in the *no eye contact* condition. This may be because people also spent less time in front of Geminoid HI-1 when it did not react at all to them. This is in line with the answers to the question of why participants recognized that Geminoid HI-1 was not human, given that most participants referred to the stiff posture and abrupt movements. Participants who encountered Geminoid HI-1 in the *no eye contact* condition were able to see only very subtle movements (blinking, breathing), if any at all, and therefore did not perceive this cue. Two participants mentioned that they recognized that the "man" sitting there was jacked up in some way.

To evaluate the Uncanny Valley effect, we assessed whether people reported unprompted (negative) feelings considered to be related to the Uncanny Valley effect, such as distress, fear, or disgust. In contrast to the study reported in [24], where it was shown that participants mentioned emotional terms related to fear and disgust, participants in the present study did not in general report any negative feelings in the interviews. Only three people mentioned that Geminoid HI-1 gave them an uneasy feeling. Against the background of Ramey's [23] thoughts on the Uncanny Valley effect and his theory that it is difficult to categorize robots into either "alive" or "not alive," it is very interesting that some participants indeed did not instantly describe Geminoid HI-1 as either a human being or a robot. They rather described their first impressions with words indicating their difficulty in categorizing the robot, e.g., "disabled person," "kind of artificial being," "extraterrestrial," or "weird person." This is in line with the findings reported in [22] that Geminoid HI-1 was perceived as an "in-between" entity. However, in terms of the participants' behavior, we found that those participants who noticed that Geminoid HI-1 was a robot showed interest rather than negative reactions. In general, the behavioral data show that, although Geminoid HI-1 should suffer from the Uncanny Valley effect, people were rather relaxed when meeting it in public in this unscripted situation.

According to related studies reported in [6, 15], we assumed that people would show huge inter-individual differences in their behavior. We therefore wished to establish which different nonverbal behaviors can be observed. Indeed, we found huge inter-individual differences for all categories, as indicated by the high standard deviations for the coded behaviors, appearance time, attention, and proximity. With respect to the occurrence of testing actions, we also observed that some people attempted very different methods to test the robot's capabilities, while others merely observed the scenery.

Furthermore, a very interesting finding was that many participants (about 13%) almost ignored the robot and quickly passed by. Although stating in the interviews that they recognized it as being a robot, they decided that buying a coffee was of higher priority than exploring it. This suggests that for a significant number of people, robots do not seem to be of any interest. They simply do not care about them being in their vicinity and proceed with their planned activities. This corresponds to the findings of Hayashi et al. [15]. In their field study with robots in a train station, they found that people differed in the amount of their interest in the robots according to the time of day and the day of the week. For example, during rush hour, people showed less interest, presumably because they wanted to arrive at work on time. It is noteworthy that our participants were not predominantly visitors to the Ars Electronica Center, but rather tourists or locals who visit the café frequently because of its good cuisine. Only three participants stated that they visited the Ars Electronica Center before they entered the café. We thus assume that the majority of our participants were not particularly interested in robots. This indicates that the samples in laboratory studies may indeed be biased if mainly participants who are already interested in robotics are recruited. Moreover, people are subject to demand characteristics and may show interest in the presented robots because they infer that they should be interested in them. Therefore, further studies should control for the general interest in robots by, for instance, including corresponding questions in their post-experiment questionnaires.

**Limitations**. The quasi-experimental setting of this study was accompanied by several problems. We were dependent on the visitors to the café agreeing to be interviewed. This caused an uneven distribution over the conditions, because, for instance, fewer people agreed to participate on the days when we installed the remote-controlled setup. Moreover, our results apply only to the android robot Geminoid HI-1 and may not be generalizable.

For the data analysis of the participants' nonverbal behavior, we had to rely on those behaviors that were easily observable in the videos. Although other studies showed that smiling is also important for investigating nonverbal behavior [6], the quality of the video material did not allow us to code smiling behavior. In addition, the participants' conversations with the remote-controlled Geminoid HI-1 were sometimes difficult to understand. Thus, it was not possible to analyze the content of all the conversations.

Furthermore, we were able to draw only implicit conclusions concerning the Uncanny Valley effect, because the questionnaire did not include specific questions asking, for example, about the participants' feelings while encountering the robot.

**Acknowledgements** This work was partially supported by a doctoral fellowship of the Studienstiftung des deutschen Volkes, by Grant-in-Aid for Scientific Research (S), KAKENHI (20220002), and a post-doctoral fellowship of the Japan Society for the Promotion of Science (JSPS).

# References

1. Rosenthal-von der Pütten, A.M., N.C. Krämer, C. Becker-Asano, K. Ogawa, S. Nishio, and H. Ishiguro. 2014. The uncanny in the wild. Analysis of unscripted human–android interaction in the field. *International Journal of Social Robotics* 6 (1): 67–83.
2. Onishi, M., Z. Luo, T. Odashima, S. Hirano, K. Tahara, and T. Mukai. 2007. Generation of human care behaviors by human-interactive robot RI-MAN. In *ICRA 2007. Proceedings of the IEEE international conference on robotics and automation*, 3128–3129. New York: IEEE Press. https://doi.org/10.1109/robot.2007.363950.
3. Matarić, M.J., J. Eriksson, D.J. Feil-Seifer, and C.J. Winstein. 2007. Socially assistive robotics for post-stroke rehabilitation. *Journal of Neuro Engineering and Rehabilitation* 4 (1): 5. https://doi.org/10.1186/1743-0003-4-5.
4. Matarić, M.J. 2006. Socially assistive robotics. *IEEE Intelligent Systems* 21 (4): 81–83.
5. Orne, M.T. 1962. On the social psychology of the psychological experiment: With particular reference to demand characteristics and their implications. *American Psychologist* 17 (11): 776–783. https://doi.org/10.1037/h0043424.
6. von der Pütten, A.M., N.C. Krämer, and S.C. Eimler. 2011. Living with a robot companion—Empirical study on the interaction with an artificial health advisor. In *ICMI'11. Proceedings of the international conference on multimodal interaction. International conference on multimodal interaction*, Alicante, Spain, November 14–18, 2011.
7. Bailenson, J.N., J. Blascovich, A.C. Beall, and J.M. Loomis. 2001. Equilibrium theory revisited: mutual gaze and personal space in virtual environments. *PRESENCE: Teleoperators and Virtual Environments* 10 (6): 583–598. https://doi.org/10.1162/105474601753272844.
8. Bailenson, J.N., J. Blascovich, A.C. Beall, and J.M. Loomis. 2003. Interpersonal distance in immersive virtual environments. *Personality and Social Psychology Bulletin* 29 (7): 819–833. https://doi.org/10.1177/0146167203029007002.
9. Krämer, N.C., N. Sommer, S. Kopp, and C. Becker-Asano. 2009. Smile and the world will smile with you—The effects of a virtual agent's smile on users' evaluation and non-conscious behavioural mimicry. In *Paper presented at ICA 2009. Conference of the international communication association*, Chicago, USA.
10. Bailenson, Jeremy N., Nick Yee, Kayur Patel, and Andrew C. Beall. 2007. Detecting digital chameleons. *Computers in Human Behavior* 24: 66–87.
11. Mori, M. 1970. The uncanny valley. *Energy* 7 (4): 33–35.
12. Gockley, R., A. Bruce, J. Forlizzi, M. Michalowski, and A. Mundell. 2005. Designing robots for long-term social interaction. *Proceedings of the IEEE/RSJ International Conference on Intelligent Robots and Systems (IROS 05)*: 2199–2204.
13. Kanda, T., T. Hirano, D. Eaton, and H. Ishiguro. 2004. Interactive robots as social partners and peer tutors for children: a field trial. *Human-Computer Interaction* 19 (1): 61–84. https://doi.org/10.1207/s15327051hci1901&2_4.

14. Kanda, T., H. Ishiguro, T. Ono, M. Imai, and R. Nakatsu. 2002. Development and evaluation of an interactive humanoid robot "Robovie". In *Proceedings of the 2002 IEEE international conference on robotics and automation (ICRA 2002)*, vol. 2, 1848–1855. https://doi.org/10. 1109/robot.2002.1014810.

15. Hayashi, K., D. Sakamoto, T. Kanda, M. Shiomi, S. Koizumi, H. Ishiguro, T. Ogasawara, and N. Hagita. 2007. Humanoid robots as a passive-social medium. In *Proceedings of the ACM/ IEEE international conference on human-robot interaction HRI '07*, ed. Breazeal, C., A.C. Schultz, T. Fong, and S. Kiesler, Arlington, VA, USA, March 08–10. https://doi.org/10.1145/ 1228716.1228735.

16. Ishiguro, H. 2006. Interactive humanoids and androids as ideal interfaces for humans. In IUI '06 Proceedings of the 11th international conference on Intelligent user interfaces, ed. *Paris, C.L., C.L. Sidner, E. Edmonds, and D. Riecken*, 2–9. New York: ACM Press. https://doi.org/ 10.1145/1111449.1111451.

17. Hara, F. 2004. Artificial emotion of face robot through learning in communicative interactions with human. In *Proceedings of the 13th IEEE international workshop on robot and human interactive communication (RO-MAN 2004)*, 7–15. https://doi.org/10.1109/roman.2004. 1374712.

18. Walters, M.L., K. Dautenhahn, R. te Boekhorst, K.L. Koay, and S.N. Woods. 2007. Exploring the design space of robot appearance and behavior in an attention-seeking 'living room' scenario for a robot companion. In *IEEE symposium on artificial life (ALIFE '07)*, 341– 347. https://doi.org/10.1109/alife.2007.367815.

19. MacDorman, K., and H. Ishiguro. 2006. The uncanny advantage of using androids in cognitive and social science research. *Interaction Studies* 7 (3): 297–337. https://doi.org/10. 1075/is.7.3.03mac.

20. MacDorman, K.F. 2006. Subjective ratings of robot video clips for human likeness, familiarity, and eeriness: An exploration of the uncanny valley. In *Proceedings of the conference on cognitive science 2006 (CogSci 06), workshop on android science*, 26–29.

21. Hanson, D., A. Olney, I. Pereira, and M. Zielke. 2005. Upending the uncanny valley. *Proceedings of the National Conference on Artificial Intelligence* 20 (4): 1728–1729.

22. Straub, I., S. Nishio, and H. Ishiguro. 2010. Incorporated identity in interaction with a teleoperated android robot: A case study. In *Proceedings of the 19th IEEE international workshop on robot and human interactive communication (RO-MAN 2010)*, 319–144. https:// doi.org/10.1109/roman.2010.5598695.

23. Ramey, C.H. 2006. An inventory of reported characteristics for home computers, robots, and human beings: Applications for android science and the uncanny valley. In *Proceedings of the ICCS: 2006 workshop: Toward social mechanisms of android science*, Vancouver, Canada, 43–47.

24. Becker-Asano, C., K. Ogawa, S. Nishio, and H. Ishiguro. 2010. Exploring the uncanny valley with Geminoid HI-1 in a real world application. In *IHCI 2010. Proceedings of the IADIS international conference on interfaces and human computer interaction*, ed. Blashki K., Freiburg, Germany, July 26–30, 121–128.

25. Ho. C., K.F. MacDorman, and Z.A.D.D. Pramono. 2008. Human emotion and the uncanny valley: a GLM, MDS, and Isomap analysis of robot video ratings. In *Proceedings of the 3rd ACM/IEEE international conference on human robot interaction (HRI'08)*, 169–176.

26. Nishio, S., H. Ishiguro, and N. Hagita. 2007. Geminoid: Teleoperated android of an existing person. In *Humanoid robots: New developments*, ed. de Pina Filho, A.C., 343–352. InTech.

27. OpenCV Wiki FaceDetection.

28. Wittenburg, P., H. Brugman, A. Russel, A. Klassmann, and H. Sloetjes. 2006. ELAN: A professional framework for multimodality research. In *LREC 2006. Proceedings of the 5th international conference on language resources and evaluation*.

# Chapter 25
# At the Café—from an Object to a Subject

Ilona Straub, Shuichi Nishio and Hiroshi Ishiguro

**Abstract** What are the characteristics that make an object appear to be a social entity? Is sociality limited to human beings? This article addresses the borders of sociality and the animation characteristics with which a physical object (here, an android robot) needs to be endowed so that it appears to be a living being. The transition of sociality is attributed during interactive encounters. We introduce the implications of an ethnomethodological analysis to show the characteristics of the transitions in the social attribution of an android robot, which is treated and perceived as gradually shifting from an object to a social entity. These characteristics should (a) fill the gap in current anthropological and sociological research, addressing the limits and characteristics of social entities and (b) contribute to the discussion of the specific characteristics of human–android interaction as compared to human–human interaction.

**Keywords** Personality · Android robot · Identity · Teleoperation

This chapter is a modified version of a previously published paper [1], edited to be comprehensive and fit with the context of this book.

Ilona Straub
Institute for Social Studies, Carl Von Ossietzky University of Oldenburg,
Oldenburg, Germany

S. Nishio (✉) · H. Ishiguro
Advanced Telecommunications Research Institute International,
Keihanna Science City, Kyoto, Japan
e-mail: nishio@botransfer.org

H. Ishiguro
Department of Systems Innovation, Graduate School of Engineering Science,
Osaka University, Osaka, Japan

## 25.1  Introduction

When dealing with social phenomena, both scientific studies and common sense refer to human beings as social entities sui generis. Social phenomena are discussed as humans' means of relating to each other as a basic unit on a micro-social level during interaction or on a macro-social level as manifestations of society. The established view of the social phenomena as well as targets of investigation is characterized by a premise that addresses social relations as genuinely attributed to human agents and is thus implicitly an *anthropocentric view* [2]. Cases such as swearing at a computer when it malfunctions, addressing plants, animals, or puppets or simply praying to spirits at a shrine, temple, or church are categorized by the established social scientific positions as de-socializing activities, that is, as "nonsocial" activities. Developments in life sciences and emerging new interaction technologies, such as social robots, encourage researchers of social theories to question the anthropocentric view and to redefine the range of social entities. This allows the analysis of (a) the features necessary for the distinction of social entities or "agency" and (b) the behavior toward a social entity as compared to that toward objects that are considered nonsocial.

Following the concept of social constructivism, beings that are capable of building relationships have dispositions on a higher cognitive level for the organization of social relationships. These complex cognitive structures constitute their own laws and own kind of social reality in addition to their given nature. There should be observable mechanisms of recognizing, accepting, and acting toward social beings. In addition to the ascription of sociality to common human-to-human relationships, this ascription to other beings or even objects is also obviously observable.

As previous studies have shown [3], embodied social robots (here, android robots) are the subjects of the personification and anthropomorphization, and even of the incorporated identity, that are ascribed to the embodied robot by human users. The acceptance of an android robot as a reference point for interactive actions raises questions about the mechanisms of acceptance and the constitution of social agency toward other entities besides human beings, and thus, about the constitution of *alter ego relations* in general. How do we assume other (nonhuman) social beings as alter egos? Which cognitive attributions must occur for the recognition and acceptance of social agency in contrast to nonsocial objects? What features make the difference in the perception, ascription, and behavior toward objects and toward social beings?

In this paper, we present a study on the transition of perception and behavior toward an android robot, which is first given the attributes of an object being, second of a reactive tool, and finally, of a subject or social being. However, first we present a quick overview of the different theoretical approaches that address human-to-nonhuman (objects) relations and of the discussions of sociality beyond anthropocentrism.

## 25.1.1  Sociality Beyond Anthropocentrism

*Sociality as sociocultural variation.* In 1970, the sociologist Thomas Luckmann published an article titled "On the boundaries of the social world" that addressed the demarcations of the anthropocentric view on sociality from a phenomenological and transcendental philosophic perspective [4]. Luckmann questioned the meta-scientific limitation of social relationships, which are defined as exclusively occurring between humans, and listed examples of relations between human and nonhumans. His analysis reflects on the ethnographic studies that demonstrate different cultural societies, with different historical backgrounds, which assert social relations with animals, plants, gods, demons, deceased persons, external powers, or objects. These social relations with external powers, objects, or deceased persons are commonly discussed in ethnographic studies as "animism," "totemism," "shamanism," or "dynamism" [4, 5]. In the view of modern western societies, the designation of relationships beyond human relations is thus considered as crossing the border of institutionally accepted social entities and therefore labeled nonsocial. Lindemann [6, 7] defined the framework of the borders of sociality in modern western societies as an *anthropological square*, which can be proclaimed as the anthropocentric focus of western societies. The "anthropological square" limits the attributions of social beings to human by following institutionally accepted borders: being alive, being dead, the human–machine difference, and the human–animal difference. Lindemann criticized anthropomorphism in social sciences and reflected on the causes for the borders of social beings in Western societies. She noted that the limitations of accepted social beings are dependent on cultural and historical legitimation in different societies and are therefore institutionalized. For a de-anthropomorphization of theories in social sciences, research is needed on the acceptance of social beings beyond the "anthropological square." By, for example, the accepted incorporation of cultural variations in social beings [8] or by the analysis of newly emerging social agents (e.g., of robots), the "anthropological square" could be overcome, redefined, and completed. To go beyond the "anthropological square" means to search for the essentials of accepted social behavior and to widen the scope of the subject of social sciences. This idea corresponds with Luckmann's approach following Scheler's [9] proclamation of phylogenetic and ontogenetic *universal projections* of the alter ego onto the surrounding world, which allows the interpretation of sociality toward each being and all objects. Luckmann noted that the given borders of sociality in modern societies are the results of social, biological, historical, and cultural conditions, which caused a de-socialization of the primary and *principal lifeworld* and became institutionalized and habituated during the secondary socialization process [10]. This means that the ascriptions of sociality and the division of social beings vs. nonsocial beings are the results of institutional manifestations and norms in different sociocultural types to different degrees [11].

Variations in sociocultural typologies of sociality and the assumption that there is no ready-made given border for sociality thus raises a question to which

Luckmann sought answers: What are the fundamental and necessary characteristics or qualities that support, affirm, and affect the apperception of social characteristics related to, for example, an animated body in contrast to an unanimated "nonsocial" object? Which features are essential for a being to be perceived as a social entity?

### 25.1.2  Excurse: Object-Related Sociality

In the following excurse, we examine concepts that suggest social agency beyond human beings and allow an innovative approach to social agency. The aim is to broaden the view of social sciences on human–nonhuman relations. First, we examine the *actor–network theory* and then we discuss the *distributed cognition approach* before we finally consider the idea of *object-related social relationships*. The idea to propose a symmetrical relation between (commonly categorized as nonsocial) objects and (human) subjects reached its summit in the *actor–network theory* as stated by its main contributors Bruno Latour and Michael Callon. The actor–network theory highlights the tight connection of humans to objects, technology, or nature in everyday activities. In this view, humans and their material surroundings are interdependent and part of a network that considers both as actors that are in a mutual relation during any activity. As part of a network, objects, technology, and nonhumans supplement the actions of humans and enable their activities in the environment [12, 13].

Although the actor–network association suspends the delineation between humans and objects, it is driven by a fallacy that is displayed in the misapplication of social features to nonsocial conditions. Objects surrounding acting entities are tools, instruments, or media for pragmatic activities that are not unconditionally considered and meant as social partners.

For providing an emphasized equation of the interdependence and relation of nonsocial environmental conditions with human beings, the *distributed cognition approach* seems to be more appropriate. According to this approach, objects, technology, or tools aid human beings to memorize, coordinate, transform, or utilize activities on a *cognitive level*. The approach indicates the integration of environmental surroundings and tools for problem-solving that cannot be restricted to intra-individual cognitive activities [14]. However, also here the complementary nature of human beings and their nonsocial surroundings is not meant as a correlation to social relations. Neither the actor–network theory nor the distributed cognition approach suits our interest in analyzing the conditions that segregate nonsocial from social beings. Our interest is in the specific characteristics nonhumans must have in order to be perceived, treated, and accepted as social beings by human beings as opposed to being treated as nonsocial objects. Therefore, ideas on the effect of the environment, experienced as a platform to enable life and as pragmatic aids, do not cover the question about the cues that suggest social behavior (toward objects) to human actors.

Knorr Cetina's [15] approach approximates in its nature to our specific question. In his article titled "Object-related Social Relationships in Post-traditional and Knowledge-based Societies" (original title: "Sozialität mit Objekten. Soziale Beziehungen in post-traditionalen Wissensgesellschaften"), he examined more closely social relations between human and nonhumans. Knorr Cetina criticized the neglect of sociality toward objects (which remains unaccepted) as a unique sociality type in social scientific research. The latter focuses solely on relationships among human beings and ignores social relations between humans and nonhumans. Knorr Cetina presented examples of human–nonhuman (object) relations in scientific communities and noted their contribution in the formation of newly emerging social structures, which are confusing the common attributions of sociality. The adaptation of sociality toward objects and the newly emerging structures are termed *creolisations of accepted social structures*. This means that human–object relations on the one hand have common features of sociality that are also observable in human–human interactions but, on the other hand, also constitute a new type of relation with different intensity and intervention means. Knorr Cetina emphasized that object relations are partially covered by interpersonal variations in sociality and should therefore be analyzed as their own discipline, or in other words, they should be integrated as human–object relations into the research area of social science studies. At this point, the argumentation needs to be extended and analyzed to cover the fundamentals that are responsible for ascriptions of sociality.

### 25.1.3    Remarks on Sociality Beyond Anthropocentrism

The current discussions imply that social sciences do not explore and observe the entire range of sociality and thus lack a proper definition of sociality and a description of the features that contribute to the acceptance of social beings. Except for research on cultural variations of accepted social beings [8], there is neither a listing of different forms of sociality nor a separation of the specifics of different social forms between humans and objects or other nonhuman social entities. Clear characteristics of attributions to define someone or something as social entity are missing and require an in-depth discussion (see also the arguments of Lindemann [16]). Furthermore, no studies exist that have addressed the transition of behavior toward nonhuman/nonsocial objects in relation to the treatment of one and the same nonhuman/nonsocial object as a social being. Thus, the study reported in this paper is the first attempt to characterize people's manners of approaching and treating a technical tool according to their gradually shifting attributions and transformations from an object to an accepted social partner.

## 25.2  Empirical Study

### 25.2.1  *Experimental Setting*

To test peoples' recognition, acceptance, and manners of approaching a humanlike robot in a (non-laboratory) open-public setting, the android robot "Geminoid HI-1" was placed in a public café in Linz for a period of three weeks. Geminoid HI-1 is an android robot that is a duplicate of its scientific originator Hiroshi Ishiguro (HI) and is designed as a human-shaped and embodied interaction tool (see Fig. 25.1). The android robot can be controlled via the Internet from a remote location by teleoperation. An interface allows the teleoperator to partially control the robots motions, to synchronize these motions with the teleoperator's motions and to transmit the teleoperator's voice (condition 3, described below). One of the robot's purposes is to replace the person controlling it in another location and to simulate that person's social presence to the interlocutor facing the robot. For a detailed description of the android robot's technical background, material, and functional applications, see [17].

**Fig. 25.1** Android robot Geminoid HI-1 (right) at the public café seated at a dinner table. Visitors of the café approached the android robot and showed features of recognizing the robot as nonhuman and furthermore treating the android according to different conditions as a social entity

To avoid attracting attention, the robot's placement in the café was not advertised. To provide the first-glance effect of a human visitor sitting in the café,
Geminoid HI-1 was seated on a chair at a dining table in a corner of the café
simulating the actions of keying in on and looking at a laptop computer (see
Fig. 25.1). The robot's hands were placed on the keyboard of the laptop computer,
and its gaze direction was focused on the laptop's monitor (except in conditions 2
and 3, described below, where the gaze direction was directed toward the visitors).
Public visitors could access and see the settings. Visitors approached the robot
spontaneously and were not instructed to communicate with or act toward the robot.
The setting of Geminoid HI-1 was recorded audiovisually by distributed microphones and video cameras covering a surround sound and view. The android robot
was periodically switched between three different mode conditions: (1) idling,
(2) face-track, and (3) interaction. We analyzed the reactions of 30 persons in each
condition qualitatively and generalized the differing behaviors related to the condition in which they occurred. The analysis thus does not focus on gender-specific
or age-related behaviors but is intended to offer a matrix of general behaviors
*toward the android robot as a starting point for further analysis.* In the following,
we provide a short introduction to the specific features of each condition and
continue with a comparison of the visitors' reactions as responses to each of the
three conditions.

The results of the comparison show significant effects of the conditions on the
visitors' relational behavior toward Geminoid HI-1 and indicate the characteristics
of sociality through transitions in treating the android robot from an object to a
subject.

## 25.3  From an Object to a Subject: Attitude to the Android Robot in Different Conditions

In this section, we compare the different conditions that show the transitions in
peoples' treatment of the android robot, starting from treating it as an object up to
relating to it as a subject. We introduce the manner in which people approach an
android robot and act differently in nonverbal, reactive, and verbal manners
according to the different nonverbal, reactive, and verbal conditions in which the
robot is set up.

### 25.3.1  Idling Condition

*Description of the android robot's activity.* The android robot shows neither
reactiveness nor motivational reference to environmental, personal, or social circumstances. It performs motions in preprogrammed patterns. At first glance, the

android robot's behavior suggests that it is writing on a laptop computer: it stares at the screen, has its hands placed on the laptop, and performs head motions in a logarithmic tonus from left to right, from looking up to looking down. The robot simulates micro-motions suggesting breathing through slight motions of the breast. Furthermore, the android robot is programmed to move its right foot up and down at certain intervals, and it sways its torso slightly back and forth. When a person approaches, the android robot continues its idling mode and stares at the laptop.

*Reactions of approaching persons.* People observe Geminoid HI-1 from a distance. If the android robot arouses their interest, they approach the table and watch Geminoid's motion patterns. There is a great interest in the manufacture of the robot, and therefore, people tend to touch the parts of Geminoid that appear to be skin (the skin is made from silicon). Beforehand, they examine the reactiveness of Geminoid through verbal utterances and motions in its immediate environmental space. A group of people (consisting of at least two persons) start to talk about the robot and describe its outer humanoid appearance, material, their perceptual and emotional experience, and their attitude toward the robot.

In the idling condition, the android robot Geminoid HI-1 shows no re-/action to environmental stimuli and social cues and is treated like an object by the people.

## 25.3.2 Face-Track Condition

*Description of the android robot's activity.* Geminoid HI-1 moves in a preprogrammed fashion as in the idling condition, except for its head motions, which show reactiveness toward appearing persons. The android robot is equipped with a face-tracking system that causes the android to lift its head in the appropriate direction in which the heads or faces of human beings appear in a range of about 4 m. Thus, the android robot's behavior suggests reactiveness and attentiveness toward the approaching persons.

*Reactions of approaching persons.* If people realize the android robot is gazing in their direction as they pass, they tend to watch the robot for a few seconds or to approach its table. Sitting at the table, the robot continues to track their faces and to gaze in their direction, offering eye contact. People tend to address the android robot with verbal utterances such as a greeting or ask about its condition or name. As soon as they receive no response, they realize the limited skills of the robot. The people most frequently check the limitation of its skills on the reactive nonverbal level, such as its range of head motions and gaze direction. As soon as they realize the robot's gaze motion is limited, they tend to talk about the robot or attempt to understand its mechanics or technical functioning. Unlike in the idling condition, a group of people (consisting of at least two persons) start to talk about the robot, addressing topics such as its humanoid appearance, material, their perceptual and emotional experience, and their attitude toward the robot.

In the face-track condition, Geminoid HI-1 shows reactiveness toward people passing it and offers limited nonverbal and visual feedback as a social cue.

People confronted with the face track condition tend to treat the android robot as an animated, reactive being with limitations in cognitive skills (reactive only on a bodily level).

### 25.3.3 Interaction Condition

*Description of the android robot's activity.* In this condition, the android robot is teleoperated by a person and has the following features in addition to the idling motions. The robots' head motions can be controlled and the robot thus seems to act according to (a) social or (b) environmental circumstances by displaying appropriate gaze behavior. For example, when people approach the robot's table, the teleoperator can control the head motions according to its gaze direction toward the visitors. Additionally, the robot can switch its gaze between different persons, the laptop, or other visual surroundings. The intervals of gaze can be controlled, and a specific person can be addressed by the gaze direction of the android robot. Furthermore, it is provided with a voice (spoken words from the teleoperator) and is thus able to perform utterances, such as a greeting or responses to questions. In this mode, the android robot is suggesting reactiveness, mutual attentiveness, and contingency in reactions to the visitors and is thus simulating higher cognitive skills than in the prior conditions.

*Reactions of approaching persons.* People look at Geminoid HI-1 and respond to its verbal address. They tend to maintain eye contact and continue to have a short talk with the android robot on topics such as its name, origin, and functioning mechanism. We can observe a limitation of the visitors' nonverbal actions to check Geminoid HI-1 s' reactiveness and observe instead that they check Geminoid HI-1 s' cognitive skill limitation. Very few persons touch the android robot, and those who do ask its permission to do so. We can also observe an increase in smiling and mimicking behavior toward the robot, together with the use of gestures for deictic references or as an addition to communication goals.

In the interaction condition, the robot shows reactiveness toward environmental circumstances by producing and reacting to social cues. Thus, the robot seems to act according to complex circumstances and offers contingencies in its verbal and nonverbal expressions. The people respond to the robot in a manner that legitimates the declaration that Geminoid HI-1 is a social being.

## 25.4 Results

Our study examined the transition of attitudes toward an android robot, starting from its perception as a mere object and shifting to its perception as a responsive and reactive tool, and highlighting its acceptance as a social being. The analysis focuses on the activities of both (a) the android robot and (b) the human visitors (exploring the robot).

(a) We could observe three reactive conditions of *the robot's activity*, which increase in complexity related to its responses to environmental conditions and to the generation of and reaction to social cues. In the first condition (idling), the robots' activity is self-centered and restricted to micro-motions. The robot shows no irritation in reaction to environmental changes or the social cues of people who explore it. In the second condition (face-track), the android robots' activity is related to reactive responses evoked by environmental changes and the stimulus of approaching people. Its actions depend on the face-tracking system, but nevertheless the robot adjusts to the occurrences in the environment. The android robot relaxes its stiffness as a self-centered object and is reactive to environmental changes. In the third condition (interaction), the android robot is teleoperated as a manipulated object and directed by an operator who adjusts its activity and acts as its senses (eyes/ears/proprioception) related to occurrences in the environment. These occurrences include social cues. We can observe that its adjustments of paraverbal and nonverbal features (in relation to the environmental occurrences) increase the acceptance of the android robot as a responsive social being. The robot's activity thus shifts from that of a self-centered object (idling condition), to that of a reactive tool (face-track condition), and its ability not merely to react to circumstances in the environment, but also to synchronize its activity with given environmental occurrences and social activities is highlighted. Furthermore, the android robot seems to initiate and cause changes in the environment on a social level. In the interaction condition, the android robot adjusts its activities related to environmental changes and causes social persons to adjust to its activities on the paraverbal, nonverbal, and verbal level.

(b) The reactions of the *human visitors* to the android robot do also depend on its activity. In the idling condition, we can observe the visitors' interest in the human appearance of the android robot and we have access to their attitudes to it. Furthermore, we can observe that the visitors talk about the robot in its presence and that they do not maintain a physical distance from the robot (e.g., they touch the robot or observe its facial area closely). The self-centered micro-motions and the fact that the robot does not give feedback to environmental changes or social cues limits the peoples' expectations of the robots skills and abilities. The data show that people do not expect cognitive skills and physical responses from a self-centered robot; thus, in the idling condition, people explore the robot as they would an object.

In the face-track condition, we can observe the visitors' irritation caused by the eye contact provided by the humanoid robot. Since the eye contact does not have vary, but rather the robot's gaze constantly follows the head positions of the visitors, they soon realize its restricted skills. By motions of their hands and by moving in the robot's close environment, the visitors put the robots reactiveness to the test and soon realize that its face detection ability is limited. This condition shows that at the beginning of the encounter the visitors are uncertain about the robot's skills and in doubt whether they are encountering a human or a robot. The restricted responsiveness of the robot to approaching visitors causes them to test the robot's abilities. When they realize its limited skills, they tend to treat the robot as an object. The synchronization of the paraverbal, nonverbal, and verbal features of the android robot and its responses according to environmental changes, together with the production of changes on social cues, influence the visitors to behave completely differently toward the android robot. They do not touch the robot without permission and their testing of skills switches from tests of physical skills to tests of cognitive ability (knowledge). The people ask the android robot about its functioning and about biographical facts. Their irritation with the robot's mismatch of body posture is thus replaced by interrogations of the robot by verbal means. We can observe that people explore the robot in a highly social manner.

From the above, we can conclude that an increase in the reaction to environmental occurrences, the production of social responses, and reactions to social activities (by, e.g., synchronization of body posture or gaze direction, in addition to verbal utterances) increases the acceptance of a technical tool as a social being. The analysis of the data offers a description of the modes of behavior toward an android robot and contributes to the deconstruction of anthropocentric theories of sociality. Follow-up studies on social theory should integrate the results of human–(android) robot encounters into current theories of sociality.

## 25.5  Discussion

As the results show, human users treated the android robot. Geminoid HI-1 differently under each of the three conditions (1) idling, (2) face-track, and (3) interaction. We can observe a shift in attitudes toward the android robot from treating it as an object to treating it as a subject, depending on the interrelation of the following three features: (a) increasing self-reference, (b) appropriate references to circumstances in the environment, and (c) reactive and initiating references to other social agents. In the discussion of theories in social sciences, the anthropocentric view should be replaced with a broader view of social agency. Through research on technical tools, which are meant to simulate social agents, Lindemann's concept of the "anthropological square" as the dominating point of reference in social studies (and in western societies) could take a turn to a concept that includes nonhuman (beings or objects) as accepted referents of social actions. The current study analyzed the acceptance of an android robot by human users and could offer a step in the direction of "creolisations of accepted social structures," (as mentioned by Knorr Cetina) as a variation of human relations, leading to the fundamentals of sociality, as mentioned by Luckmann.

For theoretical reflection on sociality, the occurrence of the three features (a) increasing self-reference, (b) appropriate references to circumstances in the environment, and (c) reactive and initiating references to other social agents should be applied to other studies on the acceptance of nonhuman social beings to verify them as general features of sociality and for the formulation of a general theory on the fundamentals of sociality.

**Acknowledgements**  The present study was enabled by the kind help of the ARS Electronica staff and the management of the Café Cubus who provided the settings for the experiments. Furthermore, we are grateful to the experimental helpers and volunteer workers during the exhibition of Geminoid HI-1 at the Café Cubus and in the ARS Electronica Building. Special thanks are due to Rafaella Gehle, Christian Dondrup, Tim Partmann (students of CITEC at University Bielefeld), Kohei Ogawa (ATR), Dr. Christian Becker-Asano (University of Freiburg), and Stefan Scherer. Furthermore, we would like to thank the AST at the Institute of Social Studies at the Carl von Ossietzky University of Oldenburg.

This work was partially supported by Grant-in-Aid for Scientific Research (S), KAKENHI (20222002). Ms. I. Straub was supported by a fellowship of the Japan Society for the Promotion of Science.

# References

1. Straub, I., S. Nishio, and H. Ishiguro. 2012. From an object to a subject—Transitions of an android robot into a social being. In *21st IEEE international symposium on robot and human interactive communication (RO-MAN 2012)*, 821–826.
2. Lüdtke, N. 2010. Sozialität und Anthropozentrik. Aktuelle Probleme der Sozialtheorie am Beispiel Mead. Arbeitsgruppe Soziologische Theorie—AST. Carl von Ossietzky Universität Oldenburg (AST-DP-4-2010). Online verfügbar unter http://www.ast.uni-oldenburg.de/download/dp/ast-dp-4-10.pdf.
3. Straub, I., S. Nishio, and H. Ishiguro. 2011. Incorporated identity creation during interaction with a teleoperated android robot: A Case Study. (submitted).
4. Luckmann, T. 2007 [1970]. Ueber die Grenzen der Sozialwelt. In Erfahrung – Wissen – Imagination. Lebenswelt, Identität und Gesellschaft. Schriften zur Wissens- und Protosoziologie, ed. T. Luckmann and J. Dreher, vol. 13, 62–90. Konstanz: UVK.
5. Tietel, E. 1995. *Das Zwischending: Die Anthropomorphisierung und Personifizierung des Computers*. Regensburg: S. Roderer.
6. Lindemann, G. 2011. Anthropologie, gesellschaftliche Grenzregime und die Grenzen des Personseins. *Ethik Med* 23 (1): 35–41.
7. Lindemann, G. 2005. The analysis of the borders of the social world: A challenge for sociological theory. *Journal for the Theory of Social Behaviour* 35 (1): 69–98.
8. Descola, P. 2011. *Jenseits von Natur und Kultur*. Suhrkamp: Frankfurt a.M.
9. Scheler, M. 1948. *Wesen und Formen der Sympathie*. Frankfurt/M.: Schulte-Bulmke.
10. Berger, P.L., and T. Luckmann. 2003 [1980]. Die gesellschaftliche Konstruktion der Wirklichkeit. Frankfurt/M.: Fischer Verlag.
11. Kelsen, H. 1982 [1941]. Vergeltung und Kausalität. Wien: Böhlau.
12. Latour, B. 1988. Mixing humans and nonhumans together. The sociology of a doorcloser. *Social Problems* 35 (3): 298–310.
13. Meyer, J.W., and R.L. Jepperson. 2000. The 'actors' of modern society. The cultural construction of social agency. *Sociological Theory* 18: 100–120.
14. Hutchins, E. 2000. Distributed cognition. http://eclectic.ss.uci.edu/~drwhite/Anthro179a/DistributedCognition.pdf.

15. Knorr, Cetina, K. 2006. Object related Social Relationships in post-traditional and Knowledge-based Societies (original title: Sozialität mit Objekten. Soziale Beziehungen in post-traditionalen Wissensgesellschaften.) In: Tänzler, D., Knoblauch, H., und Soeffner, H. G. (Hg.): Zur Kritik der Wissensgesellschaft. Konstanz: UVK-Verl.-Ges (Erfahrung – Wissen – Imagination, 12), S. 83–120.
16. Lindemann, G. 2010. *Das Soziale von seinen Grenzen her Denken*. Weilerswist: Velbrück.
17. Ishiguro, H., and S. Nishio. 2007. Building artificial humans to under-stand humans. *The International Journal of Artificial Organs* 10: 133–142.

# Chapter 26
# At the Hospital

**Eri Takano, Yoshio Matsumoto, Yutaka Nakamura, Hiroshi Ishiguro and Kazuomi Sugamoto**

**Abstract** Recently, many humanoid robots have been developed and actively investigated all over the world in order to realize partner robots which coexist in the human environment. For such robots, high communication ability is essential in order to naturally interact with human. However, even with the state-of-the-art interaction technology, it is still difficult for human to interact with humanoids without conscious efforts. In this chapter, we apply the android robot, which has a quite similar appearance to a human, to a bystander in human-human communication. The android is not explicitly involved in the conversation, however makes small reactions to the behaviors of the humans, and the psychological effects on human subjects are investigated. Through the experiments, it is shown that mimicry behavior of the subject by the android is quite effective to harmonize the human-human communication.

This chapter is a modified version of a previously published paper [1], edited to be comprehensive and fit with the context of this book.

E. Takano · Y. Matsumoto (✉) · Y. Nakamura · H. Ishiguro
Graduate School of Engineering, Osaka University, Osaka, Japan
e-mail: matsumoto@ams.eng.osaka-u.ac.jp; yoshio.matsumoto@aist.go.jp

E. Takano
e-mail: eri.takano@ams.eng.osaka-u.ac.jp

Y. Nakamura
e-mail: nakamura@ams.eng.osaka-u.ac.jp

H. Ishiguro
e-mail: ishiguro@ams.eng.osaka-u.ac.jp

Y. Matsumoto
Robot Innovation Research Center, National Institute of Advanced Industrial Science and Technology (AIST), Tsukuba Central 2, 1-1-1 Umezono, Tsukuba, Ibaraki 305-8568, Japan

K. Sugamoto
Graduate School of Medicine, Osaka University, Osaka, Japan
e-mail: sugamoto@ort.med.osaka-u.ac.jp

**Keywords**  Android robot · Clinical examination · Trilateral communication
Synchronized behavior · Impression

## 26.1  Motivation and Related Work

Recently, many humanoid robots have been developed and actively investigated
worldwide in order to realize partner robots that can exist in the human environment.
High-level communication ability is essential for such robots so that they can naturally interact with humans. Among these robots is an android robot developed by
Ishiguro et al. [2], the appearance of which is quite similar to that of a human. It is
very important that the generated motions and behaviors of an android appear natural since an android displaying unnatural motions may give a worse impression than
other robots, as the "uncanny valley" theory [3] insists. Noma et al. showed that 70%
of human subjects were unable to distinguish an android from a human by observing it for two sec when small humanlike fluctuations that correspond to breathing
and blinking were added to its static posture [4]. They also showed that 77% of subjects recognized the android as a robot when it was standing still. This indicates that
small differences in an androids behavior can result in a large difference in people's
impression of it.

However, no psychological studies on androids have thus far been conducted in
which the influence of their behavior on humans in the context of communication
was examined. Thus, the aim of this study was to investigate the psychological effects
of the presence and the behavior of an android on real communication. In research in
the field of psychology, it has been shown that the nod and facial expressions are two
major channels in nonverbal communication by which humans judge the degree of
intimacy [5]. In addition, the "chameleon effect," which constitutes the mimicking
behavior of the conversation partner in bilateral communication, is known to make
human–human communication smoother [6]. We hypothesize that the behavior of
an android positively affects human communication when nodding and a smiling
expression are appropriately generated.

## 26.2  Android as a Bystander in Trilateral Communication

### 26.2.1  Android Robot

Figure 26.1 shows the android robot Repliee Q2 utilized in this research study. The
main features of the android are its appearance and motions, which resemble those
of a human. The face of the android is constructed from soft silicon rubber and
copied from a real human face. The height of the android is approximately 150 cm,
and the upper body including the face has 42 degrees of freedoms. The lower body

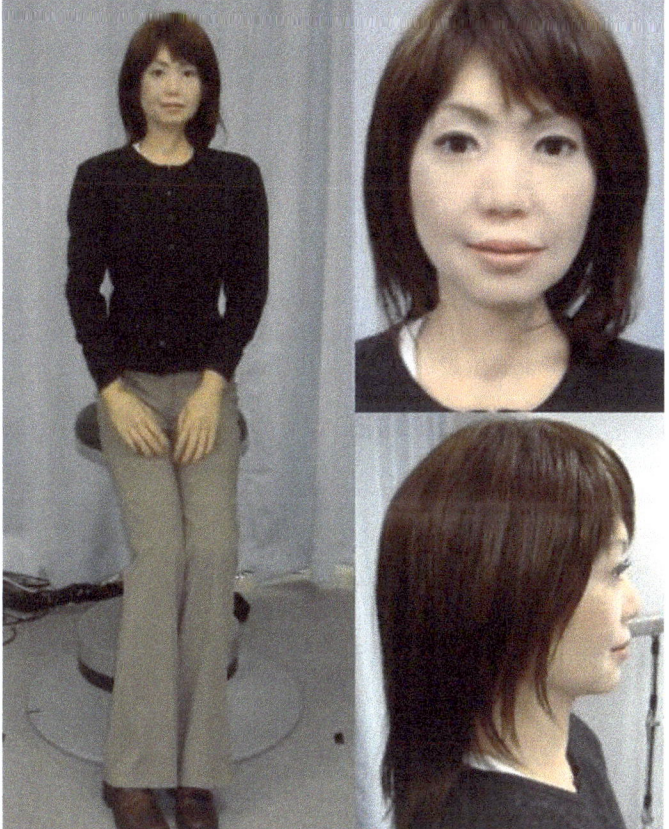

**Fig. 26.1**   Android robot Repliee Q2

(i.e., the legs) cannot be moved. All of the degrees of freedom are driven by pneumatic cylinders using an air servo system. Therefore, an air compressor is necessary for controlling the android. The face has 13 degrees of freedom, which allows various facial expressions. The position of all the joints is controlled using an external PC. The use of air actuators provides the robot with the physical compliance that is required to realize smooth motion and safe interaction with humans.

## 26.2.2   Android as a Bystander

Many robotic systems that can interact with humans have been designed. For example, a humanoid robot, ROBITA, can communicate with two persons [7] by recognizing speech and vision information. However, the conversation partners have to take into account the ability of the robot when interacting with it, and thus, the communication is quite limited. Even state-of-the-art robots cannot easily maintain a long direct interaction with a human since their communication abilities, such as speech recognition, speech synthesis, dialog processing, and gesture generation, are far from those of humans.

In order to compensate such disabilities, we took an approach in which the android is utilized as a bystander in trilateral communication. Several research studies have addressed trilateral communication in robotics, where the role and the relationship of three subjects, each of which is a human or a robot, can be variously determined (e.g., [7, 8]). In this study, we addressed a situation comprising a real clinical examination at a hospital, where a doctor and a patient seriously communicate with each other in an examination room. The android acts as a bystander, like a nurse or a medical student, and does not explicitly participate in the conversation. That is, it directly interacts with neither the doctor nor the patient. Our hypothesis was that, despite the lack of interaction, the nodding and a smiling behavior of the android bystander, when appropriately generated, would positively affect human communication.

## 26.3 Preliminary Experiment with Human Bystander

### 26.3.1 Experimental Condition

We first conducted a preliminary experiment using a human bystander in order to confirm how the behavior of a bystander psychologically affects patients in an examination room. This experiment was performed in an orthopedics examination room in the outpatient department of Osaka University Medical Hospital. Forty-four patients aged in their thirties to seventies (on average, in their fifties) participated in the experiment. In this experiment, a female graduate student, wearing a white coat so that she would resemble a nurse or a medical student, played the role of a bystander observing the examination. She behaved in one of two conditions as follows:

- **Condition 1: Synchronized with patients**: The bystander nods and smiles in synchrony with the patient,
- **Condition 2: Without expressions**: The bystander does not either nod or smile.

The behavior in Condition 1 was determined based on the chameleon effect found in bilateral conversation. We first asked the subjects about their opinion on the presence of the bystander, and they selected one response to the question from among the following three alternatives:

- **Prefers presence**: I preferred the presence of the bystander,
- **Indifferent**: I was indifferent to the presence of the bystander,
- **Prefers absence**: I didn't like the presence of the bystander.

The experimental results for this question are shown in Table 26.1. The subject's impression on the bystander seem to be better in Condition 1; however, the dominant

**Table 26.1** Subjects' opinion on the presence of the human bystander

| | Condition 1 (%) | Condition 2 (%) |
|---|---|---|
| Prefers presence | 33.3 | 4.3 |
| Indifferent | 61.9 | 91.3 |
| Prefers absence | 4.8 | 4.3 |

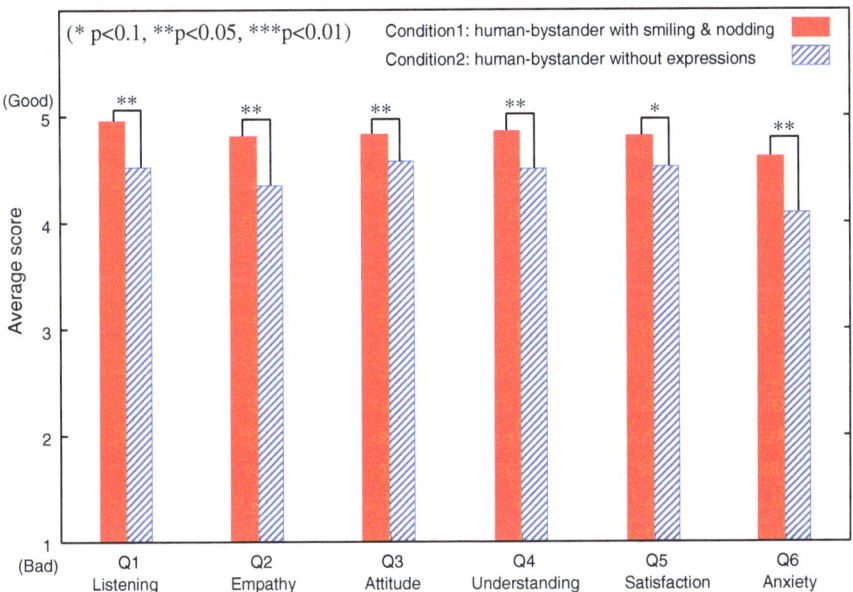

**Fig. 26.2** Experimental results of preliminary experiment with human bystander

opinion of the subjects was that they were indifferent to the presence of the bystander. Most of the subjects also mentioned that the bystander was irrelevant for the clinical examination. Then, the subjects were asked the following six questions about the examination.

- **Q1**. Did the doctor listen kindly to your opinion?
- **Q2**. Did the doctor empathize with you?
- **Q3**. Was the doctor's attitude good?
- **Q4**. Did you understand the doctor's explanation?
- **Q5**. Was the doctor's explanation satisfactory?
- **Q6**. Was your anxiety decreased by the clinical examination?

Figure 26.2 shows the responses of the subjects to the questions. It can be seen in this figure that the responses to all questions show that the smiling and nodding behavior of the bystander in synchrony with the subjects positively affected them at a statistically significant level. It should be noted that the questions were related not to the bystander but to the clinical examination. It should also be noted that the subject's conscious impression of the bystander in both conditions was not different at a significant level, as shown in Table 26.1; however, the existence of the human bystander clearly affected the impression of the subjects unconsciously.

## 26.4  Clinical Experiment with Android

In order to confirm the hypothesis that the appropriate behavior of an android bystander can positively affect human communication, we conducted a clinical experiment at Osaka University Hospital with the authorization of the Ethical Review Board. This experiment was also performed in an orthopedics examination room in the outpatient department; its duration was approximately one month.

### 26.4.1  Experimental Setup

The android was placed behind the doctor, as shown in Fig. 26.3, and a PC terminal, together with an air compressor to drive it, was placed in the neighboring room. The patients could observe the bystander in their peripheral visual field, while it was not within the doctor's field of view. The doctor and patients were asked to conduct an examination as usual; most of the subjects did not consciously pay attention to the bystander android. The behavior of the bystander android during the clinical examination was generated in the following four conditions.

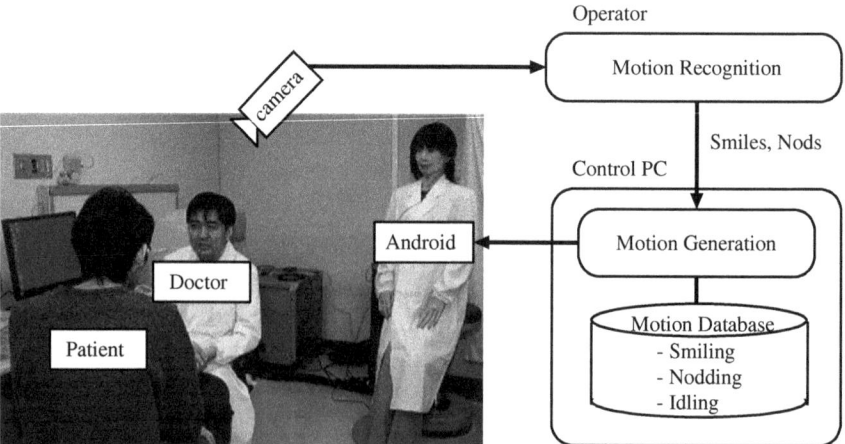

**Fig. 26.3**  System setup for clinical experiment

- **Condition 1: Synchronized with patients**: The bystander nods and smiles in synchrony with the patient,
- **Condition 2: Without expressions**: The bystander neither nods nor smiles,
- **Condition 3: Random**: The bystander nods and smiles randomly,
- **Condition 4: Synchronized with doctor**: The bystander nods and smiles in synchrony with the doctor.

One of these conditions was assigned to each subject. The bystander androids' behavior was controlled according to the condition during the clinical examination. Condition 1 corresponds to the chameleon effect, and we expected the psychological effect on the patient to be good. In Conditions 1 and 4, an operator in the neighboring room observed the examination room through a monitor and determined the timing of the androids nodding and smiling behavior. In all conditions, the android was constantly moving slightly to decrease the unnaturalness of its appearance [4]. Sixty-four patients (24 males and 40 females) took part in the experiment as subjects. The same doctor took part during the entire experiment, but was not informed of the experimental conditions. In addition, we also asked some patients who experienced a normal examination without the android in the room to fill the same questionnaire to provide a comparison.

## 26.4.2   Experimental Results

The most frequent response of the subjects concerning the presence of the bystander robot in all conditions was that they were indifferent to its presence, which was the same result as that of the preliminary experiment. However, the android's presence tended to be preferred when the its behavior was synchronized with that of the patient in Condition 1. In fact, many subjects also commented that the presence was comfortable or relaxing in this condition. Another frequent comment was that they were interested in the android, and therefore, the subjects seem to have regarded it as an object of interest rather than of communication.

Figures 26.4 and 26.5 show the responses of the subjects to the questions about the clinical examination. The questionnaire was the same as that utilized in the preliminary experiment. In Fig. 26.4, the impressions of the subjects in Condition 1 (android synchronous with subjects) and Condition 2 (android without expressions) are compared with those in the case where no android was present in the examination room. When the android did not express smiling and nodding behaviors, the impression of the subjects as expressed by their answers to all the questions was worse than when the android smiled and nodded in synchrony with the subjects. These are the same results as those for the preliminary experiment. In addition, it became clear that the subject impressions were even worse than in the case where the android was not present in the examination room.

As shown in Fig. 26.5, the timing of the smiling and nodding behaviors was varied in order to investigate the effect of the synchronization of the android

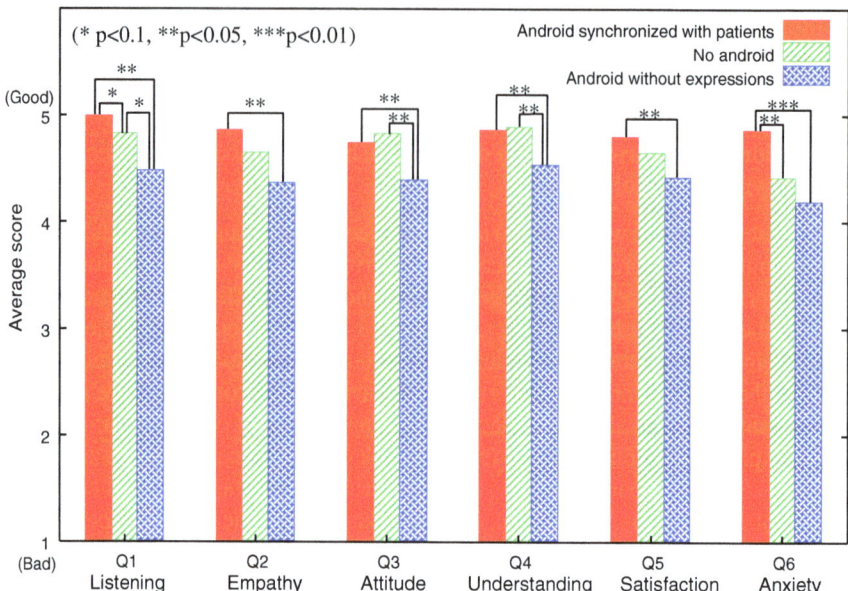

**Fig. 26.4** Experimental results (1): Impression of patients of clinical examination with/without android and synchronization

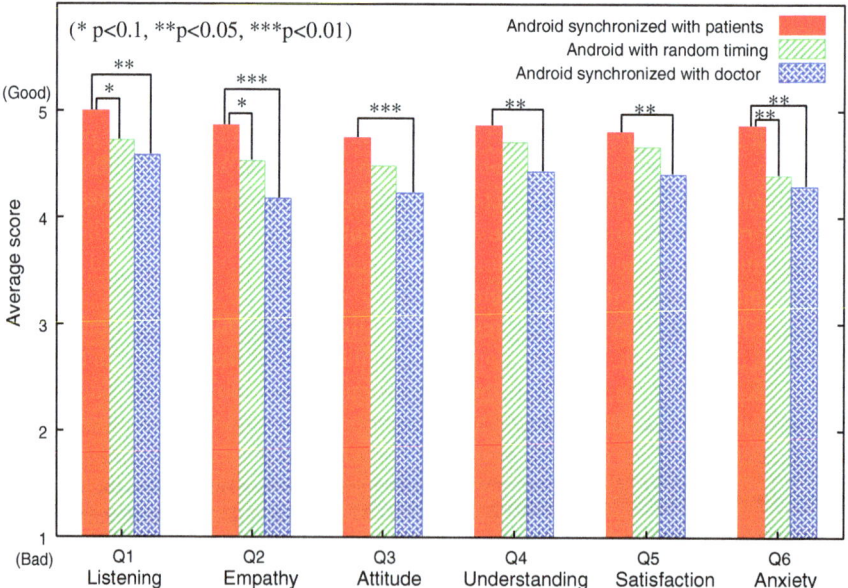

**Fig. 26.5** Experimental results (2): Impression of patients of clinical examination with different timing of behaviors

with the subjects. The best timing of the android's behavior was that in which it was synchronized with the subjects, which yielded a significantly higher score than the random timing condition. An unexpected result was that the condition where the android was synchronized with the doctor yielded a worse impression than the other two conditions, although the number of nodding and smiling behaviors was similar. This is probably because the subjects unconsciously recognized that the android was empathizing not with them but with the doctor.

## 26.5   Conclusion

In this study, we investigated the psychological effects of the presence and the behavior of an android on trilateral communication during a clinical examination at a hospital. The experimental results indicate that the behavior of the bystander android affected the subjects impression of the communication and that the synchronization of the android with the subject was important for yielding positive effects. The best results were achieved when the android smiled and nodded in synchrony with the patient, and our hypothesis was thus supported. We also obtained the result that both the android without expressions and the android synchronized with the doctor produced an even worse impression on the subjects than the case where no android was present. It was also shown that even the current android, with its low interaction ability, could be applied as a bystander in the practical clinical situation to improve communication.

Our future work will include further investigation of the effect of the appearance of the robot in order to confirm that other humanoids with a mechanical appearance and a CG agent on the monitor have a similar psychological effect. Currently, the behavior of the android is most effective when it is synchronized with the subject; however, we should also test other timings, such as changing the synchronization target according to the context of the conversation. We are also planning to automate the recognition process of the behavior generation system of the android by utilizing vision technology.

**Acknowledgements**  This research was supported by "Special Coordination Funds for Promoting Science and Technology: the Yuragi Project" of the Ministry of Education, Culture, Sports, Science, and Technology, Japan.

## References

1. Takano, Eri, Yoshio, Matsumoto, Yutaka, Nakamura, Hiroshi, Ishiguro, and Kazuomi, Sugamoto. 2009. The psychological effects of attendance of an android on communication. *Experimental Robotics*, 221–228.
2. Ishiguro, H. 2007. Scientific issues concerning androids. *The International Journal of Robotics Research* 26 (1): 105–117.

3. Mori, Masahiro. 2005. On the Uncanny Valley. In *Proceedings of the humanoids 2005 workshop: Views of the Uncanny Valley*.
4. Noma, Motoko, Naoki, Saiwaki, Shoji, Itakura, and Hiroshi, Ishiguro. 2006. Composition and evaluation of the humanlike motions of an android. In *Proceedings of the IEEE/RAS international conference on humanoid robots (Humanoids 2006)*, 163–168.
5. Daibo, I. 2003. Communication studies by a social psychological approach: Interpreting interpersonal relationships. *Japanese Journal of Language in Society* 6.
6. Chartrand, Tanya L., and J.A. Bargh. 1999. The chameleon effect—The perception-behavior link and social interaction. *Journal of Personality and Social Psychology* 76: 893–910.
7. Matsusaka, Yosuke, Tsuyoshi, Tojo, and Tetsunori, Kobayashi. 2003. Conversation robot participating in group conversation. *Trans. IEICE* J84-D-II (6): 898–908.
8. Hayashi, K., D. Sakamoto, T. Kanda, M. Shiomi, S. Koizumi, H. Ishiguro, T. Ogasawawra, and N. Hagita. 2007. Humanoid robots as a passive-social medium—a field experiment at a train station. In *Proceedings of the ACM 2nd annual conference on human-robot interaction*, 137–144.

# Chapter 27
# At the Department Store—Can Androids Be a Social Entity in the Real World?

## Miki Watanabe, Kohei Ogawa and Hiroshi Ishiguro

**Abstract** In this paper, we discuss an autonomous android robot that is recognized as a social entity by observers in the real world. We conducted field experiments to investigate the type of function with which an android should be provided so that it is can be recognized as a social and humanlike entity by observers. In the field experiment, the android conversed with multiple visitors through several touch displays. The results show that the visitors evaluated the android as humanlike, although this type of interaction differs from normal human–human interaction. Moreover, the results of the experiment suggest that the android exerts a social influence-advertisement effect on visitors.

**Keywords** Android · Human–robot interaction · Conversation system

## 27.1 Introduction

The number of partner robots [2] present in our daily lives has been gradually increasing, and the roles of robots have become more diverse with the increase in the variety of robots available on the market. In the future, android robots having an appearance that closely resembles that of a human could be employed instead of humans in locations where the presence of a receptionist or a guard, for example, is required. An android robot called Repliee Q1 was developed in 2005. Repliee Q1 can reply to a human autonomously according to the input into the system from a tactile

This chapter is a modified version of a previously published paper [1], edited to be comprehensive and fit with the context of this book.

M. Watanabe · K. Ogawa (✉) · H. Ishiguro
Department of Engineering Science, Osaka University, Toyonaka, Japan
e-mail: ogawa@irl.sys.es.osaka-u.ac.jp

M. Watanabe
e-mail: watanabe.miki@irl.sys.es.osaka-u.ac.jp

H. Ishiguro
e-mail: ishiguro@sys.es.osaka-u.ac.jp

© Springer Nature Singapore Pte Ltd. 2018
H. Ishiguro and F. Dalla Libera (eds.), *Geminoid Studies*,
https://doi.org/10.1007/978-981-10-8702-8_27

sensor, a floor sensor, and a microphone. Unfortunately, the quality of the input data is insufficient for providing a humanlike response. This mismatch between the anthropomorphic appearance and the machinelike response may lead to a problem known as the Uncanny Valley effect [3]. In order to overcome this problem, Geminoid HI-1, which is copy of a real person, was developed with tele-operational functionality, which enables a human operator to provide responses instead of an autonomous system [4]. Another Geminoid version, called Geminoid F, is being used as an actress in the "Android Theater." In the limited situation of a theater play, Geminoid F can be perceived as a real human.

As mentioned above, several questions on natural language processing, computer vision, and reasoning systems remain to be answered before an autonomous conversation system for real-world usage can be created for robots. However, in specific situations, an android can be perceived as a humanlike entity by people. Furthermore, in these situations, an android that can interact with people in the real world can affect them positively.

In this paper, we describe an autonomous conversation system that allows an android to interact with people in limited interaction situations. We performed a field experiment using this system to evaluate how people perceive Geminoid F and how Geminoid F affects people when it interacts with them. We analyzed the human behavior using video observation and questionnaires.

## 27.2  Field Experiment

The purpose of this experiment was to verify how people perceive Geminoid F and how it affects people during an interaction by generating conversations between Geminoid F and multiple visitors via touch displays. In this study, Geminoid F conversed with the visitors in a department store via three touch displays (Fig. 27.1). Four questions were shown on the touch displays. When a visitor touched one of these questions, Geminoid F answered making eye contact with the person. The questions were chosen randomly from 50 questions related to the robot and the department store. In the case of one-to-one conversation, it is difficult to create an autonomous system using our system, because deeper content is required. However, conversation with multiple people enables every interacting user to independently initiate a one-turn conversation with Geminoid F. We expected that these simple interactions would give users the feeling that they were more engaged in the conversation. We evaluated human likeness and how people perceived Geminoid F through questionnaires and by analyzing the video data recorded at the two touch displays (duration, 78 h and 7 min). We also observed the interactions between people and the android that occurred during the experiment.

**Fig. 27.1**  Experimental situation

## 27.3  Results

Out of 2159 people who talked with Geminoid F using the touch displays, we randomly interviewed 60 (39 female, 21 male). Furthermore, we interviewed 40 people (24 female, 16 male) who only observed the interaction. We analyzed the data of the two groups separately. We can state that the visitors understood how to interact with Geminoid F without any instructions, as most of them attempted to talk to Geminoid F by using the touch display. In the interviews, we asked questions addressing the naturalness of the conversation with Geminoid F, its behavior, how much time the interviewees felt they could spend talking with it, and what kind of questions they would like to ask it. In their responses to the question addressing the naturalness of the conversation, we found that 68.3% of the interviewees who talked with Geminoid F and 77.5% of those who only observed it evaluated the robot positively; i.e., they rated it with a score of 5 or higher on a 7-point Likert scale (Fig. 27.2). Furthermore, the naturalness of the behavior was evaluated positively by 61.7% of interviewees who talked with Geminoid F and 80.0% of those who only observed it (Fig. 27.3). This means that the interviewees accepted this situation and recognized Geminoid F as a humanlike entity, although the situation was different from normal human conversation. In response to the question about the length of time that they felt that could spend talking with Geminoid F, 41% of the interviewees answered that they could spend more than one hour, and some answered that they wanted to live with Geminoid F eternally (Fig. 27.4 (Left)). Furthermore, 64% were interested in its daily life or private matters (Fig. 27.4 (Right)). This means that Geminoid F

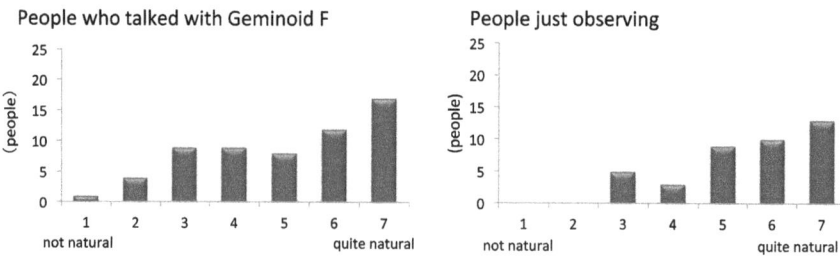

**Fig. 27.2**   Naturalness of the conversation with Geminoid F

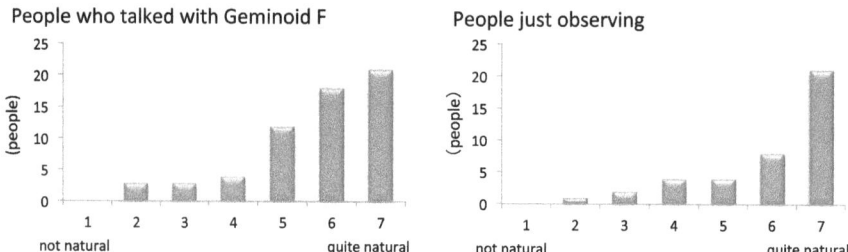

**Fig. 27.3**   Naturalness of the behavior of Geminoid F

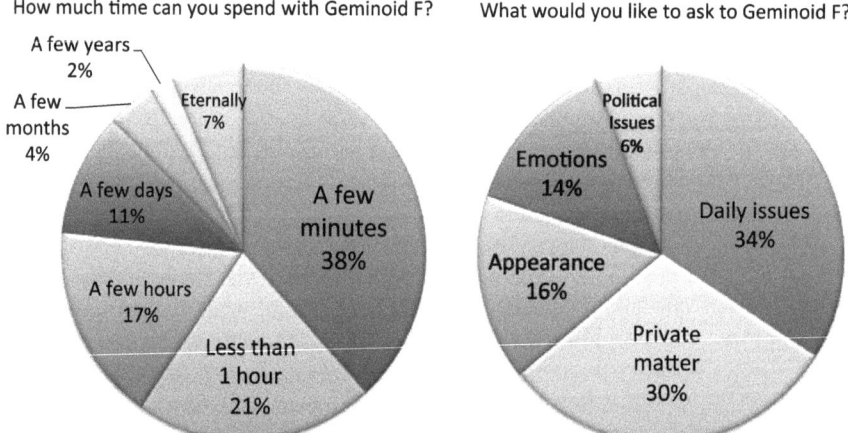

**Fig. 27.4**   Left: How much time can you spend talking with Geminoid F? Right: What would you like to ask to Geminoid F?

aroused the visitors' interest, although they interacted with it in a limited situation, and that it was perceived as a humanlike entity.

In addition, the sweater and skirt that Geminoid F was wearing were afterward sold out, suggesting that Geminoid F could be an effective advertising medium. The female visitors said that the costumes that Geminoid F wore were very beautiful, and they asked where they could buy them. Sales staff said that the sales were considerably higher than usual. Based on these facts, we assume that Geminoid F could produce the above-mentioned incidental effect because the visitors perceived Geminoid F as a humanlike and social entity and desired to own it.

## 27.4   Conclusions

We developed an autonomous conversation system that is able to interact with people in a limited situation. In our experiment, we focused on a situation where an android was working. Our experimental results showed that the android can interact with people autonomously and they perceive it as a humanlike and social entity. The results of this study can be applied to design an autonomous android robot that will be recognized as a social entity by observers in the real world. In the future, by improving the content of the conversation according to various other practical situations, it may be possible to start using androids in the real world.

## References

1. Watanabe, M., K. Ogawa, and H. Ishiguro. 2014. Field study: Can androids be a social entity in the real world? In *2014 ACM/IEEE international conference on Human-robot interaction*, 316–317.
2. Fujita, M. 2000. Development of robot entertainment system AIBO (Special features human interface out of box). *Information Processing Society of Japan* 41 (2): 146–150.
3. Mori, M., K.F. McDorman, and N. Kageki. 2012. The uncanny valley. *IEEE Robotisc and Automation Magazine* 19 (2): 98–100.
4. Becker-Asano, C., K. Ogawa, S. Nishio, and H. Ishiguro. 2010. Exploring the uncanny valley with Geminoid HI-1 in a real-world application. In *IADIS international conference on interfaces and human computer interaction*, 121–128, Freiburg, Germany.

# Chapter 28
# At the Department Store—Can Androids Be Salespeople in the Real World?

**Miki Watanabe, Kohei Ogawa and Hiroshi Ishiguro**

**Abstract** The roles of robots in the real world have become more diverse according to the bodily properties with which they are endowed. In this study, we aimed to determine the roles that androids, the bodily properties of which resemble those of humans, could serve in the real world. Selling and purchasing are common human activities. Therefore, we propose the use of an android as a salesperson utilizing cognitive and affective strategies that exploit the advantages of online- and counter-selling methods. We conducted a field study to investigate whether androids can sell goods in a department store. The results show that the sales strategies were effective and that over 10 days the android sold 43 sweaters that cost approximately $100 each. These results provide important knowledge for determining how androids may fill new roles and communicate with humans in the real world.

**Keywords** Android robot · Field study · Sales · Conversation · Affection Human–robot interaction

## 28.1 Introduction

The number of robots present in our daily lives has been gradually increasing, and their roles have become more diverse according to the bodily properties with which they are provided. In particular, humanoid robots that have bodily properties similar

---

This chapter is a modified version of a previously published paper [1], edited to be comprehensive and fit with the context of this book.

M. Watanabe · K. Ogawa (✉) · H. Ishiguro
Department of Systems Innovation, Osaka University, 1-3 Machikaneyame, Toyonaka, Osaka 560-8531, Japan
e-mail: ogawa@irl.sys.es.osaka-u.ac.jp

M. Watanabe
e-mail: watanabe.miki@irl.sys.es.osaka-u.ac.jp

H. Ishiguro
e-mail: ishiguro@sys.es.osaka-u.ac.jp

© Springer Nature Singapore Pte Ltd. 2018
H. Ishiguro and F. Dalla Libera (eds.), *Geminoid Studies*,
https://doi.org/10.1007/978-981-10-8702-8_28

**Fig. 28.1** Android called
Geminoid F

to those of humans have been used in real-world applications. For example, Kanda et al. reported that humanoid robots can serve as guides in a shopping mall [2]. Shiomi et al. also used humanoid robots for advertisement, and they developed a robot system for distributing coupons in a shopping mall [3]. In addition, a robot with an animal appearance, Paro, has been used in the real world. Paro's baby-seal appearance and its interaction function are considered to contribute to its effective use as a therapeutic robot in care facilities for older persons [4]. This example implies that the outer appearance of a robot is tightly connected to the situation in which it is applied in the real world.

An android (Fig. 28.1) that resembles a human can serve in a situation where its application was not previously expected. The android utilized in this research is called Geminoid F and was developed in 2010. Its facial expressions are sufficiently rich to convey emotions to an interlocutor. Several studies demonstrated certain unique characteristics of this android. For example, Nishio et al. reported that this android could be a medium for transferring an actual human's presence [5]. Further, Watanabe et al. showed that Geminoid F can conduct a natural conversation with customers as a receptionist, and it was perceived as a social entity in the real world [6].

Selling and purchasing are common activities in humans daily lives. Recently, the types of sales activities have been changing. One type, online shopping, has spread widely with the increase in Internet use. In addition, the method of counter selling in a department store is still popular, because customers desire to experience a social relationship with a salesperson when purchasing goods [7]. These two types of sales activity have different advantages. The advantage of online shopping is the availability to the customer of specific information, such as the specifications of electrical devices (e.g., a laptop PC, a smartphone, or an audio surround system), whereas the advantage of counter selling is the salespersons ability to affect customers when they purchase goods that have less specific criteria that influence their choice, such as a dress or accessories. An android is clearly an artifact; however, it is still recognized as a human-like entity. This unique feature allows the android to utilize both the online- and counter-selling methods.

In this paper, we propose using an android as a salesperson that provides the customer with detailed information and affects them through conversation. We conducted a field study to investigate whether the android could sell goods to humans through conversation in a real sales situation. Moreover, we discuss the reasons why customers purchased from the android and what types of customer accepted the sales activity of the android.

## 28.2  Related Work

As mentioned previously, online shopping has become widespread with the increase in Internet use [8]. Many technologies have been proposed. Sekozawa et al. developed a recommendation system for online shopping that can extract customers' tastes on the basis of big data [9]. This system has already been implemented in Web services, such as Amazon and e-Bay. As an aspect of psychology, the reasons why humans use online shopping have been discussed. For example, Ward et al. noted that a person who shops online feels less annoyed than he/she would when purchasing at a counter, where communication with salespeople is involved [10]. The results of these studies imply that the availability of more information online is sufficient for persuading customers to purchase goods.

However, counter selling is still popular because humans regard social relationships with a salesperson as important for purchasing goods. The selection of goods that have less specific choice criteria may be rather difficult for customers if they lack confidence. Therefore, customers desire to ask the subjective opinions of salespeople to facilitate their purchasing decision. The results of several studies indicated that the attribution of the source of a message significantly influences how the message is perceived by the recipients [11, 12]. Hence, affecting customers such that they form a positive relationship with a salesperson is important, because they will then accept a subjective opinion that consequently leads to the purchase of goods.

In this study, we implemented a conversation system that can cognize information and affect customers in a sales situation. Through an application of the system, we validated the performance of an android as a salesperson that sells sweaters in a counter-selling situation.

## 28.3  Field Study

In this study, we investigated whether androids could serve as salespeople in the real world through a field study in a department store. The situation was a counter-selling situation in which Geminoid F conducted a conversation with a customer via a touch display (Fig. 28.2).

**Fig. 28.2**  Sales situation: conversation with Geminoid F through a touch display

### 28.3.1  Commodity

The androids sold cashmere sweaters that are manufactured in a variety of types and color, a total of 105 different sweaters. For males, there were 4 types (U-neck, V-neck, cardigan, and turtleneck sweater) and 8 colors for each type, whereas for females, there were 4 types (the same as for males) and 18 colors for each type.

### 28.3.2  Strategy

In this study, the android utilized two strategies for selling goods. As mentioned previously, the perception and availability of specific information, as well as the salespersons ability to influence customers, are important for achieving the sale of a product to a customer. Therefore, we expected that utilizing both cognitive and affective strategies might be effective for selling goods. Specifically, the android first provided a color consultation to the customers to determine a suitable color for them. Second, after informing the customers which color was suitable for them, the android complimented them and gave them a subjective impression to evoke their affection and persuade them to purchase sweaters.

### 28.3.3  Sensor System

In this system, we use a Kinect sensor to detect the situation in front of the android. When no customer was sitting in the chair in front of the android, the android called out to customers, saying "Welcome to the shop," "Would you like me to analyze your personal color?" "I can recommend sweaters that suit you." When the sensor determined that a customer was sitting in the chair, the android started to converse with the customer. During the conversation, the sensor detected the face position of the customer, and the android made eye contact with and nodded back to the customer.

### 28.3.4   Conversation System

The android conversed with customers through a touch display (Fig. 28.3). A touch display was employed because a precise voice-recognition system is required for a conversation. Recently, the quality of voice-recognition systems has been improved, but it is still not satisfactory for androids, in particular in a noisy environment, because they are expected to have the same quality of voice-recognition ability as humans owing to their human-like appearance. We installed a touch display in front of the android, which showed several questions and replies that could be addressed to the android. When a customer touched one of these questions or replies, a recorded voice was output from the display according to their choice. The voice was altered according to the gender of the customer to make them feel that they were actually talking to the android. Then, the android produced utterances regarding the question or reply that the customer had selected. Moreover, as an advantage, the customer had to choose from the inductive selections on the touch display to communicate

**Fig. 28.3**  Using a touch display to talk with Geminoid F

with the android. For example, during the color-consultation phase, the fact that only one selection option was displayed, "Please provide me with a color consultation," caused the customers to feel that they were willing to ask the android for this service.

### 28.3.5   Conversation Scenario

We prepared a conversation scenario in advance, which was divided into four parts. The duration of the conversation was approximately 15 min. In the first part of this scenario, the android greets and builds a social relationship with a customer. The android asks about the customers gender, hobbies, and so on. In the second part, a color consultation is provided to the customer. The android does not ask the customer questions directly but derives information by using "cold reading," which can derive information from a person without prior knowledge. For example, the android says "Your eyes are really beautiful. Are your eyes brown?" The customer replies "Yes, my eyes are brown," or "No, my eyes are black." In the third part, the android recommends several sweaters according to the results of the color consultation. The customer then selects one of the sweaters that the android recommends, and a human salesperson brings the sweater to the customer. A customer examines his/her appearance with the sweater and asks the android whether the sweater is suitable or not. Then, the android compliments the customer in an exaggerated manner. If a customer wants to look at another color or type of sweater, the android repeats this part. During the fourth part, the android persuades the customer to purchase a sweater. The android asks whether the customer wants to purchase the sweater. Even if a customer's answer is no, the android continues the conversation to further persuade the customer (e.g., Android: Was my recommendation bad? Customer: Yes, your recommendation was bad. Android: I'm so sorry about that, but I did my best to help you. Would you reconsider purchasing a sweater from me?). The android attempts to persuade the customer three times at most. Then, the android says good-bye. As a strategy in all parts, the android forces customers to select a negative response with the expectation of deriving a positive attitude toward the android.

## 28.4   Results

### 28.4.1   Sales Results

We conducted a two-week field study at a department store in Osaka, Japan. The android served 515 customers over this period. Note that it took four days to stabilize the androids conversation system; therefore, we used the data from the fifth day

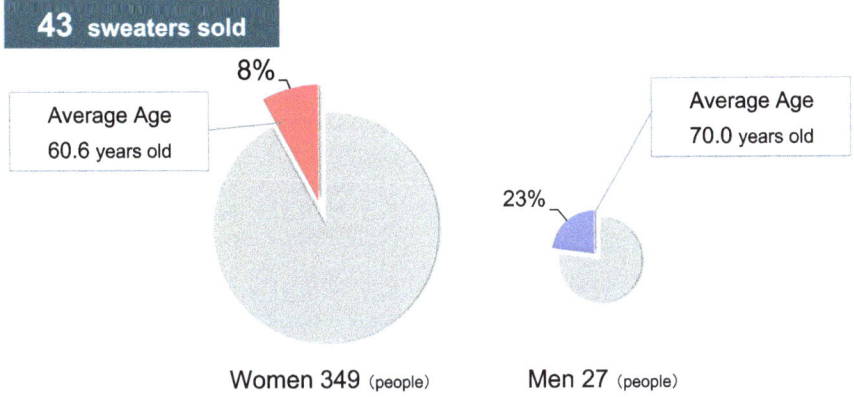

**Fig. 28.4** Breakdown of customers who bought sweaters by gender

onward. According to the sales results for these 10 days, the android served 349 customers and sold 43 sweaters. The gender breakdown of the customers is 349 women and 27 men. Of these customers, 8% of women and 23% of men purchased sweaters (Fig. 28.4).

The average age of the female purchasers was 60.6 years old and of male purchasers 70.0 years old. A comparison of the number of customers who were served by an actual human clerk and the android shows that the human clerk served approximately 15 customers per day and the android served 39.6 customers per day. Note that 76% of the customers answered that they previously knew about our android and its function, because this was the fourth field study conducted at this department store. Therefore, we assume that the novelty of the android's existence in a public environment was not the central factor for purchasing a sweater.

## 28.4.2   Questionnaire Results

We randomly chose 242 customers who experienced a conversation with the android for the purpose of interviewing them about their demographics and impressions of the android. We asked them the following four questions. Question 1 (Q1): Did you trust the results of the color consultation? Question 2 (Q2): Did you accept the complimentary words from the android? Question 3 (Q3): Did you feel hesitation when you refused the androids multiple offers of a sweater? and Question 4 (Q4): Did you feel as if you were talking to the android through the touch display? The customers ranked their responses on a seven-point Likert scale. The age distribution of the

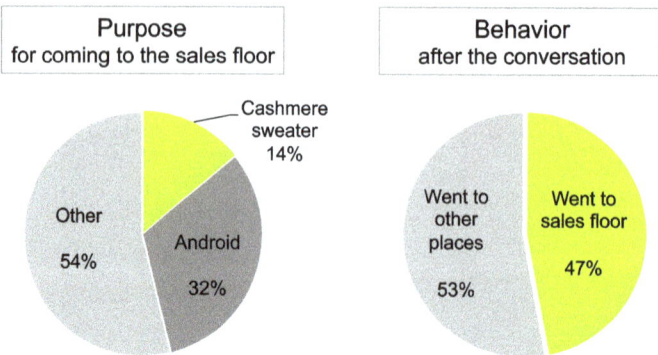

**Fig. 28.5** Purpose for coming to the sales floor and behavior after the conversation

customers was as follows: 3% under 20 years old, 22% 20–39 years old, 39% 40–59 years old, and 36% over 60 years old. The reasons for the customers presence in the store were as follows: 14% "to look for a sweater," 32% "to see the android," and 54% "other." The results of observing the behavior of customers after the conversation with the android are as follows: 47% of the customers visited the selling space (Fig. 28.5).

To examine the aspects of the cognitive strategy, in Q1, we asked the customers about the extent to which they could trust the results of the color consultation. The results of a t test comparing the scores for this question and the chance level (score = 4) showed a significant difference (Fig. 28.6 (Q1)). Moreover, one customer reported "I can trust the android because she recognized the color of my skin and eyes." To examine aspects of the affective strategy, in Q2 we asked the customers about the extent to which they could accept the exaggerated compliments of the android. The results of a t test comparing the scores for this question with the chance level (score = 4) showed a significant difference (Fig. 28.6 (Q2)). One customer reported "When I hear this type of compliment from actual humans, I feel their strong intention to sell the sweater, but in a case of the android, this impression was suppressed." To examine the difficulty that the customer experienced when refusing the offers of the android, we asked Q3, "Did you feel hesitation when you refused the androids multiple offers of the sweater?" The results of the t test comparing the scores for this question with the chance level (score = 4) showed a significant difference (Fig. 28.6 (Q3)). Finally, in Q4 we asked customers to evaluate the performance of proposed conversation system with the touch display. The results of the t test comparing the scores for this questions with the chance level (score = 4) showed a significant difference (Fig. 28.6 (Q4)).

|     | N   | DF  | Mean(SD)    | P-value  |
| --- | --- | --- | ----------- | -------- |
| Q1  | 235 | 234 | 5.6(1.41)   | <.03 **  |
| Q2  | 242 | 241 | 4.94(1.77)  | <.03 **  |
| Q3  | 158 | 157 | 3.15(2.07)  | <.03 **  |
| Q4  | 242 | 241 | 4.32(1.72)  | <.05 *   |

**Fig. 28.6**  Result of the t tests

## 28.5  Discussion

### 28.5.1  Sales

The result that the android sold 43 sweaters indicates that it has the ability to persuade customers to purchase clothes through a touch display conversation in a department store. This result indicates that the android can serve a social role involving persuasion in the real world. We now discuss the distribution of customers who purchased sweaters. First, the price and conservative design of the sweaters explain the average age of the customers who bought the sweater. Second, the difference between the genders in the ratio of purchasing indicates that male customers are more open to persuasion than female customers. In fact, for example, one male customer reported "She was so helpful because I didnt know which would look nicer on me." In contrast, a female customer reported "I just wanted to buy the one I prefer. It's more fun just looking." These comments also indicated the gender differences in attitudes toward shopping. Therefore, when using androids as salespersons, the strategy and the goods should be carefully considered according to the age and gender of customers. The number of customers served by an actual human and the android shows that the android served nearly twice as many customers as the actual human, which is one of the advantages of the android.

### 28.5.2  Strategy

In this section, we discuss the cognitive and affective strategies mentioned above. First, the results pertaining to the acceptance of the color consultation show that the android could provide customers with information that they felt they could trust. After cognizing the information, the android gave compliments and subjective impressions to the customers. The results regarding the acceptance of compliments show that the android could influence customers and persuade them to purchase sweaters. Meanwhile, the results regarding the customers hesitation to refuse the androids repeated offers show that its communication in a sales situation did not annoy them. From the sales results showing that the android sold 43 sweaters, we can conclude that using both strategies was effective for persuading customers to buy goods.

### 28.5.3 System

Although our proposed conversation system produces a conversation style that is different from that between two humans, we found that people could understand how to talk with the android without an explanation of the conversation system. Furthermore, people purchased sweaters through this conversation style, and thus, a conversation through a touch display that limits selections could be an effective persuasion tool.

## 28.6 Limitation

In this field study, the android utilized several strategies during the sales conversation. Thus, we cannot determine the specific factor that caused customers to purchase sweaters. However, we offer some hypotheses: (1) The combination of cognitive and affective persuasion is effective in a sales situation, and (2) conversation through a touch display could persuade people effectively. We are currently verifying these issues through comparative experiments in laboratory studies.

## 28.7 Conclusion

In this research study, our aim was to determine the roles androids could serve in the real world. We conducted a field study to investigate whether an android can fill the role of a salesperson and persuade customers to buy goods in sales situation by using the proposed cognitive and affective strategies that utilize the advantages of both online- and counter-selling methods. We found that the android could sell goods and that our sales strategy was effective in a real environment. We believe that these results provide important knowledge for determining how androids may fill new roles and communicate with humans in the real world.

## References

1. Watanabe, M., K. Ogawa, and H. Ishiguro. 2015. Can androids be salespeople in the real world? In *Proceedings of the 33rd annual ACM conference extended abstracts on human factors in computing systems*, 781–788.

2. Kanda, T., M. Shiomi, Z. Miyashita, H. Ishiguro, and N. Hagita. 2009. An affective guide robot in a shopping mall. In *Proceedings of the 4th ACM/IEEE international conference on Human robot interaction*, 173–180.

3. Shiomi, M., K. Shinozawa, Y. Nakagawa, T. Miyashita, T. Sakamoto, T. Terakubo, and N. Hagita. 2013. Recommendation effects of a social robot for advertisement-use context in a shopping mall. *International Journal of Social Robotics* 251–262.

4. Wada, K., and T. Shibata. 2006. Robot therapy in a care house-its sociopsychological and physiological effects on the residents. In *Proceedings 2006 IEEE international conference on robotics and automation ICRA 2006*, 3966–3971.

5. Nishio, S., H. Ishiguro, and N. Hagita. 2007. Geminoid: Teleoperated android of an existing person. In *Humanoid robots: New developments*, 343–352. Vienna, Austria: I-Tech Education and Publishing.

6. Watanabe, M., K. Ogawa, and H. Ishiguro. 2014. Field study: Can androids be a social entity in the real world? In *Proceedings of the 2014 ACM/IEEE international conference on Human-robot interaction*, 316–317.

7. Rintamaki, T., A. Kanto, H. Kuusela, and M. T. Spence. 2006. Decomposing the value of department store shopping into utilitarian, hedonic and social dimensions: Evidence from Finland. *International Journal of Retail and Distribution Management* 6–24.

8. Limayem, M., M. Khalifa, and A. Frini. 2000. What makes consumers buy from Internet? A longitudinal study of online shopping. *IEEE Transactions on Systems, Man and Cybernetics, Part A: Systems and Humans* 421–432.

9. Sekozawa, T. 2010. Recommendation system for apparel online shopping. *WSEAS Transactions on Systems* 488–497.

10. Ward, M. R. 2001. Will online shopping compete more with traditional retailing or catalog shopping? *Netnomics* 103–117.

11. Petty, R.E., and J. T. Cacioppo. 1986. The elaboration likelihood model of persuasion. In *Advances in experimental social psychology* 123–205.

12. McLuhan, M., and Q. Fiore. 1967. *The medium is the message*. New York

# Chapter 29
# At the Theater—Designing Robot Behavior in Conversations Based on Contemporary Colloquial Theatre Theory

Kohei Ogawa, Takenobu Chikaraishi, Yuichiro Yoshikawa, Oriza Hirata and Hiroshi Ishiguro

**Abstract** The design of humanlike behavior for a robot that interacts with humans remains a central issue in the human–robot interaction (HRI) field because of humans sensitivity to humanlike objects. This issue is very challenging, because humanlikeness is an important factor in designing better interactions, since an imperfect design can easily cause negative impressions, as reflected by the uncanny valley phenomenon. This paper addresses this issue using a novel approach that utilizes implicit know-how for performing on stage dedicated to the stage representation of human beings. Contemporary colloquial theatre theory (CCTT), which is a theory applied in a method of directing plays, is appropriate for this purpose, since its reality-oriented instructions are directly applicable to improving robot behavior. In this paper, we report a case study involving the performance of a play in which both humans and a robot played roles. The play in our study was evaluated by the audiences of public performances in Japan. We also report a detailed analysis of HRI or human–human–robot interaction in comparable short plays. Our analysis implies that the robots utterances and motion timings should be tuned according to the situation. In future work, a motion capture system will be applied to obtain more precise data and more useful knowledge.

**Keywords** HRI · Robot · Theatre play

---

This chapter is a modified version of a previously published paper [1], edited to be comprehensive and fit with the context of this book.

---

K. Ogawa (✉) · T. Chikaraishi · O. Hirata
Center for Study of Communication Design, Osaka University,
1-16 Machikaneyama-cho, Toyonaka, Osaka 560-0043, Japan
e-mail: ogawa@is.sys.es.osaka-u.ac.jp; ogawa@irl.sys.es.osaka-u.ac.jp

Y. Yoshikawa · H. Ishiguro
Graduate School of Engineering Science, Osaka University, 1-3 Machikaneyama, Toyonaka, Osaka 560-8531, Japan

## 29.1 Introduction

Human-like robots, which have the same body structure as humans, as well as androids that have an appearance identical to that of a real person, are expected to assume human roles in society. Many previous studies have demonstrated successful interactions between many types of robots endowed with various humanlikeness aspects and humans in real-world applications [2–4]. However, to design the robots behavior such that is appears natural is not a trivial task, since the uncanny valley effect sometimes appears [5], probably because of human sensitivity to the whether the coordination of gestures and speech in humanlike objects appears natural.

The naturalness of human behavior on a theater stage has been addressed since the time of ancient Greece. The implicit know-how of playwrights and directors about staging, i.e., making actors appear fascinating, may be effectively utilized for designing robots. However, much of the know-how about acting that reproduces natural human behavior cannot be directly applied to the design of robot behavior, because frequently actions are unnaturally over-expressed, and the directors instructions to modify the actions of actors are frequently given through suggestions about human emotions that are too ambiguous to be applied to robots. As compared with the traditional direction described above, Hirata proposed Contemporary Colloquial Theater Theory (CCTT) to reproduce natural daily conversations in stage plays [6]. The CCTT guidelines, which indicate the physical aspects of actions that should be changed, are expected to be directly applicable to developing robots [7]. This implies that we can create a performance involving humanoid robot actors based on CCTT and construct an effective theory to provide robots with humanlike behaviors in daily conversations.

In this paper, we report the results of our first important step toward realizing this approach, which is aimed to provide an android actor with humanlike behaviors based on CCTT, and present an evaluation of the audience impressions of a performance that included an android actor. We argue that implicit know-how for designing robot behavior can be extracted by constructing and analyzing comparable short plays that are also directed based on CCTT precepts.

To demonstrate the feasibility of our approach, we composed a 20 min play, titled *Sayonara*, based on CCTT, in which an android actor interacts with a human actor. The multimodal actions of the androids operator, such as those for producing the robots speech or its gaze at its interlocutors, were captured by a tele-operating system, transferred to the androids performance in real time, and recorded. In the rehearsals using the recorded performance, the instructions focused on the physical aspects of the android's actions that were expected to be provided by the director based on CCTT, such as the onset timing of motion and speech, rate of speech, gaze direction, standing position. The androids performance shaped by our proposed process was evaluated by the audiences of public performances in Japan. The same method was then applied for creating relatively short plays involving human–robot interactions, in which almost the same script was presented but elaborated for slightly different situations. We analyzed the instructions given by the director during the acting performances based on the instruction terms for the physical aspects. This analysis is expected to facilitate the extraction of the distinctive features of robot behaviors that are required in different situations. In practice, based on CCTT, we modeled conversations between a person and a humanoid robot named Robovie-R3 and between a robot and two persons, and analyzed their differences. We show that the order relation of the timings of the onsets of body motions and utterances appeared to be tuned based on the number of people participating in the conversation. Future work will be based on the expectations from the successful case trials in our current study.

## 29.2  Creation of Android Theater

Geminoid F, an android that resembles an actual living woman, was cast in a play called *Sayonara*. In this section, we describe our system that provided the android with an acting performance based on a tele-operating system, as well as the process of producing a play based on CCTT (Fig. 29.1).

**Fig. 29.1**  Android theater: *Sayonara*

## 29.2.1 Acting Performance System for the Android

We created the androids acting performance based on a tele-operating system [8] (Fig. 29.2). The performances are given by a human actor using the tele-operating system, instead of the human actor being present on stage himself. They are captured by image and voice processing techniques. The captured data are transformed to the androids motion to reproduce the same acting performance. If actor's skills that respond to the directors instruction are implicitly acquired, they can be derived only when the actor gives a normal performance in a similar situation. Therefore, a tele-operating system must be designed such that the operator of the robot can produce behavior that is as natural as possible. In this study, we adopted a method that estimates the posture of the operator's head by image processing of face tracking results [9], instead of placing a recording device on the operator, to avoid constraining his/her performance. The data are then transferred to the androids head and coordinating back motions. To detect the mouths opening and closing motions, we do not apply an image processing technique, since this might fail because mouth movements are very rapid. Instead, we use a sound processing method to estimate the shape of the robots mouth according to the formant-based acoustic features of the utterance to be produced [10]. Since such subtle unconscious behaviors as blinking or breathing, which are important for representing a humanlike presence, are considered important but less distinguishable, we generate them using a random process independent of the operator's performance as implemented in different applications [11].

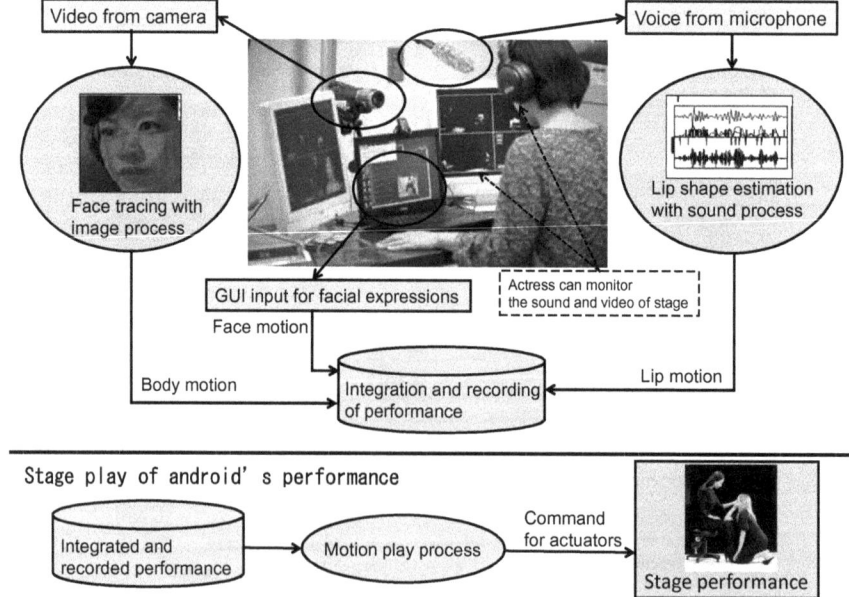

**Fig. 29.2** Tele-operating system

To produce facial expressions, such as smiles and frowns, the actuators must be precisely controlled to deform the androids facial skin or move its eyeballs. Therefore, we previously tuned the command sequences of the actuators to produce the appropriate facial expressions and prepared an interface function to reproduce them in arbitrary timing when the operator wanted them.

Note that during the androids operation by the actor to produce an acting performance, vivid vision and sound sensory feedback are given to the operator so that he/she can improve the naturalness of its acting performance. The operator can see and hear the motions and utterances of the human actor who is the androids interlocutor, as well as any changes on stage, through video cameras and microphones installed on the stage. To reproduce the performance in the playing phase, the recorded sequence of the commands for all of the androids actuators and the utterances synchronized with them are sent to the actuators at the same sampling rate as that in the recording phase.

## 29.2.2  Acting and Directing

The recorded motions and utterances given to the android must be elaborated before being used in the actual play. The instructions given by the director based on CCTT are used for this purpose in the following process.

(1) **Read-through**: The script is read aloud by an actor and a second actor who takes the role of the androids operator to learn the script and the timing of their lines.

(2) **Initial directing**: A performance is produced by the human actor and the android tele-operated by the operator actor. In this process, the director instructs both the human actor and the android in order to modify their performances. Instructions are given based on the physical aspects of the last performance, such as the timing of the utterances, the standing or sitting positions, and body movements. When the director is satisfied with the performance to a certain extent, the voices of both the actress playing the android and the opposite actress are recorded.

(3) **Final utterance recording**: The androids utterances are recorded in a soundproof room. The actor who operates the android produces the same utterances while listening to the sound recorded during the last session in the initial directing phase. This recording style helps the operator reproduce the utterances at the same timing and speed as in the initial directing phase.

(4) **Final motion recording**: The operator tele-operates the android while listening to the sound recorded during the final utterance recording phase. The operator focuses only on repeating the utterances at the same time as the motions so that the motion is recorded in synchrony with the sound obtained in the phase of the final utterance recording. The recorded motions are transformed to the command sequence for the android and recorded.

(5) **Final performance**: The human actor tunes the timing of her performances for synchronization with the androids performance produced by the data recorded in phases (3) and (4).

In this studys trial, the director concentrated on fine-tuning the androids motions, in particular its gaze directions and the timing of its utterances. He also guided the human actor who was the androids interlocutor. In the direction method based on CCTT [12], actors are not required to have the same feelings as the characters in the story [13]. According to the reports of previous studies that analyzed plays based on CCTT, the instructions given for this play also included those mentioning the onset timing of utterances, their speed, the gaze directions, and the standing/sitting positions of the actors. Interestingly, the instructions given to the android and the human actor basically mentioned identical aspects.

## 29.3  Android Drama: *Sayonara*

We applied the system and procedure proposed in Sect. 29.2 to create a play titled *Sayonara*. It was written and directed based on CCTT by Hirata, a world-famous director and a coauthor of this study. It was produced by the Seinendan theater company, which provided all the technical staff necessary for a professional production. The actors, technicians, lighting and acoustic designers, and scenographer all belonged to this company, since their technological knowledge was necessary for the human actor and the androids satisfactory performances.

### 29.3.1  Script

*Sayonara* represents a scene where a woman dying from a serious illness is talking to a communication partner robot. The woman is played by a human actress, and the robot is played by Geminoid F. In the stage play, the android reads poetry written by Arthur Rimbaud and Shuntaro Tanikawa to the dying woman. Through the serene dialogue, the play raises issues of life and death.

### 29.3.2  Acting Performance

Figure 29.3 illustrates a sample scene from the play. The story is advanced by the conversation between the android and the actress. In this scene, the actress approaches the android, snuggles against its knee, takes its hand, and talks to it. In this play, the androids emotional expressions are limited to moderate ones. The lack of exaggerated actions reflects a distinctive feature of CCTT by which the director reproduces natural styles of behavior from daily life.

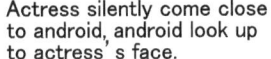
Actress silently come close to android, android look up to actress' s face.

Android look down to her face and recite Rimboud.

**Fig. 29.3**  Scene from *Sayonara*

## 29.4 Evaluation of Play Through Public Performances

To investigate the type of android performance that is elicited by the proposed method, we introduced it into public performances of the play in Japan during the summer of 2012 and distributed questionnaires to all the members of the audiences. Through the questionnaire, we obtained the personal information of the audience members, such as their age and gender, and their reactions to the play and to the android itself.

We obtained 291 filled questionnaires from the audiences of three performances. The data of 18 audience members younger than 16 years old were removed from the analysis. None of the audience members expressed problems related to seeing or hearing the events on the stage. All the questionnaire results used in the analysis were obtained from Japanese subjects. Some of these data exclusion criteria overlapped.

### 29.4.1  Questions About the Android

In this play, since we implemented an acting performance of the android based on CCTT, we expected that it would be endowed with a high level of humanlikeness. However, since humanlikeness is not an easily understandable index for audiences, we examined it from several viewpoints.

In the questionnaires, we asked the audience members to identify which actor was the android in order to determine whether they had difficulty distinguishing between a human and an android endowed with humanlikeness. At the beginning of the performance, 5.0% of the audience and after seeing the play 5.7% thought the actress on the right side was the android, although in fact the android was on the left. Although we failed to build a humanlike android that completely deceived the entire audience, some subjects were indeed confused.

Another interesting point is whether the humanlike aspects represented in the android were successfully recognized or enhanced more than the human's. In this study, we investigated the following three aspects in the audiences evaluation of the android in the play: mind (M), attractiveness (A), and weight of the words (W). These attributions will probably be reduced if the android lacks sufficient humanlikeness.

We asked the audience members to indicate the extent to which they responded positively/negatively to the following seven questions on a 7-point Likert scale. Note that we adopted exaggerated expressions in some questions to avoid a ceiling effect. The list of questions is as follows.

1. **Mind**
   [M1] Did you feel that the android expressed emotion?
   [M2] Did you occasionally feel that the android had a purity that humans lack?
2. **Attractiveness**
   [A1] Did you find the android very beautiful?
   [A2] Did you find the androids voice very impressive.
   [A3] Did you occasionally feel that the android was more attractive than other women?
3. **Weight of words**
   [W1] Did you regard the poetry recited by the android as a message from the android?
   [W2] Did the android enhance the original impressions of the poems more than any human could?

### 29.4.2  Results and Discussions

Table 29.1 shows the t-test results comparing the median value (i.e., 4) of the scale we used with the value given by members of the audience. The audience recognized humanlike emotions in the android to a significant extent (M1) and felt that its inner purity (M2) was greater than that of a human. The audience evaluated the androids beauty (A1) and its voice (A2) at a level significantly higher than the median but did not evaluate its attractiveness as higher than that of other women (A3). The audience

**Table 29.1**  Questionnaire results: t-test results comparing the median value (i.e., 4) of the scale with the value of audience responses

| Q. | Ave | M | SD | t | p | Effect size |
|----|-----|---|----|----|----|-------------|
| M1 | 4.6 | 5.0 | 1.6 | $t(267) = 7.0$ | <0.01 | $r = 0.40$ |
| M2 | 5.1 | 6.0 | 1.6 | $t(264) = 11.4$ | <0.01 | $r = 0.57$ |
| A1 | 5.1 | 6.0 | 1.6 | $t(270) = 12.2$ | <0.01 | $r = 0.60$ |
| A2 | 5.0 | 6.0 | 1.7 | $t(269) = 10.6$ | <0.01 | $r = 0.54$ |
| A3 | 3.7 | 4.0 | 1.7 | $t(265) = 3.1$ | <0.01 | $r = 0.18$ |
| W1 | 4.5 | 5.0 | 1.8 | $t(264) = 5.2$ | <0.01 | $r = 0.30$ |
| W2 | 3.9 | 4.0 | 1.6 | | n.s. | |

also regarded the poetry recited by the android as a message from the android (W1) with a small effect size; however, there was no significant difference in the androids enhancement of the impression of poems as compared to that of a human (W2).

Our questionnaire results indicate that the androids that performed in plays were perceived positively by the audiences in terms of the mind, attractiveness, and weight of words aspects, although we adopted exaggerated expressions in our questionnaire. However, as compared with humans, the android underperformed in the attractiveness and weight of words aspects.

People who face androids in general feel that they are in fact humanlike. However, they also experience them as being uncanny. In our play, the audiences reacted positively to the android, although it appeared on stage for at least 20 min. This means that adopting CCTT for designing androids behavior is an effective way to make them more humanlike entities.

## 29.5   Analysis of Short Plays

The results of our investigation of the audiences impressions indicate that our proposed method for designing robot behavior based on CCTT is sufficiently effective to allow us to design a robot having a humanlike entity. Thus far, robots have been deemed mere sideshows in the entertainment field. However, if professional directors create plays including robots based on CCTT, the robots can be perceived as entities that can exist harmoniously with humans. A play based on CCTT contains the essential knowledge that can be applied to endow a robot with greater humanlikeness. Therefore, the analysis of a play that includes robots among the actors may contribute to the design of robots behavior in conversations. Hence, in this experiment, we extracted beneficial knowledge by analyzing plays based on CCTT and interviewing their director.

Using robots in existing plays to precisely derive beneficial knowledge is futile, because they were created for situations that simultaneously engage many people. In this experiment, therefore, we prepared a short play using communication primitives that can be easily analyzed to help us derive beneficial knowledge for designing robots behavior in conversations.

### 29.5.1   Short Plays

In this experiment, we prepared two types of scene (Fig. 29.4). A director created plays for each of them. The first play presents a scene where a robot, and human are interacting with each other (Scene 1) to represent two-person interactions and the second presents a scene where a robot and two humans are interacting (Scene 2) to represent three-person interactions. In both, the robot asks Human A about the progress of his/her research. Note that the scripts used for the two scenes are identical

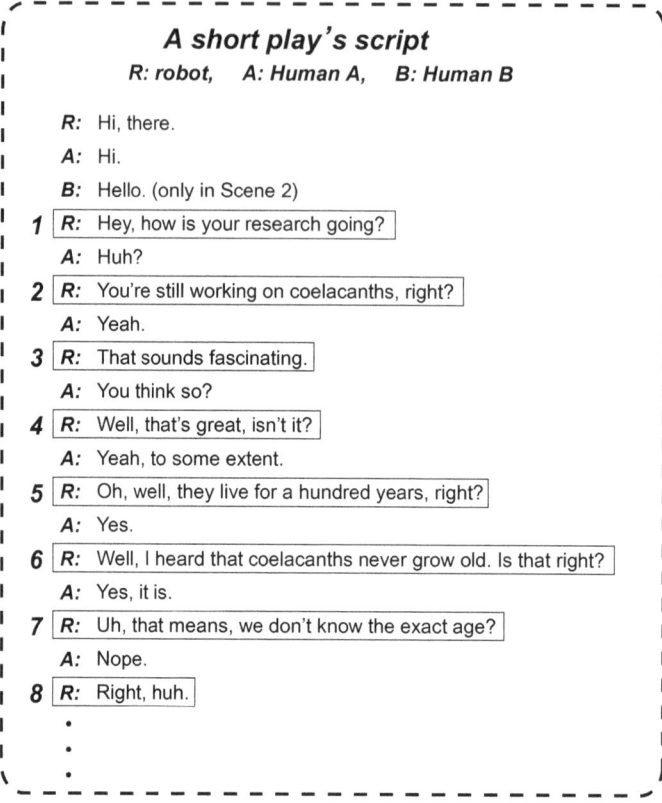

**Fig. 29.4** A short play's script: **A** and **B** are the humans lines and **R** are those of the robot. Sentences surrounded by rectangles were analyzed for the time gaps described in Sect. 3.2.2. The values on the left of each rectangle correspond to the first row in Table 29.2

except that a greeting by the robot to Human B is added at the beginning of Scene 2. For our experiment, we used Robovie R3, a humanoid robot, produced by VStone, Japan, because it can produce active arm movements that androids cannot.

### 29.5.2 Analysis Methods

We conducted two types of analysis: interviews with the director and a quantitative analysis of the sequential robot motion as directed by the director. We asked the director why he directed the robot's behavior as he did. We analyzed the relationship between the robots sequential motions and its utterances from the aspect of the time between them.

## 29.5.3  Analysis Results

### 29.5.3.1  Interviews with Director

As a result of interviewing the director, we identified the following knowledge that can be applied for designing a robots behavior in conversations and assigned it to three categories: (A) overall interaction, (B) three-person interaction, and (C) general knowledge.

A. **Overall interaction**

(1) When a robot greets its interlocutor, it must move its arms before it talks.
(2) When a robot raises its arms, it has to keep them raised for a certain period of time. They should not be returned to their original position too soon.
(3) While the interlocutor is talking, the robot should sometimes break eye contact. It should not continuously look at the interlocutors eyes.

B. **Three-person interaction**

(1) The frequency of looking away from the interlocutor in the two-person interaction is the same as in the three-person interaction.
(2) When the robot starts talking during the three-person interaction, it must primarily greet the person to whom it wants to talk.

C. **General knowledge**

(1) When the robot stands in front of the interlocutor, it must slightly change its direction for added naturalness.
(2) When the robot talks, it has to move its body occasionally.
(3) When the robot moves both its arms, their movements have to be slightly different.
(4) When the robot makes a big motion with one of its arms, it must slightly move the other arm.
(5) After the robot makes a big motion with its arms, they have to be slightly swung at the end point of the motion because of the kickback from stopping their movements.
(6) When the robot turns its body, the opposite arm has to move; i.e., its left arm should move when the robot turns clockwise.

The knowledge we found by interviewing the director can be directly applied to robot implementation.

This knowledge came from the short play that we prepared for our experiment. Therefore, we need to be careful about applying it according to the situation. Nevertheless, it can contribute to the design of robots behavior in conversations since the short play was created using communication primitives.

Note that some of the statements obtained from the interviews reflect the facts ambiguously because of the limitations of language for representing implicit knowledge. Therefore, further analysis is necessary for verifying or quantifying the facts.

### 29.5.3.2    Timing Between Utterances and Motions

Next, we analyzed the timing between the robot's motion data and its utterances. Tables 29.2A and B show the analysis results. Table 29.2A shows the time gap data between the onsets of the robots arm and head movements and utterances in Scene 1. Table 29.2B shows the time gap data between the robots arm and head movements and utterances in Scene 2. The values in the table represent the time gaps between the start of the utterances and the start of the movement of each of the robot's features, i.e., its arms and head. The unit of the value is seconds. A negative value indicates that the movement starts before the robot talks: utterance-first behavior. Conversely, the positive value indicates that the movement starts after the robot talks: movement-first behavior.

Concerning the ARM results in Scene 1 (Table 29.2A, first row), six of the seven data items are movement-first behaviors. Concerning the HEAD results in Scene 1 (Table 29.2A, second row), all the data items are movement-first behaviors. Concerning the ARM results in Scene 2 (Table 29.2B, first row), six of the seven data items are utterance-first behaviors. Concerning the HEAD results in Scene 2 (Table 29.2B, second row), two of the five data items are utterance-first behaviors.

This result indicates that even when a third person is not directly participating in the conversation, robots have to adjust the timing of their utterances. In two-person interactions, movement-first behaviors should be used more frequently than utterance-first behaviors.

Our quantitative analysis of the robots movements reveals important knowledge. Interestingly, the director did not mention the timing between the robots movements and its utterances during our interviews. The source of his aesthetic sense is his intuition inside his mind. It was quite difficult to extract this knowledge from him during our in-depth interviews. However, our quantitative analysis that uses a short play using communication primitives can demonstrate his implicit knowledge.

**Table 29.2**    Upper part (A): the results of Scene 1. Lower part (B): the results of Scene 2. The time gap between the robots utterance and its movement are shown for eight types of sentence. The values in the table represent the amount of time before or after its utterance that the robot moves. The unit of the values is seconds. A negative value indicates the movement starts before the robot talks, and a positive value indicates the movement starts after it talks. The actual sentences of all eight utterances are shown in Fig. 29.4

A: Scene 1

|        | 1    | 2    | 3   | 4    | 5   | 6    | 7   | 8    |
|--------|------|------|-----|------|-----|------|-----|------|
| ARM    | 1.5  | 1.7  | 1.8 | 1.8  | 1.7 | 1.8  |     | −1.4 |
| HEAD   | 0.4  | 1.2  |     |      | 4.2 | 5.7  | 8.5 |      |

B: Scene 2

|        | 1    | 2    | 3    | 4    | 5   | 6    | 7   | 8    |
|--------|------|------|------|------|-----|------|-----|------|
| ARM    | −1.2 | −0.9 | −0.9 | −1.0 | 4.9 | −1.0 |     | −1.4 |
| HEAD   | −0.4 | −0.8 |      |      | 4.3 | 3.1  | 5.1 |      |

## 29.6  Conclusion

In this paper, we described the first play that included a robot in its cast and presented the evaluation of the audience impressions of it, which were in general positive. The members of the audience felt that the android was a humanlike entity when its behaviors were designed based on the CCTT concepts. We extracted knowledge for designing robots behavior in conversations from two short plays based on CCTT. Our analysis of the plays revealed several useful items of knowledge, including the timing of utterances and body movements.

Designing robots that can communicate with humans is a shared purpose in the HRI field. In this study, we found that plays provide important knowledge that can be extracted through interviews with their directors and by the quantitative analysis of the directed robots motions. We believe our approach is an effective way to identify useful knowledge for designing robots behavior during conversations. Future work will use a motion capture system to obtain more precise data and more useful knowledge.

## References

1. Ogawa, K., T. Chikaraishi, Y. Yoshikawa, S. Nishiguchi, O. Hirata, and H. Ishiguro. 2014. "Designing robot behavior in conversations based on contemporary colloquial theatre theory," In *23rd IEEE international symposium on robot and human interactive communication (RO-MAN 2014)*, 168–173.
2. Kobayashi, Y., Y. Hoshi, G. Hoshino, T. Kasuya, M. Fueki, and Y. Kuno. 2008. Museum guide robot with three communication modes, In *IEEE/RSJ international conference on intelligent robots and systems, IROS 2008*, 3224–3229, Sept 2008.
3. Kamashima, M., T. Kanda, M. Imai, T. Ono, D. Sakamoto, H. Ishiguro, and Y. Anzai, Embodied cooperative behaviors by an autonomous humanoid robot, In *IROS*, 2004, 2506–2513.
4. Ogawa , H., and T. Watanabe. 2000. Interrobot: A speech driven embodied interaction robot. In *Proceedings 9th IEEE international workshop on robot and human interactive communication, RO-MAN 2000*, 322–327.
5. Ishiguro, H. 2007. Scientific issues concerning androids. *International Journal of Robotics Research* 26 (1): 105–117.
6. Hirata, O. 1995. *Gendai Kogo Engeki no tameni (For Contemporary Colloquial Theater)*. Banseisha (in Japanese).
7. Ishiguro, H., and O. Hirata. 2011. Robot theater. *The Robotics Society Japan* 29 (1): 35–38. (in Japanese).
8. Sakamoto, D., T. Kanda, T. Ono, H. Ishiguro, and N. Hagita. 2007. Android as a telecommunication medium with a human-like presence. In *2007 2nd ACM/IEEE international conference on human-robot interaction (HRI)*, 193–200.
9. S. M. Inc. Faceapi. http://agesage.co.jp/.
10. Ishi, C., C. Liu, H. Ishiguro, and N. Hagita. 2011. Speech-drivenlip motion generation for teleoperated humanoid robots. In *International conference on auditory-visual speech processing*, 131–135.
11. Chikaraishi, T., T. Minato, and H. Ishiguro. 2008. Development of an android system integrated with sensor networks. In *The proceedings of the 2008 IEEE international conference on intelligent robots and systems (IROS 2008)*, 326–333.

12. Hirata, O. 2004. *Engi to Enshutsu (Acting and Directing)*. Kodansha.
13. Goan, M., T. Fukaya, and K. Tsujita. 2008. Two succeeding stages in acquisition process of a rehearsed drama: Applying system dynamics to human collaborative behavior,. In *The 2nd international conference on knowledge generation, communication and management (KGCM 2008), Proceedings of the 12th world multi-conference on systemics, cybernetics and informatics (WMSCI 2008)*, vol. 7, 126–131.

# Chapter 30
# At the Theater—Possibilities of Androids as Poetry-Reciting Agents

Kohei Ogawa and Hiroshi Ishiguro

**Abstract** In recent years, research on very humanlike androids has increased, in general investigating the following: (1) the manner in which people treat these very humanlike androids and (2) whether it is possible to replace existing communication media, such as telephones or TV conference systems, with androids as a communication medium. We found that androids have advantages over humans in specific contexts. For example, in a collaborative theatrical project between artists and androids, audiences were impressed by the androids that read poetry. We therefore experimentally compared androids and humans in a poetry-reciting context by conducting an experiment to illustrate the influence of an android who recited poetry. Participants listened to poetry that was read by three poetry-reciting agents: an android, a human model on which the android was based, and a box. The experimental results show that the participants scored their enjoyment of the poetry highest under the android condition, indicating that the android has an advantage for communicating the meaning of poetry.

**Keywords** Robot · Android · Art · Geminoid · Poetry

## 30.1 Introduction

Over the past several years, many humanoid robots have been developed that have arms, a head, and legs. In the human–robot interaction (HRI) field, researchers focus

---

This chapter is a modified version of a previously published paper [1], edited to be comprehensive and fit with the context of this book.

---

K. Ogawa (✉) · H. Ishiguro
Hiroshi Ishiguro Laboratory, ATR, 2-2-2 Hikaridai Seikacho Sorakugun,
Kyoto 6190288, Japan
e-mail: ogawa@irl.sys.es.osaka-u.ac.jp

H. Ishiguro
Graduate School of Engineering Science, Osaka University, 1-1 Machikaneyama,
Toyonaka 5600043, Japan

© Springer Nature Singapore Pte Ltd. 2018
H. Ishiguro and F. Dalla Libera (eds.), *Geminoid Studies*,
https://doi.org/10.1007/978-981-10-8702-8_30

on natural interaction between humans and humanoid robots. For example, ATR developed Robovie, which is used for route guidance or as a shop advisor agent in such real environments as shopping malls [2]. Robovie is also used for educational purposes in elementary schools [3]. Wakamaru, developed by Mitsubishi Heavy Industry [4], is based on Robovie and is used not only as a communication robot but also as a stage actor.

Against this background, we have been developing very humanlike androids called Geminoids, which resemble real persons (Fig. 30.1). The effects of an androids appearance and body movements have thus far been investigated mainly in empirical studies conducted in laboratory environments.

Noma et al. showed that a Geminoid is indistinguishable from a real human if the observer sees the robot only for three seconds [5]. Minato et al. used a Geminoid to investigate whether the uncanny feeling it sometimes induces is diminished when it exhibits increasingly complex behavior [6].

Several researchers focused on issues involving tele-operated androids by treating an android not as an autonomous being but as a tele-operated communication medium in order to compare Geminoids and such traditional communication media

**Fig. 30.1** Geminoid HI-2

as TV conference systems and phones. Sakamoto et al. investigated whether a Geminoid could be treated in the same manner as its human model by comparing them [7]. Ogawa et al. and Nishio et al. concluded that a Geminoid was as persuasive as its human model [8, 9]. Moreover, its social influences on visitors were tested in the real world [10, 11].

The influences of Geminoids on their operators have also been investigated. Watanabe et al. reported that when Geminoid operators were deeply concentrating on manipulating the robot, they sometimes felt the perceptual illusion that the Geminoid was part of their own body [12]. They experienced Chandrans Rubber Hand Illusion (RHI). In this experiment, while participants were operating the Geminoid, its hand received an injection from a needle. The results showed that the RHI was illustrated during Geminoid tele-operation. Such past research shows that as a tele-operated android, a Geminoid influences not only its communication partners but also its operators.

As described above, androids have a potential to influence humans that has never been examined. In other words, androids have abilities with which traditional robots or communication media cannot be endowed. Previous android studies focused on research platforms for examining humans recognition of robots or investigated only whether an android can replace traditional communication media. However, androids have specific features that outperform those of humans. Some android features that surpass those of humans have emerged through our activities with them. In this study, we conducted a psychological experiment to identify those functions that surpass those of humans in order to explore androids potential.

## 30.2   Theater Using an Android

We are currently collaborating with dramatic artists using androids to discover additional potential that has been overlooked because only traditional engineering or science aspects were examined. In one collaboration, an android and an actor perform live on stage (Fig. 30.2). In the play, an android reads a poem called "Sayonarar" to a dying person. The play has been performed over ten times to favorable worldwide reviews. We asked the audience members for their impressions of the play. Many gave the following impressive comments:

- I couldn't distinguish between the android and the actor.
- I was much more impressed by this android performance than by a normal play performed by humans.

These two comments contain important indications. Previous android research approaches in general substituted androids for traditional communication media or assumed a similar presence of humans in the context of the interaction. These comments suggest a potential advantage that androids may have over humans in a theater context. In other words, androids have the potential to emotionally affect humans more than an actual human can.

Photo by Tatsuo Nambu

**Fig. 30.2** Theater using an android

In this paper, we conducted a psychological experiment to investigate two issues. (1) In what kind of situations are humans moved/impressed by androids? (2) Why were the audiences impressed by the acting android? Through this experiment, we investigated the further potential of androids. The following three questions describe the purposes of this study.

(1) Is there a situation where humans are moved/impressed more by androids than by humans?
(2) If so, in what type of situation?
(3) Why are humans moved/impressed in such situations?

## 30.3   Geminoid F and Tele-Operation System

In this section, we describe the features of Geminoid F, which is the android that we used for the experiment, and the tele-operation system.

**Fig. 30.3**   Geminoid F

### 30.3.1   Geminoid F

We developed Geminoid F (Fig. 30.3), a female android that is modeled on a real person, to convey the feeling that a specific individual is present at the robots location using a tele-operation system. The body of Geminoids source model was duplicated using a 3D scanner system. Her face features were precisely duplicated by using a plaster cast. Geminoid Fs actuators are not electronic but pneumatic motors. Because pneumatic motors are very quiet, Geminoid F is perceived as moving very naturally. Geminoid F has 12 degrees of freedom (DoFs), 7 of which are installed in its face to produce such rich facial expressions as anger, sadness, or laughter.

### 30.3.2   Tele-operation System

Figure 30.4 shows an overview of the tele-operation system that transfers an operators movements and voice to an android in real time. The operators face direction, mouth movements, and facial expressions are captured by a face recognition system, the video stream of which is obtained using a Web camera attached to a laptop

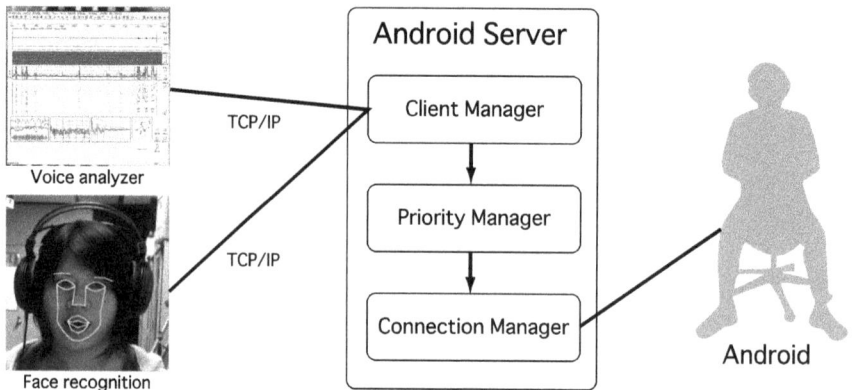

**Fig. 30.4** Tele-operation system for an android

computer. These face-tracking results are used to create commands that are sent to the Android server by TCP/IP.

The face recognition system and other clients also send control commands to the Android server by TCP/IP. The Android server processes the priority of the commands sent from multiple clients. After this process, the Android server sends the signals to the android.

The Android server has several modules, including Client, Priority, and Connection Managers, rendering it more flexible. For example, if a user wants to add sensor information to the tele-operation system, the user simply develops the client module and establishes the priority of the command. If a user wants to operate different androids, he/she simply replaces the Connection Manager, which is developed for specific androids.

## 30.4 Experiment

In this experiment, we established whether humans were impressed by an android. If so, we also investigated the reasons. The android, a human, and a box read poetry to the participants. After the reading, we asked the participants for their impressions.

The experiment was inspired by the use of androids in theater productions. In a play, the android read a poem to a dying person. An actor read it through the android using a tele-operation system. After the play, some members of the audience mentioned one poem, "Sayonara," written by Shuntaro Tanigawa, a famous Japanese poet. Since the audience was impressed by the androids reading of the poem in the theater, we used it in our experiment.

### 30.4.1   Hypothesis

Since poetry is expressed using aesthetic and evocative qualities, its full impact and meaning are conveyed by elements outside the actual text; room for interpretation remains. In this section, we discuss why the audiences of a play that included an android as an actor were impressed by the androids poetry reading as a message that allows room for interpretation. We also describe our experiments hypothesis based on the discussion.

It is not always true that a human recitation of a poem provides the medium that allows a listener to appreciate it to the full. A reader's presence may constrain the freedom of interpretation. For example, poetry readers never act as actors on stage do when they read poetry. Rather, they seem to suppress their bodily gestures. Moreover the readers' eyes tend to be on the book from which they are reading. This behavior could be interpreted as a strategy for communicating a poems implicit meanings by suppressing the readers natural presence. The androids appearance resembles that of a real human. Its behaviors are also humanlike. However, observers realize it is not an actual human. Androids seem to exist on the borderline between being human and non-human. Therefore, an android may have an advantage over a human reader when communicating poetry-like messages because of its weak presence.

Assuming that the humans presence impairs the freedom of the interpretation of the poem, a medium with less presence than that of an android may be more appropriate for poetry readings. In the experiment, we employed a simple box as the condition involving the least presence. We investigated the following two hypotheses by comparing the results for the Human, Android, and Box conditions.

(1)   An android has an advantage for communicating poetry as compared with a human because of its neutral presence.
(2)   An android has an advantage for communicating poetry as compared with a non-humanoid object because of its humanlike features.

### 30.4.2   Method

The experiment was conducted to verify our hypotheses. It included the following three conditions: (1) Human Condition (Condition H): android model, (2) Android Condition (Condition A): Geminoid F, and (3) Box Condition (Condition B): box.

Seventeen university students (12 men and 5 women) whose average age was 21.2 years participated in the experiment. They experienced all three conditions, the order of which was chosen randomly as a counterbalance. The post-experiment questionnaires showed that they had no previous knowledge of Geminoid F.

In Condition A, the androids model operated the android such that its movements were as synchronized as possible. The model also memorized the poem. The box was a 25-cm wide, 25-cm high, 40-cm deep cube placed on a chair used in Conditions H and A. We installed a speaker inside the box that was used for Condition A.

### 30.4.3  Procedures

The details of the procedures are as follows.

(1) First, we explained the experiments purpose to the participants. (1) "You are going to listen to three poetry-reciting agents: Human, Android, and Box"; (2) "After listening to the poetry readings, you will be asked about your impressions".
(2) Before the experiment, we distributed copies of the poem to the participants so that they could read it. This decreased the order effect of the poems novelty.
(3) The participants entered the experiment room and listened to the poem.
(4) After listening, they filled questionnaires. After filling the questionnaires, they started the next condition.
(5) After completing the three conditions, they were asked to choose under which condition they would prefer to listen to the poem again.

### 30.4.4  Environment

Participants listened to the poem alone sitting in chairs 1.5 m from the poetry-reciting agent.

### 30.4.5  Analysis

We established the participants impressions toward the poetry-reciting agent and the agent-recited poetry for each condition by asking them the following six questions. Question 1 (Q1) and question 2 (Q2) address the poetry-reciting agent, and question 3 (Q3) to question 6 (Q6) address the agent-recited poetry. The participants ranked their responses to Q1–Q6 on a one-to-seven point scale, where seven denoted the most positive point on the scale. Finally, the participants were asked which condition they would prefer if they listened to the poetry again.

Q1.  I felt an affinity to the [Human/Android/Box].
Q2.  I felt the [Human/Android/Box] was uncanny.
Q3.  I felt that I was able to understand the poem read by the [Human/Android/Box].
Q4.  I thought the poem read by the [Human/Android/Box] was the [Human/Android/Box]'s own messages.
Q5.  The poems implicit meanings were impaired by the [Human/Android/Box].
Q6.  The poems impact was intensified by the [Human/Android/Box]'s poetry reading.

## 30.5  Results

The participants subjective impressions of the agent-recited poetry and the poetry-reciting agent were evaluated by an ANOVA. The results showed a significant difference between the conditions in the responses to Q1 to Q6.: ($Q1$: $F(2, 51) = 27.9$, $Q2$: $F(2, 51) = 5.0$, $Q3$: $F(2, 51) = 27.8$, $Q4$: $F(2, 51) = 4.1$, $Q5$: $F(2, 51) = 7.7$, $Q6$: $F(2, 51) = 6.8$)

We conducted multiple comparisons among the conditions for each question using the Bonferroni method. The mean values, standard deviations, and significant differences are shown in Fig. 30.5. The results for Q1 showed that the score of Condition B was less than that of Conditions A and H with a significant difference ($p < 0.01$). Hereafter, we write these results as ($A > B$ **, $H > B$ **). The results for Q2 were $A > H$ *, $B > H$ **. The results for Q3 were $A > B$ **, $A > H$ **, $H > B$ **. The results for Q4 were $H > B$ **. The results for Q5 were $B > A$ **, $A > B$ *. The results for Q6 were $A > B$ *, $H > B$ **.

After the experiment, we asked the participants which condition they preferred for listening to the poem again. We also asked why they preferred this condition. The results are shown in Fig. 30.6. Nine out of 17 (53%) participants chose Condition A, 6 (35%) chose Condition H, and 2 (12%) chose Condition B.

The results for Q1 showed that the box was considered a less-attractive agent than the human and the android, whereas the results of Q2 showed that the android and the box caused the participants to experience more uncanny feelings than did the human. The results for Q1 and Q2 indicated that the android was an attractive poetry-reciting agent. At the same time, it gave rise to uncanny feelings. These results seem inconsistent. However, we believe that they simply indicate one of the differences between an android and an object: The participants did not feel attractiveness in parallel with uncanniness in the case of the box. These results suggest that an android may exist on the borderline between being human and not human.

The Q3 results showed that participants were more attracted to the poetry when it was read by the android than when it was read by the box or the human. These results verify hypothesis 1: An android has an advantage over a human for communicating poetry because of its neutral presence. Poetry has room for interpretation. Therefore,

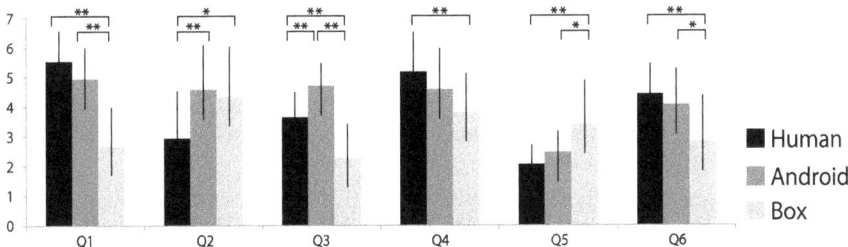

**Fig. 30.5**  Results for questions 1–6

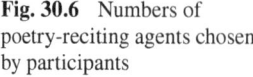

**Fig. 30.6** Numbers of poetry-reciting agents chosen by participants

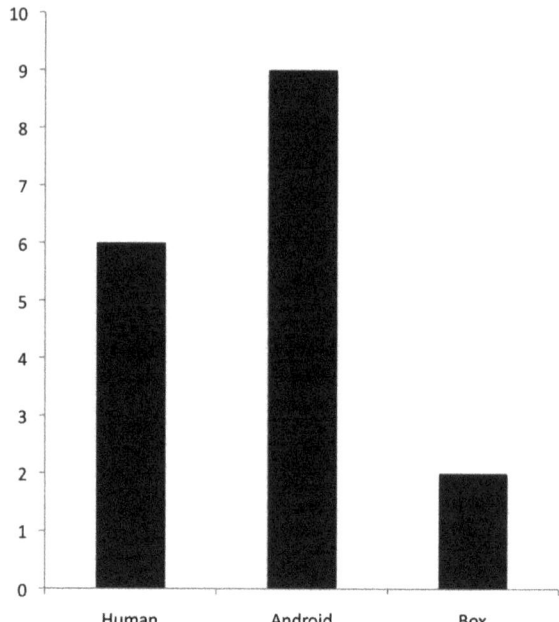

the android encouraged participants to experience the poetry and intensified their interpretations.

The results for Q4 showed that when the poem was read by a human, the participants felt as if the poem was the agents own message more than when it was read by the box. The Q3 and Q4 results indicate that receiving the poetry as a message influenced by the agents own intentions impaired the participants enjoyment of the poem. However, we cannot clearly draw such a conclusion because there was no significant difference between the results for Conditions H and A.

The results for Q5 showed that the box impaired the communication of the poems implicit meanings more than the human and the android. Hypothesis 2 is as follows: Because of its humanlike features, an android has an advantage for communicating poetry over a non-humanoid object. The box may have less influence on the poems interpretation space because of its weak presence as compared with the human or the android. However, the Q5 results showed that the box was the worst poetry-reciting agent in terms of freedom of interpretation. The poetry-reciting agent needs a humanlike appearance to communicate the message of a poem that was created by a human. Moreover, this result does not contradict the Q6 results that showed that the android intensified the poems original meaning more than the box.

After the experiment, we asked participants which condition they preferred for listening to the poem again. Fifty-three percent chose Condition A, and 35% chose Condition H, indicating that in this experiments context, the android was the most attractive poetry-reciting agent. We asked the participants to explain their responses to this question. The following is a typical answer from a participant who chose the android. "I could concentrate on listening to the poem because the android was not human." The following is a typical answer of the participants who chose the human: "I was bothered by the android, because its movements were a bit different from a humans." These opinions indicated that the android has an advantage as a poetry-reciting agent over humans. At the same time, however, its strange behaviors sometimes impaired the participants enjoyment.

## 30.6   Discussions

The experimental results showed that the android was the most effective poetry-reciting agent in that it provided a sense of immersion into the poetry. Such immersion deeply facilitates appreciation. This implies that the androids neutral presence positively affected the communication of the poetry because it allowed room for interpretation. Moreover, the responses to the questionnaires revealed that the box as an object having less presence could not compete with the human or the android.

We employed the box for the less-presence condition for the following reasons. (1) Participants cannot imagine that it has some type of character, and (2) its appearance and function are not directly linked, as in traditional communication media, such as telephone, radio, and TV. The experimental purpose was to compare the presence of the poetry-reciting agents. If participants regarded the box as a radio-like object, the experimental results might have shown higher scores because most humans have such preconceptions as "a radio must use recorded sound." If this were the case, the participants would have positively answered the questions about the box. However, participants scored the box lower than the android. This shows that participants did not regard the box as a radio-like object. Poetry-reciting agents have to have a humanlike presence, even if they do not impair the space for interpretation of the poetry.

Over half of the participants preferred the android for listening to the poetry again. This result may provide collateral evidence supporting hypothesis 1.

## 30.7   Conclusion

In this paper, we described the advantages androids may have as poetry-reciting agents. The results indicated that an android may have an advantage in terms of providing a sense of immersion when people listen to poetry. Moreover, it may also be a better poetry-reciting agent than a human.

Thus far, many robots have been proposed for supporting humans or as substitutes for traditional communication media. However, it has never been proposed that a robot can cause a human to consider a poems implicit meaning by facilitating a sense of immersion in the poem better than a human. We believe that androids can be used not only as a practical medium but also as interfaces for communicating poetry-like messages.

**Acknowledgements** This research was partially supported by KAKEN-20220002.

# References

1. Ogawa, Kohei, Koichi Taura, Hiroshi Ishiguro. 2012. Possibilities of androids as poetry-reciting agent. In *21st IEEE international symposium on robot and human interactive communication (RO-MAN 2012)*, 565–570.
2. Shiomi, M., T. Kanda, D. Glas, and S. Satake. 2009. Field trial of networked social robots in a shopping mall. In *International conference on intelligent robots and systems* 2846–2853.
3. Kanda, Takayuki, Rumi Sato, Naoki Saiwaki and Hiroshi Ishiguro. 2007. A two-month field trial in an elementary school for long-term human-robot interaction. *IEEE Transactions on Robotics (Special Issue on Human-Robot Interaction)* 23 (5): 962–971.
4. Shiotani, S., T. Tomonaka, K. Kemmotsu, S. Asano, K. Oonishi, and R. Hiura. 2006. World's first full-fledged communication robot "wakamaru" capable of living with family and supporting person. *Mitsubishi Juko Giho* 43 (1): 44–45.
5. Noma, Motoko, Naoki Saiwaki, Shoji Itakura and Hiroshi Ishiguro. 2006. Composition and evaluation of the humanlike motions of an android. In *Proceedings of the the IEEE-RAS international conference on humanoid robots (Humanoids'06)*, 163–168.
6. Minato, Takashi, Michihiro Shimada, Shoji Itakura, Kang Lee, and Hiroshi Ishiguro. 2006. Evaluating the human likeness of an android by comparing gaze behaviors elicited by the android and a person. Advanced Robotics 20 (10): 1147–1163.
7. Sakamoto, D., T. Kanda, T. Ono, H. Ishiguro, and N. Hagita. 2007. Android as a telecommunication medium with a human-like presence. In *Proceedings of the ACM/IEEE international conference on human-robot interaction*, 200.
8. Nishio, S., and H. Ishiguro. 2011. Attitude change induced by different appearances of interaction agents. *International Journal of Machine Consciousness*. 3 (01): 115–126.
9. Ogawa, K., C. Bartneck, D. Sakamoto, T. Kanda, T. Ono and H. Ishiguro. 2009. Can an android persuade you?. In *Robot and human interactive communication (RO-MAN2009)*, 516–521.
10. Becker-Asano C., K. Ogawa , S. Nishio and H. Ishiguro. 2010. Exploring the uncanny valley with Geminoid HI-1 in a real-world application. In *IADIS international conference on interfaces and human computer interaction*, 121–128, Freiburg, Germany.
11. Straub, I., S. Nishio, and H. Ishiguro. 2010. Incorporated identity in interaction with a teleoperated android robot: A case study. In *19th IEEE international symposium in robot and human interactive communication*, 119–124, Viareggio, Italy.
12. Watanabe, Tetsuya, Shuichi Nishio, Kohei Ogawa, and Hiroshi Ishiguro. 2011. Body ownership transfer to an android by using tele-operation system. *Transaction of IEICE* 94 (1): 86–93 (Japanese).